Progress in Inflammation Research

For further volumes:
http://www.springer.com/series/4983

Nathalie Vergnolle • Michel Chignard
Editors

Proteases and Their Receptors in Inflammation

Springer

Editors
Nathalie Vergnolle
INSERM U 1043
CHU Purpan
BP 3028
31024 Toulouse Cedex
France
nathalie.vergnolle@inserm.fr

Michel Chignard
INSERM U 874
Institut Pasteur
Unite de Defense Innee et Inflammation
Rue du Dr. Roux 25
75724 Paris Cedex 15
France
chignard@pasteur.fr

Series Editors
Prof. Michael J. Parnham, Ph.D.
Visiting Scientist
Research & Clinical Immunology Unit
University Hospital for Infectious Diseases
"Dr. Fran Mihaljević"
Mirogojska 8
HR-10000 Zagreb, Croatia

Prof. Eugen Faist, MD, FACS
Ludwig-Maximilians-University Munich
Klinikum Grosshadern
Department of Surgery
Marchioninistr. 15
81377 Munich
Germany

ISBN 978-3-0348-0156-0 e-ISBN 978-3-0348-0157-7
DOI 10.1007/978-3-0348-0157-7

Library of Congress Control Number: 2011934026

Printed on acid-free paper

Springer Basel AG is part of Springer Science + Business Media (www.springer.com)

Preface

For some time, we had in mind to write together on the role of proteases and anti-proteases in inflammation. We had considered different options over the years, but concretization was still a long way from our busy schedules. The idea of this book arose in 2009, at the nineth World Congress on Inflammation in Tokyo. We met there at the booth held by Springer Verlag (at that time Birkhenhouse Verlag) with Pr. Vincent Lagente, who had just published in the same series and was presenting his most recent volume. The enthusiasm of Pr. Lagente and the persuasiveness of the publisher representative convinced us that a book on Proteases and their Receptors in Inflammation was the best way to shed some light on the crucial role the protease-anti-protease balance appears to play in inflammatory diseases. We thus embarked on this project, associating colleagues and friends to contribute to the 13 chapters of this book. Each contributor has been a key scientific player in raising new knowledge on the role of proteases, and we want to express here our most sincere gratefulness for the time and energy they spend on their chapters, providing this volume with their invaluable expertise.

The existence of proteases has been known for centuries. Their control and the use of proteolytic activity have occupied a place of choice very early in the everyday life of human beings. Back to the Antiquity, the properties of proteases were exploited by humans for their food processing. As such, rennet, a natural complex of enzymes mainly composed of proteases produced in the stomach of many mammals to digest their mother's milk, was used in the production of cheese. A reference to this enzymatic activity can be found in Homer's classic, the Iliad, and likewise the philosopher Aristotle wrote several times about the process of milk curdling [1]. Along the same lines, wheat flour, a major component of bread, contains gluten, an insoluble protein indigestible by a number of individuals and that affects loaf processing yield. Proteinases from *Aspergillus oryzae* have been identified very early on and used to modify wheat gluten, inducing a limited proteolysis. The proteolytic treatment of the dough facilitated its handling and machining, and resulted in increased loaf volumes [2]. Proteases have also been used by other civilizations. For instance, people from the Pacific Islands have used for centuries the juice of the papaya fruit as a tenderizer for meat. It is now known that the protease papain is the active component accounting for this effect [1]. Papain is nowadays merchandized as a powder, and sold as meat tenderizer. This protease is also recommended as a home remedy treatment for insect stings or bites, because of its ability to degrade insect protein toxins. This story of papain has led two scientists: the French Réaumur (1683–1757) and the Italian Spallanzani (1729–1799), to hypothesize and demonstrate that gastric juice (full of proteases) are responsible for food digestion, through a complex process, now attributed to the protease pepsin.

Proteases are everywhere from prokaryotes to eukaryotes, from virus to bacteria and in all human tissues, playing a role in many biological functions ranging from digestion, fertilization, development, to senescence and death. The innate immune response to all types of aggression and tissue damage constitutes one of the major function in which proteases play a role. Almost 14,000 entries in PubMed are reported to the keywords "proteases and inflammation". While the role of proteases in inflammation-associated tissue damage was considered for years merely as a degradative role, where proteases would serve as "cleaners" or "spoilers" of the inflammatory site, the discovery of receptors for proteases has highlighted proteases as true signaling molecules that actively participate to inflammatory signals.

In the present book, the first two chapters are devoted to resume the type of signals proteases might send to cells and how those signals might be modulated by protease inhibitors in the context of inflammation. Then, the role and expression of different types of proteases: Kallikreins, proteases from inflammatory cells, and Matrix Metalloproteases at sites of inflammation is addressed in the following three chapters. Six chapters are devoted to discuss the role of proteases in specific organ inflammatory diseases: the lungs, the gastro-intestinal tract, the skin, the joints, or specific inflammation-associated events: fibrosis, coagulation, pain. Because proteases are not only produced by the host, but as stated before, also by microorganisms, it appeared important to have a special chapter focusing on the role of microbial proteases in the inflammatory response to infection. Finally, the last chapter discusses the different pathways by which protease receptor signaling terminates, thereby ending protease signaling events.

As editors, we are profoundly indebted to the chapter authors, and we want, once again, to express our gratitude for the contribution they made to this volume, providing this book with the most advance knowledge in this field.

We also want to thank Hans Detlef Kluber of Springer Verlag, for his enthusiasm, his patience, and his expert assistance in the preparation of this volume. A special thank to Ursula Gramm for her editorial and coordinator help.

Toulouse Cedex, France — Nathalie Vergnolle
Paris Cedex 15, France — Michel Chignard

References

1. Copeland R.A. (2000) Enzymes: a practical introduction to structure, mechanism, and data analysis. Wiley-VCH, Inc., New York
2. Rao M.B. et al. (1998) Molecular and biotechnological aspects of microbial proteases. Microbiol Mol Biol Rev, 62(3):597–635

Contents

Contributors

Elisabeth Boichot INSERM UMR 991, Université de Rennes 1, 2 avenue du professeur Léon Bernard, 35043 Rennes cedex, France

Mariarosaria Bucci Department of Experimental Pharmacology, University of Naples "Federico II", via Domenico Montesano 49, 80131 Naples, Italy

Eleonora Cavarra Department of Physiopathology and Experimental Medicine, University of Siena, via Aldo Moro n.6, 53100 Siena, Italy

Nicolas Cenac Inserm, U1043, Toulouse, 31300, France; CNRS, U5282, Toulouse, 31300, France; Université de Toulouse, UPS, Centre de Physiopathologie de Toulouse, Purpan (CPTP), Toulouse 31300, France, nicolas.cenac@inserm.fr

Giuseppe Cirino Department of Experimental Pharmacology, University of Naples "Federico II", via Domenico Montesano 49, 80131 Naples, Italy, cirino@unina.it

Lawrence de Garavilla Johnson and Johnson Pharmaceutical Research and Development, L.L.C., Welsh & McKean Roads, Spring House, PA 19477-0776, USA

Kathryn A. DeFea Biomedical Sciences Division, University of California, Riverside, CA 92521, USA, katied@ucr.edu

Eleftherios P. Diamandis Department of Laboratory Medicine and Pathobiology, University of Toronto, Toronto, ON, Canada M5S 1A8; Department of Pathology and Laboratory Medicine, Mount Sinai Hospital, Joseph and Wolf Lebovic Centre, 60 Murray Street, Toronto, ON, Canada M5T 3L9; Department of Clinical Biochemistry, University Health Network, Toronto, ON, Canada M5G 1X5, ediamandis@mtsinai.on.ca

Azza Eissa Department of Laboratory Medicine and Pathobiology, University of Toronto, Toronto, ON, Canada M5S 1A8; Department of Pathology and Laboratory Medicine, Mount Sinai Hospital, Joseph and Wolf Lebovic Centre, 60 Murray Street, Toronto, ON, Canada M5T 3L9

Silvia Fineschi Department of Physiopathology and Experimental Medicine, University of Siena, via Aldo Moro n.6, 53100 Siena, Italy

Julie Gabillet Immunology-Hematology Department, INSERM U1016, Institut Cochin, Université Paris Descartes Paris, 27, rue du faubourg Saint Jacques, 75014 Paris, France

Kristina K. Hansen Department of Physiology and Pharmacology, University of Calgary, Calgary, AB, Canada T2N 4N1

Melissa Heightman Centre for Respiratory Research, University College London, 5 University Street, London, WC1E 6BT, UK

Morley D. Hollenberg Department of Physiology and Pharmacology, University of Calgary, Calgary, AB, Canada T2N 4N1; Department of Medicine, University of Calgary, Calgary, AB, Canada T2N 4N1, mhollenb@ucalgary.ca

Takahisa Imamura Department of Molecular Pathology, Kumamoto University School of Medicine, Kumamoto, 860-8556, Japan

Ken Kilgore Immunology Research, Centocor R&D Inc, 145 King of Prussia Road, Radnor, PA 19087, USA

Vincent Lagente INSERM UMR 991, Université de Rennes 1, 2 avenue du professeur Léon Bernard, 35043 Rennes cedex, France, vincent.lagente@univ-rennes1.fr

Geoffrey J. Laurent Centre for Respiratory Research, University College London, 5 University Street, London, WC1E 6BT, UK, g.laurent@ucl.ac.uk

Monica Lucattelli Department of Physiopathology and Experimental Medicine, University of Siena, via Aldo Moro n.6, 53100 Siena, Italy

Giuseppe Lungarella Department of Physiopathology and Experimental Medicine, University of Siena, via Aldo Moro n.6, 53100 Siena, Italy, lungarella@unisi.it

Bénédicte Manoury Institut Curie U932, 24 rue d'Ulm, 75005 Paris, France

Laurence Martin Inserm, U1043, Toulouse, 31300, France; CNRS, U5282, Toulouse 31300, France; Université de Toulouse, UPS, Centre de Physiopathologie de Toulouse Purpan (CPTP), Toulouse 31300, France

Jason J. McDougall Department of Physiology and Pharmacology, University of Calgary, 3330 Hospital Drive NW, Calgary, AB, Canada T2N 4N1, mcdougaj@ucalgary.ca

Koichiro Mihara Department of Physiology and Pharmacology, University of Calgary, Calgary, AB, Canada T2N 4N1

Jean-Paul Motta Inserm, U1043, Toulouse, 31300, France; CNRS, U5282, Toulouse 31300, France; Université de Toulouse, UPS, Centre de Physiopathologie de Toulouse Purpan (CPTP), Toulouse 31300, France

Tatiana Ort Immunology Research, Centocor R&D Inc, 145 King of Prussia Road, Radnor, PA 19087, USA

Magali Pederzoli-Ribeil Immunology-Hematology Department, INSERM U1016, Institut Cochin, Université Paris Descartes Paris, 27, rue du faubourg Saint Jacques, 75014 Paris, France

Jan Potempa Department of Microbiology, Jagiellonian University, 30-387 Kraków, Poland; Department of Oral Health and Rehabilitation, University of Louisville School of Dentistry, Louisville 40202, KY, USA, potempa@uga.edu

Rithwik Ramachandran Department of Physiology and Pharmacology, University of Calgary, Calgary, AB, Canada T2N 4N1

Ali Roghanian Cancer Sciences Division, University of Southampton School of Medicine, Southampton General Hospital, Southampton SO16 6YD, UK

Fiona A. Russell Department of Physiology and Pharmacology, University of Calgary, 3330 Hospital Drive NW, Calgary, AB, Canada T2N 4N1

Jean-Michel Sallenave Institut Pasteur, Unité de Défense Innée et Inflammation et Inserm U874, 25 rue du Dr. Roux, 75724 Paris Cedex 15, France; Inserm U874, 25 rue du Dr. Roux, 75724 Paris Cedex 15, France; Université Paris, 7-Denis Diderot, Paris, France, jms@pasteur.fr

Nathalie Vergnolle Inserm, U1043, Toulouse 31300, France; CNRS, U5282, Toulouse 31300, France; Université de Toulouse, UPS, Centre de Physiopathologie de Toulouse Purpan (CPTP), Toulouse 31300, France, nathalie.vergnolle@inserm.fr

Tatiana Victoni INSERM UMR 991, Université de Rennes 1, 2 avenue du professeur Léon Bernard, 35043 Rennes cedex, France

Véronique Witko-Sarsat Immunology-Hematology Department, INSERM U1016, Institut Cochin, Université Paris Descartes Paris, 27, rue du faubourg Saint Jacques, 75014 Paris, France, veronique.witko@inserm.fr

Proteolytic Enzymes and Cell Signaling: Pharmacological Lessons

Morley D. Hollenberg, Kristina K. Hansen, Koichiro Mihara, and Rithwik Ramachandran

Abstract The innate immune response to invading organisms or inflammatory injury involves the activation of proteolytic enzyme cascades that in turn trigger the host defense signaling pathways. Although it has been known for over 40 years that proteinases such as thrombin, trypsin, and chymotrypsin can trigger hormone-like signal transduction pathways in target tissues, the mechanisms for signaling have only recently come into focus. Thus, enzymes of the coagulation cascade (thrombin, factor VIIa/Xa, activated protein C) are now known to signal to cells by cleaving and activating so-called proteinase-activated receptors (PARs). This cleavage unmasks a PAR "tethered ligand" sequence that triggers receptor signaling. Not only can endogenous proteinases activate the PARs by unmasking a "TL" sequence, but they can also "disarm" PAR signaling. Receptor "disarming" results from cleaving and removing the TL sequence entirely, thus preventing its subsequent activation by other proteinases. This chapter provides an overview of the multiple mechanisms whereby proteinases can affect tissue function. This overview emphasizes the key role that a "pharmacological approach" using PAR-activating peptides, receptor antagonists, bioassays of tissue responses both in vitro and in vivo and employing receptor-null mice ($PAR1^{-/-}$/$PAR2^{-/-}$) has played in providing insight into the physiological roles that proteinases can play as hormone-like mediators of inflammation and pain.

M.D. Hollenberg (✉)
Department of Physiology and Pharmacology, University of Calgary, Calgary, AB, Canada, T2N 4N1

Department of Medicine, University of Calgary, Calgary, AB, Canada, T2N 4N1
e-mail: mhollenb@ucalgary.ca

K.K. Hansen • K. Mihara • R. Ramachandran
Department of Physiology and Pharmacology, University of Calgary, Calgary, AB, Canada, T2N 4N1

N. Vergnolle and M. Chignard (eds.), *Proteases and Their Receptors in Inflammation*,
Progress in Inflammation Research, DOI 10.1007/978-3-0348-0157-7_1,

Keywords Activity-based proteinase probes • Biased signaling • Bioassay • Inflammation • PARs • PAR antagonists • PAR-activating peptides • Proteases • Proteinases • Receptors • Signal transduction

1 Introduction

In dealing with injury, the body's very first responses that mobilize its "innate defense system" include (1) triggering of the coagulation and complement cascades, (2) activation of pain pathways, and (3) the rapid recruitment of neutrophils to the site of damage. All of these immediate responses that generate the hallmark signs of inflammation (pain, redness, swelling, heat, and decreased function) involve the production and activation of proteinases. As a testament to the importance of proteolysis for regulating body processes, more than 2% of the human genome has been found to code for either proteinases (also colloquially referred to as "proteases") or their inhibitors [1]. Understandably, early work with proteinases was focused on the biochemical properties of the distinct proteinase families and on their unique catalytic activities, as summarized later. However, the discovery of hypotensive peptide principles in urine that have contractile activity in uterine smooth muscle [2] rapidly focused attention on the ability of proteinases to generate inflammatory "kinins" from their precursors (summarized by [3]). Further, the mechanisms for the processing of proinsulin to insulin were just coming into view [4]. Thus, by the early 1970s, the role of proteinases to generate physiologically active peptides from polypeptide precursors was well established. What was not fully appreciated at the time was that the proteinases themselves could also generate tissue responses that in many ways mirror the actions of peptide hormones. For example, in the mid-1960s, pepsin and chymotrypsin were shown by the Riesers to mimic the ability of insulin to promote glycogen formation in a rat diaphragm preparation [5, 6]. This insulin-like action of proteinases was also observed in isolated fat cells, wherein trypsin, like insulin, can stimulate glucose oxidation and inhibit lipolysis [7]. These actions of trypsin have been attributed to its ability to activate the insulin receptor via the tryptic cleavage of a regulatory domain of the receptor α-subunit [8]. In another context, thrombin and trypsin have been shown, like insulin and epidermal growth factor (EGF), to stimulate mitogenesis in cultured cell systems by acting at the cell surface [9–13]. Given these pharmacological actions of the proteinases, it was therefore reasonable to anticipate that the triggering of the inflammatory innate immune response by the enzymes might involve the activation of hormone-like signals in target tissues via "receptor" mechanisms. Stimulated by this hypothesis, it was the search for the mechanism of action whereby thrombin activates platelets and stimulates fibroblast mitogenesis that has revealed the G-protein-coupled membrane receptor family responsible for many of the inflammation-related actions of proteinases. These "proteinase-activated receptors" or "PARs," like the one activated by thrombin via a unique "tethered ligand" mechanism, have become a key area of interest for signaling by a number of serine proteinases. The proteinases that can participate in signaling, their biochemical mechanisms of catalysis, and their receptor-related mechanisms of action are summarized in the following sections. In part, progress in this field has been stimulated by a pharmacological approach to understand the effects of proteinases on their target tissues. The aim of this chapter is to provide an overview of the multiple mechanisms whereby proteinases can affect tissue function and to illustrate how a "pharmacological approach" has provided insight into the physiological roles that the proteinases can play as hormone-like mediators.

2 The Proteolytic Enzymes: Their Families and Mechanisms of Catalysis

Before going on to discuss the mechanisms whereby proteinases can signal to cells, it is of value to understand the catalytic properties and potential targets of the many proteinase families. Following, is a brief overview of the so-called degradome that comprises not only the proteinases and their substrates, but also their potential endogenous inhibitors (e.g., serine proteinase inhibitors or "serpins") [14, 15]. In principle, any of these proteinases can stimulate cell signaling via a number of mechanisms to be discussed. For purposes of this chapter, it will be important to keep the "degradome" concept in mind; but our focus will be on providing a succinct overview of the enzymes and their potential targets that can result in inflammation-related signaling. To understand the broader context, the reader is encouraged to consult some excellent reviews in this area [1, 14, 15]. Further, for more detailed information about individual proteinases, the reader is referred to the MEROPS peptidase database (http://merops.sanger.ac.uk/).

The "degradome" of the human genome codes for about 560 proteinases, including 178 serine-, 28 threonine-, 148 cysteine-, 21 aspartic-, and 186 metalloproteinases. Over 150 proteinase inhibitor genes are also present. Of note, due to the expansion of gene families clustered in selected gene regions, the rodent genome codes for even more proteolytic enzymes than does the human genome, making extrapolations from rodent models of disease to human diseases that involve proteinases a challenge [1, 16]. Despite this diversity, the basic mechanisms of catalysis of the different proteinase families are few in number, as outlined in the following paragraph.

2.1 Proteinase Mechanisms of Catalysis

Proteinases catalyze peptide bond hydrolysis in a peptide sequence-selective manner, which is dependent on having an enzyme binding site that is complementary to one or more substrate residue(s). In one class of proteinases, this cleavage is accomplished by having an enzyme-bound activated water molecule directly

Table 1 Human PARs 1–4: The sequence accession numbers, tethered ligand sequences, and the receptor-selective activating peptides and inactive control peptides are shown for human PARs 1–4

Receptor	NCBI accession number	Cleavage site (/) and TL (italicized)	Selective activating peptide	Inactive control peptide
PAR_1	NM_001992	TLDPR/*SFLLRN*	TFLLR-NH_2	FTLLR-NH_2
PAR_2	NM_005242	SSKGR/*SLIGKV*	SLIGKV-NH_2, SLIGRL-NH_2, 2-furoyl-LIGRLO-NH_2	LSIGKV-NH_2 LSIGRL-NH_2 2-furoyl-OLIGRL-NH_2
PAR_3	NM_004101	TLPIK/*TFRGAP*	No selective activating peptides; TL peptide activates PAR_1 and PAR_2	
PAR_4	NM_003950	LPAPR/*GYPGQV*	AYPGQV-NH_2	YAPGQV-NH_2

attacking the amide carbonyl of the peptide bond that is to be cleaved. This water molecule is held in place by a zinc cation in metalloproteinases or two aspartic acid residues in the active site of aspartic proteinases. Renin, an aspartic proteinase, and angiotensin-converting enzyme (ACE), a metalloproteinase, are both involved in the renin-angiotensin system which mediates intracellular volume and arterial vasoconstriction. Matrix metalloproteinases (MMPs) regulate a number of signaling pathways that control cell growth, inflammation, or angiogenesis and may even work in a nonproteolytic manner [17]. In the second class of proteinases, an amino acid residue of the enzyme initiates cleavage of the peptide bond in the first step and a water molecule subsequently cleaves the enzyme-bound intermediate. For serine and threonine proteinases, the hydroxyl group of an active site serine or threonine residue, respectively, is responsible for the attack of the peptide bond. Thrombin, trypsin, tryptase, elastase, and cathepsin G are all examples of serine proteinases that have been shown to be key regulators of proteinase-activated receptors (PARs) (Table 1). In the cysteine proteinases, an active site thiol group of a cysteine residue initiates cleavage. Cathepsins B, K, and L and the caspases are all members of this enzyme class. Cathepsin B, in particular, has been shown to be important in inflammation and cancer, and the caspases are involved in apoptosis, necrosis, and inflammation (the caspase acronym comes from the mechanism of action of this cysteine proteinase: *c*ysteine-*asp*artic prote*ase* or *c*ysteine-dependent *asp*artate-directed prote*ase*). As already mentioned earlier, the "MEROPS" database (http://merops.sanger.ac.uk/) can provide a wealth of further information for any individual proteinase of interest to the reader.

2.2 Identifying PAR-Regulating Proteinases by Activity-Based Probe Labeling

Knowledge of the proteinase catalytic mechanism provides an avenue to identify proteinase families that may potentially regulate PAR activity in an in vivo setting.

Thus, the "active serine" of a serine proteinase can be covalently labeled using a Biotin-linker-Pro-Lys-diphenylphosphonate "activity-based" probe (ABP). The ABP we have used was designed in accord with the strategy for tagging serine proteinases with a biotinylated diphenylphosphonate probe [18]. The phosphonate reactive group, which alkylates the active-serine residue within the active site of any trypsin-like proteinase, is preceded with a proline-lysine sequence in the P2/P1 enzyme target site. This enzyme target is separated from the biotin tag with an N-terminal-attached spacer (Bio-PK-DPP4: [19–21]). The interaction between the probe and the enzyme is characterized as a suicide inhibitor mechanism, such that the inhibitor is covalently and irreversibly bound to the proteinase. In a similar way, tagged activity-based probes have been designed to form covalent bonds selectively with the active-site thiol of a cysteine protease, allowing its identification in cell and tissue systems [22]. Unfortunately, for the aspartic acid and metalloproteinases the catalytic moiety is a water molecule that unlike the serine or cysteine residues of the serine and cysteine proteinases is not a substrate for covalent labeling. Other strategies that use fluorescent resonance energy transfer (FRET) are being developed to identify these enzymes, like the metalloproteinases, as "active" proteinases in tissues [17, 21, 22].

Using activity-based probes, the presence of an active proteinase in a biological sample can be identified and matched to its pharmacology for regulating PAR activity. With such a reagent, we were able to identify mammalian PAR_2-activating serine proteinases in colonic washings from mice infected with *Citrobacter rodentium* to provide a rationale for the colitis generated by this bacterium in mice [23]. Thus, the pharmacology of the proteinase in terms of PAR regulation can be matched with the chemical identity of a proteinase that is produced in a pathophysiological setting.

2.3 *Identification of Proteinase Targets*

It is often difficult to determine substrate specificity of newly identified proteinases and their in vivo targets. The most straightforward approach is to use high-throughput screening methods with libraries of chromogenic substrates (*p*-nitroanilide-conjugated peptides) or fluorogenic substrates (7-amino-4-methycoumarin-conjugated peptides), either in solution or on membranes/chips. A more efficient and unbiased method to identify proteinase substrates is with phage display techniques, since approximately 10^7 small peptides can be tested at once [24, 25]. However, these techniques only give the "theoretical" substrate specificity of small peptides and not whole proteins so the assayed proteinase may seem more promiscuous than it is in vivo, where the three-dimensional structure of the substrate plays an important role. Thus, proteomics approaches such as: (1) conventional DNA microarray chips, (2) proteinase-specific protein chips, (3) proteinase-activity chips, or (4) substrate chips may be more useful [14]. Another novel approach uses a system in which enzyme substrates are determined in the setting of a cell that does or does not express an active proteinase of interest. For each condition (i.e., in the presence or absence of active proteinase), cells are stably labeled metabolically with amino acids of different isotopic masses which become incorporated into the proteome of the cells for each distinct condition (acronym: "SILAC," for *s*table *i*sotope *l*abeled *a*mino acids in living *c*ells) [26]. The unique "natural" proteolysis products that are found by mass spectral analysis only in the "active proteinase condition" are used to identify potential endogenous cellular proteinase substrates. When focusing on the PARs, cells that do or do not express a PAR of interest can be used to assess the ability of a given enzyme to affect PAR function (or not) by a process of activating, disarming, or disabling receptor function [27]. Thus, the response to a proteinase of a cell that does not express a functional PAR can be compared with the response of the same cell that is PAR transfected so as to express the PAR target. PAR activation can be assessed by monitoring increases in cellular calcium or increases in MAPkinase activity [27, 28]. A proteomic approach can also be used to identify PAR-related peptides released from the cell surface by comparing enzyme-exposed versus nonexposed PAR-expressing cells [29, 30], the ability of an enzyme of interest to cleave synthetic PAR-related sequences can provide evidence for or against the PAR-regulating property of an individual proteinase (e.g., [31]). Thus, an evaluation of the biochemical pharmacology of a specific enzyme for its action on PAR-expressing cells (signal activation or disarming) compared with the ability of the enzyme to release PAR-related sequences from the cell surface and to cleave synthetic PAR-derived polypeptides can provide insight about the impact that the proteinase might have on PARs in vivo.

3 Targets for Proteinase-Mediated Signaling: Proteinase-Activated Receptors and More

3.1 Pharmacological Actions of Proteinases That Mimic Hormone Action

As outlined briefly earlier, the ability of proteinases such as trypsin, thrombin, and other serine proteinases to stimulate hormone-like tissue and cell responses was recognized by the mid-1960s. Further, the proteolytic generation of active peptides from precursors and the proteolytic degradation of these peptide agonists to terminate their action were also understood. However, the receptor signaling mechanisms involved in the action of proteinases on tissues were not known until relatively recently. The discovery that proteinases can regulate receptors for growth factors like insulin either by activation or "disarming," mentioned earlier, provided the first evidence for "insulin-like" signaling by proteinases via a "receptor." Thus, a very brief exposure of fat cells to insulin (15 s) triggers an insulin-like response (uptake of glucose or inhibition of lipolysis: [7]), whereas a longer exposure (minutes) "disarms" the insulin receptor, so as to reduce or abolish its response to insulin [32, 33]. By the early 1970s, cell culture studies had also shown that, like insulin and other "growth factors," serine proteinases in general and specifically, trypsin and thrombin can stimulate mitogenesis [9–11]. It was the pharmacological analysis of the mitogenic action of trypsin, thrombin, and other serine proteinases that led ultimately to the cloning of a receptor for thrombin.

3.2 Discovery of a Proteinase-Activated G-Protein-Coupled Receptor for Thrombin

One strategy used to isolate the "thrombin" receptor made a link between (1) the pharmacology and mitogenic action of thrombin in hamster fibroblasts and (2) the regulation by thrombin of G-protein coupled signaling including inhibition of adenylyl cyclase [34] and a stimulation of phosphoinositide-specific phospholipase C. This link led Van Obberghen-Schilling and colleagues to isolate a G-protein-coupled receptor-related cDNA clone from thrombin-responsive hamster CCL39 cells. When expressed in *Xenopus* oocytes the clone resulted in thrombin-triggered calcium signaling [35]. It was thus an understanding of the pharmacology of thrombin action in a cell culture system that facilitated the isolation of the hamster cDNA that coded for the receptor target for thrombin action. A complementary strategy using a comparable oocyte expression system in which thrombin-mediated calcium signaling was monitored led to the isolation of the thrombin-responsive "receptor" from human megakaryocyte cell lines that also responded to thrombin via calcium signaling [36]. Thus, the cloning of the "thrombin receptor" relied on the pharmacological match between the enzyme's ability to trigger calcium signaling in cultured cells and its ability to activate calcium signaling via the oocyte-expressed receptor cDNA. Not only did these two cloning efforts establish the essential role for the catalytic activity of thrombin to trigger signaling, but the work of Vu and colleagues discovered in addition, an entirely unique mechanism whereby proteolytic cleavage of the receptor N-terminal domain unmasks a "tethered ligand" that self-activates the receptor [36] (Fig. 1). Moreover, it was found that a relatively short synthetic peptide 14-mer representing the N-terminal sequence of the proteolytically revealed "tethered ligand" could activate the receptor in the absence of proteolysis (Fig. 1) [36]. This work created a paradigm shift in understanding the novel mechanism whereby proteinases can regulate G-protein receptor function and opened the door, via the synthetic "receptor-activating peptides," to understand in depth the kinds of pharmacological responses that the "PARs" could stimulate without the need for using the enzymes as physiological probes. The search for other members of the PAR family and the analysis of the potential roles that the PARs could play in vivo followed shortly after the publication of the two seminal manuscripts describing the cloning of "the thrombin receptor."

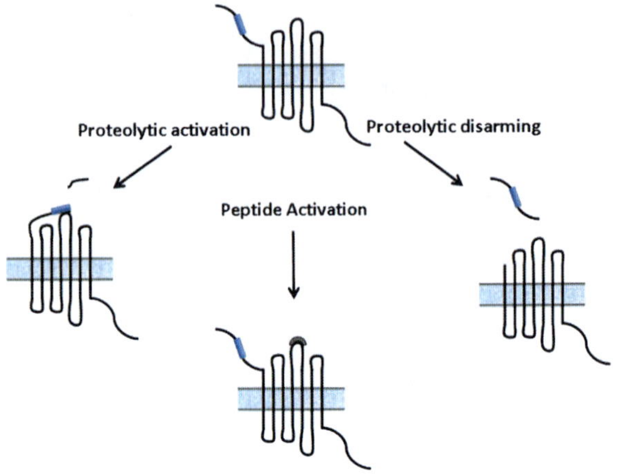

Fig. 1 Scheme for the activation of PARs by either enzyme or PAR-activating peptide, and proteolytic "disarming" of PARs to "silence" signaling by other PAR-targeted proteinases

3.3 Activation Versus Inactivation-Disarming of PARs by Proteinases

In addition to activating PARs via the unmasking of a "tethered ligand" sequence, proteolysis can also "silence" the PARs by "disarming" or removing the tethered ligand sequence, so that the receptor cannot be activated subsequently by a second proteinase (Fig. 1). There are many examples of such disarming responses in a number of cell types and physiological settings. Thrombin signaling via PAR_1 is abolished by prior exposure of cells to neutrophil proteinases such as cathepsin-G [37], elastase, and proteinase-3 (PR3) [38]. Chymase, a mast cell proteinase, can desensitize keratinocytes to thrombin signaling [39]. In endothelial cells, it has been shown that trypsin challenge can render PAR_1 unresponsive to subsequent thrombin stimuli [40, 41], while plasmin, a proteinase of the coagulation cascade, is also reported to desensitize PAR_1 by truncation of the TL [42]. Elastase and cathepsin-G have also been revealed to inactivate PAR_2 by cleaving amino terminal domains distinct from the activation site, rendering the receptors unresponsive to the activating proteinases [43]. Cathepsin-G has also been shown to abolish signaling by thrombin in PAR_3 transfected cells [44]. Interestingly, a number of proteinases can cleave PARs both at their activation sites and at other disabling sites. For example, cathepsin-G cleaves PAR_1 at the $Arg^{41}Ser^{42}$ receptor activation site, albeit with much lower efficiency than cleavage at the Phe^{55}–Trp^{56} site which results in the removal of the TL [37]. Thus, depending on the enzyme concentration and the rates of protein hydrolysis, an individual proteinase may be able to both activate and inactivate PARs. On human platelets, cathepsin-G disarms and silences PAR_1 for thrombin activation, but can activate PAR_4 to regulate platelet function [45]. In contrast, on fibroblasts or endothelial cells which do not express PAR_4, this proteinase would be solely responsible for disabling PAR_1 [46]. Similarly, tryptase a PAR_2 activating proteinase [47] can also cleave PAR_2 at the Lys^{41}–Val^{42} site, which would inactivate the receptor [37]. In murine platelets, cathepsin-G through inactivating PAR_3 could abolish its co-receptor function for PAR_4 thus preventing platelet responses to thrombin [44].

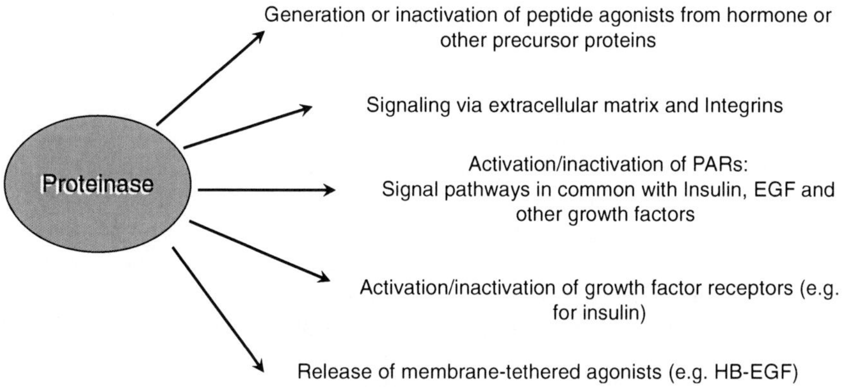

Fig. 2 Multiple mechanisms of proteinase-mediated signaling. The scheme points to the many ways that proteinases can trigger hormone-like signals, as outlined in the text

3.4 Multiple Mechanisms of Proteinase-Mediated Signaling

Although the PARs rapidly became a central focus for work related to signaling by proteinases, it must be emphasized that, as shown in Fig. 2, this receptor family represents only one of multiple mechanisms whereby proteinases affect tissue function. In Fig. 2, in addition to regulating PARs, these proteolytic mechanisms range from the generation and degradation of active peptides (including the release of agonists from the cell surface, e.g., EGF receptor ligands) to the modulation of matrix/integrin receptor interactions, to the activation or inactivation of growth factor receptors. These multiple ways whereby proteinases can affect cell function via cell surface and extracellular proteolysis are matched by complex intracellular proteolytic mechanisms (e.g., caspase activation and apoptotic signaling; proteasomal disposal of intracellular mediators) that also regulate tissue signaling. These intracellular signaling mechanisms involving proteinases will not be dealt with in this chapter, but must be kept in mind. Thus, overall, the complexity of the proteinase families themselves is matched equally by the complexity of mechanisms whereby the enzymes can affect cell signaling.

4 PAR-Activating Peptides: Their Pharmacology and Lessons Learned

4.1 PAR-Activating Peptide Pharmacology and the Discovery of Multiple Members of the PAR Family

The landmark manuscript from the Coughlin laboratory already cited [36] demonstrated that a 14-mer synthetic peptide (SFLLRNPNDKYEPF) could mimic the ability of thrombin to activate human platelets. Work by us and by others rapidly took advantage of this "mimicry" to synthesize shorter analogs of this sequence that could also regulate PAR activity [48–51]. We and others quickly established that the pentapeptide, SFLLR retained the ability to activate human platelets and that the second and fifth amino acids played key roles in receptor activation. Thus, the sequences SALLR or SFLLA were either inactive (SALLR) or very low in activity (SFLLA). Our own approach was to use the pharmacological principles established by Ahlquist [52], who identified α and β adrenoceptors by observing the distinct relative activities of a number of catecholamine agonists in several bioassay systems. Thus, we measured the relative contractile and relaxant activities of several different PAR-activating peptide analogs in vascular and gastric smooth muscle preparations. We observed significant differences in the relative potencies of these peptides for triggering endothelium-dependent NO-mediated vascular relaxation compared with the activation of smooth muscle contraction [53–55]. The pharmacological data provided convincing evidence that the receptor-activating peptides were acting via distinct receptors in these tissues, whereas at that time, the restriction fragment DNA mapping suggested that there was only a single "thrombin receptor" gene. In keeping with our evidence based on peptide structure–activity data pointing to multiple thrombin receptor "subtypes," it was found that the receptor-activating peptide that mimicked thrombin action in human platelets (SFLLRNPNDKYEPF) was not able to affect rodent platelets, which were otherwise thrombin responsive [56]. Thus, the pharmacological data identified distinct PAR subtypes well in advance of the cloning of the other three members of the PAR family (reviewed by [57, 58]). Four members of this unique G-protein-coupled receptor family are now known (PARs 1–4), each of which has its distinct proteinase-revealed "tethered ligand" sequence (Table 1). A "tethered ligand sequence" in PARs 1, 3, and 4 can be unmasked by thrombin and trypsin, whereas the tethered ligand sequence in PAR_2 is revealed by trypsin, but not by thrombin cleavage. Although proteolytic cleavage/activation of PARs 1, 2, and 4 generate intracellular signals, surprisingly, PAR_3 does not cause cell signaling on its own, but in general, appears to act as a synergistic receptor facilitating thrombin-triggered activation of PAR_4 and possibly functioning as a PAR heterodimer [59, 60]. In select circumstances, however, data suggest that PAR3 can signal on its own to activate MAPkinase and to trigger IL-8 release [60]. Thus, the signaling repertoire for PAR_3 acting "autonomously" remains to be fully established. Importantly, all of PARs 1, 2, and 4 can be activated by synthetic peptides that mirror their proteinase-revealed tethered ligands, whereas the synthetic peptide based on the tethered ligand sequence of PAR_3 can activate PARs 1 and 2, but not PAR_3 [62]. Based on the distinct tethered ligand sequences of PARs 1, 2, and 4, it was possible, using cell signaling bioassays as a pharmacological "readout," to develop receptor-selective activating peptides that at appropriate concentrations could activate each of PARs 1, 2, and 4 without affecting the other family members. Similarly, partial or complete reverse-sequence peptides based on the PAR-activating peptides were developed to serve as "control" peptides that cannot activate the PARs (Table 2). The PAR-activating peptides, along with the PAR-inactive "control" peptides, in

Table 2 Proteinases that target PARs for activation or disarming inactivation. The table lists the proteinases that have been evaluated for their ability to activate or disarm PARs 1, 2, and 4. The ability of the dust mite proteinase to activate PAR_2 may depend on the receptor environment, and thus a question mark is thus shown in the table

	Action on		
Proteinase	PAR_1	PAR_2	PAR_4
Thrombin	Activates		Activates
APC	Activates		
TF/VIIa		Activates	
Factor Xa		Activates	
Complement proteinase, MASP-1			Activates
Plasmin	Inactivates	Inactivates	Activates
Tryptase		Activates	
Neutrophil elastase	Inactivates	Inactivates	
Cathepsin-G	Activates	Inactivates	Activates
Proteinase 3	Inactivates	Inactivates	
Derp1	Inactivates	Activates??	
Derp3		Activates	
Derp9		Activates	
KLK1	Inactivates		
KLK5		Activates	
KLK6		Activates	
KLK14	Activates/inactivates	Activates	Activates/inactivates
MMP1	Activates		
Trypsin		Activates	Activates

conjunction with the use of tissues from PAR-null mice, became key tools to use as pharmacological probes for determining the physiological roles that PARs can play either in vitro or in vivo. This pharmacological approach will be outlined in the following section. The effective use of the PAR-activating peptides as probes for PAR function relies heavily on understanding the pharmacology of their actions and employing the use of several PAR-activating peptides along with their PAR-inactive peptide control analogs to rule in or rule out a role for a given PAR in a physiological process.

4.2 *Physiological Roles for PARs: Use of PAR-Activating Peptides, PAR-Null Animals, and PAR siRNA to Assess PAR Function In Vitro and In Vivo*

One of the first studies we did to evaluate a potential physiological role for PAR_2 focused on the ability of trypsin to stimulate intestinal ion transport using an Ussing chamber model assay with intact rat colon tissue. Perhaps ironically for this first use of our PAR-activating peptides in intestinal tissue, our peptide structure–activity data (relative potencies of PAR-activating peptides) indicated that the trypsin-triggered increase in chloride current, mimicked by the PAR_2-activating peptides, involved a "proteinase"-activated receptor that appeared pharmacologically distinct from either PAR_2 or PAR_1 [63]. The identity of that receptor with an unusual structure–activity profile for the PAR_2-activating peptides has yet to be identified. Notwithstanding, with the use of a murine colon test tissue in an Ussing chamber model, our subsequent efforts were able to establish a role for PAR_1 in inhibiting neurally evoked (electrical field stimulation) chloride secretion [64]. This work benefitted from the use of PAR_1 null mice in addition to the PAR_1-selective peptide agonist to establish the role for PAR_1 unequivocally. Thus, this work illustrated the complementary use of the pharmacological and genetic approach to explore a physiological role for the PARs in intact tissue. To establish a role for PARs 1 and 2 in the inflammatory response, we first employed a paw edema model to monitor a "physiological" inflammatory response triggered by thrombin, trypsin, and the PAR-activating peptides [65, 66]. The data showing parallels between the inflammatory actions of the PAR-activating peptides (but not the reverse-sequence PAR-inactive peptides) and the inflammation caused by trypsin and thrombin strongly supported the proposal that the proteinases were causing inflammation via the PARs. Additionally, the work pointed to an anti-inflammatory role for thrombin that could not be attributed to PAR activation at that time [66]. Of interest, the inflammation caused by PAR_2 was found to be in part "neurogenic" [67], implicating a physiological role for neuronal PARs. An inflammatory role for PAR_2 in colonic inflammation was established with the combined use of the PAR_2-activating peptides, the control PAR-inactive peptides, and PAR_2 null mice [68]. This kind of approach has been used in general to validate a physiological role for PARs 1, 2, and 4 in a variety of settings, as outlined by the following paragraphs. Of note, the PAR-selective activating peptides can have actions that are PAR independent [69, 70]. Thus, a careful analysis of the pharmacology of the activating peptides in each system of interest is necessary to validate or not a role for the PARs using the peptide agonists. Other settings in which proteinases, via PAR regulation can modulate tissue function are described in more detail in subsequent chapters. In addition to the lessons that have been learned from a pharmacological analysis of the actions of PAR agonists in vitro or in vivo, studies of the molecular pharmacology of the PARs, in terms of their diverse signaling pathways has provided another clue for understanding the complex effects that PAR activation can have in vivo. The differential regulation of PAR signaling via distinct mechanisms is outlined in the following section.

4.3 Biased Signaling Through the PARs

Early on it was found that PARs can be activated either by proteolytic cleavage of the receptor or by short synthetic tethered-ligand-derived peptides. These peptides have proven to be very valuable tools in studying the signal transduction of individual PAR subtypes. Recently, a number of GPCRs have been shown to exhibit biased signaling, a phenomenon where a receptor selectively couples to different signaling pathways following activation by different ligands [71, 72]. In the case of the PARs, a number of studies have now shown distinct signaling response to PAR activation by peptide agonists or proteinase-mediated TL signaling. Both thrombin and the PAR_1 peptide agonist, SFLLRNP have been shown to stimulate a transient ERK and MAPkinase activation, with only thrombin-activated PAR_1 leading to a sustained activation of MAPkinase [73]. More recently, a study showed that PAR_1 activation with peptide agonists biased the signaling response towards $G\alpha_q$ coupling and calcium signaling, while thrombin stimulated PAR_1 coupled preferentially to $G\alpha_{12/13}$ causing changes in endothelial barrier permeability [74]. These divergent signal transduction responses through PAR_1 are not just limited to the difference between enzyme and peptide activation of the receptors. A number of different proteinases now have been shown to promote PAR_1 coupling to distinct signaling pathways.

Activated Protein-C (APC) and thrombin have divergent effects on endothelial barrier permeability [75–77]. While thrombin acts as a proinflammatory mediator and disrupts endothelial barrier integrity, APC signaling is cytoprotective and enhances the barrier integrity. The exact mechanism by which PAR_1 mediates these divergent effects on barrier integrity is still poorly understood, however recent studies have thrown some light on the mechanisms underlying the differential signaling through PAR_1. Thrombin activation of PAR_1 results in RhoA activation but does not stimulate Rac1 while APC activated PAR_1 strongly increases Rac1 but not RhoA [78]. The location of the signaling events also appears to be important, with caveolae and lipid raft localization shown to be critical for APC activation of PAR_1, but not affecting thrombin-mediated receptor signaling [79–80]. Thus, PAR_1 co-localization with the APC accessory receptor, endothelial protein-C receptor (EPCR), and $G\alpha_{i/o}$ in a membrane microdomain may sequester it from thrombin or stabilize a receptor conformation that favors signaling through APC. The protective signaling through PAR_1 in the setting of sepsis has also been shown to require the expression and transactivation of PAR_2, a receptor that is upregulated by inflammatory stimuli in endothelial cells [81].

Recent reports have also indicated that biased signaling through PAR_1 is triggered by the metalloproteinase, MMP1. MMP1-activated PAR_1 stimulates transcription of different subsets of proangiogenic genes compared to thrombin activation of PAR_1 [82, 83]. MMP1 cleaves PAR_1 at a distinct site upstream of the thrombin cleavage site and strongly activates Rho-GTP pathways, cell shape change and motility, and MAPkinase signaling. Blockade of MMP1-dependent PAR_1 activation suppresses thrombogenesis and prevents thrombosis in animal models of disease [84].

Some more evidence for possible determinants of differential signaling through PAR_1 has come from a series of experiments using resonance energy transfer techniques that have shown that PAR_1 can exist in multiple receptor states, forming preassembled complexes with $G\alpha_{i1}$ but not with $G\alpha 12$. These different receptor states result in distinct kinetics of $G\alpha_{i1}$ activation and $G\alpha_{12}$ recruitment to PAR_1

[85]. These multiple receptor populations coupled to distinct signaling partners could be stabilized by different modes of activation resulting in biased signaling.

Like PAR_1, PAR_2 can also be activated by numerous proteinases including the coagulation proteinases TF-VIIa or TF-VIIa-Xa, mast cell tryptase, tissue kallikreins, and trypsin to name a few. It has, however, not been established whether PAR_2 activation by these different proteinases can result in biased signaling. Recent molecular pharmacological studies have shown that PAR_2 has different determinants for receptor coupling to signaling when the agonists are presented to the receptor either as proteolytically revealed "tethered ligand" sequences or as soluble peptides. The differences in signaling via these two modes of PAR_2 activation indicate that the receptor can exhibit biased signaling [86]. Further, it has been shown that mutations within the proteolytically revealed tethered ligand for PAR_2 disable the receptor for coupling to the $G\alpha_q$-calcium-signaling pathway. Intriguingly, the mutated receptors that cannot trigger elevations in intracellular calcium retain the ability to trigger the ERK-MAPKinase pathway. These data show that, depending on the activating ligand, it is possible for PAR_2 to couple selectively to different signaling pathways downstream of receptor activation. The activation of MAPKinase by the mutated receptors is independent of β-arrestin recruitment but was dependent on Rho-Kinase and represented coupling to $G\alpha_{12/13}$ [28].

PAR signaling can also occur without coupling to a G-protein via receptor interactions with β-arrestins. β-arrestins were first identified as proteins that interact with agonist-activated GPCRs to desensitize signaling [87]. More recently it has emerged that β-arrestins can also act as signaling scaffolds and can mediate signaling through GPCRs independently of receptor interactions with G-proteins [88]. PAR_2 activation by trypsin-like enzymes and by TL-activating peptides recruits β-arrestins to the receptor. This interaction is an important determinant of proper subcellular localization of ERK-MAPkinase [89] and these spatial differences result in the activation of different transcriptional targets. It has also emerged that the PAR_2 stimulated signaling responsible for regulating cell migration and actin assembly through dephosphorylating and activating cofilin, an actin filament-severing protein, is G-protein independent and meditated by β-arrestins [90, 91].

The exact determinants of differential PAR_2 signaling are still unclear, but like PAR_1, the site of membrane localization of PAR_2 could be an important factor. Disruption of lipid rafts or silencing of caveolin-1 can disrupt the ability of TF/VIIa to activate signaling through PAR_2 [92]. This requirement for sequestration in lipid rafts to result in "biased" signaling via PAR_2 may hold true for other proteinases as well.

Thus, emerging evidence for two of the PAR family members (PAR_1 and PAR_2) shows that these receptors are able of coupling to multiple G-proteins as well as signaling in a G-protein-independent manner through interactions with β-arrestin. The ability of PAR_3 or PAR_4 to exhibit biased signaling has not been examined. Biased signaling is thus one of the mechanisms by which these receptors regulate the numerous cellular responses. It is entirely likely that the lessons about biased signaling revealed by the molecular pharmacology studies will lead to an understanding of the diverse effects that an individual PAR can have in vivo, ranging from the stimulation to inhibition of inflammation as has been observed for both PARs 1 and 2.

5 Developing PAR Antagonists: Small Molecule Peptidomimetic Antagonists and PAR-Targeted Antibodies

5.1 *Antagonists for PAR_1*

As for all studies of receptor function, antagonists for the PARs are key reagents for understanding the potential physiological roles the PARs can play. Developing an antagonist for PAR_1 was particularly attractive for therapeutic purposes, since the impact of thrombin on platelets and endothelial cells could in principle be blocked, without affecting the essential coagulation properties of thrombin. Even before the "thrombin receptor" (PAR_1) had been cloned, synthetic peptide analogs were synthesized that were able to block the action of thrombin on platelets, without inhibiting thrombin's proteolytic activity [93, 94]. Although not very potent (K_i values in the 140–450 μM range) these compounds pointed the way for developing PAR antagonists. It was the pharmacology of the PAR-activating peptides, including a substantial amount of structure–activity work in this area [48–51, 53–55, 95], that led to the synthesis of the first thrombin antagonists, which were also of low potency, but like the dipeptides synthesized by Ruda and coworkers [93, 94], these compounds did set the stage for antagonist development [50, 95]. Studies of the amino acid "side-chain" requirements for thrombin receptor agonists [96] led to the synthesis of relatively potent peptide antagonists [97]. It was not fully appreciated at that time, however, that the PAR_1-targeted peptide antagonists could also be PAR_2 agonists [27]. The real success in developing potent and selective PAR_1 antagonists came not from "rational drug design" based on the agonist peptides, but from a high-throughput screening approach looking for lead compounds that blocked either thrombin or PAR-activating peptide-stimulated calcium signaling or platelet aggregation [98–100]. The development of clinically useful PAR_1 antagonists is outlined in the following paragraphs.

The first promising PAR_1 antagonist reported was RJW-56110 (Fig. 3), which was developed by Andrade-Gordon and colleagues [98]. This indole peptidomimetic was shown to be a potent, selective PAR_1 antagonist, devoid of PAR_1 agonist and thrombin inhibitory activity. It binds to PAR_1, interferes with PAR_1 calcium mobilization and cellular function (platelet aggregation; cell proliferation), and has no effect on PAR_2, PAR_3, or PAR_4. The starting point for the design of this compound was based on computational modeling of a range of distances between the key ammonium, phenyl, and guanidinium groups in the agonist peptide SFLLRN-NH_2 (doubly protonated form). The next generation of PAR_1 antagonists

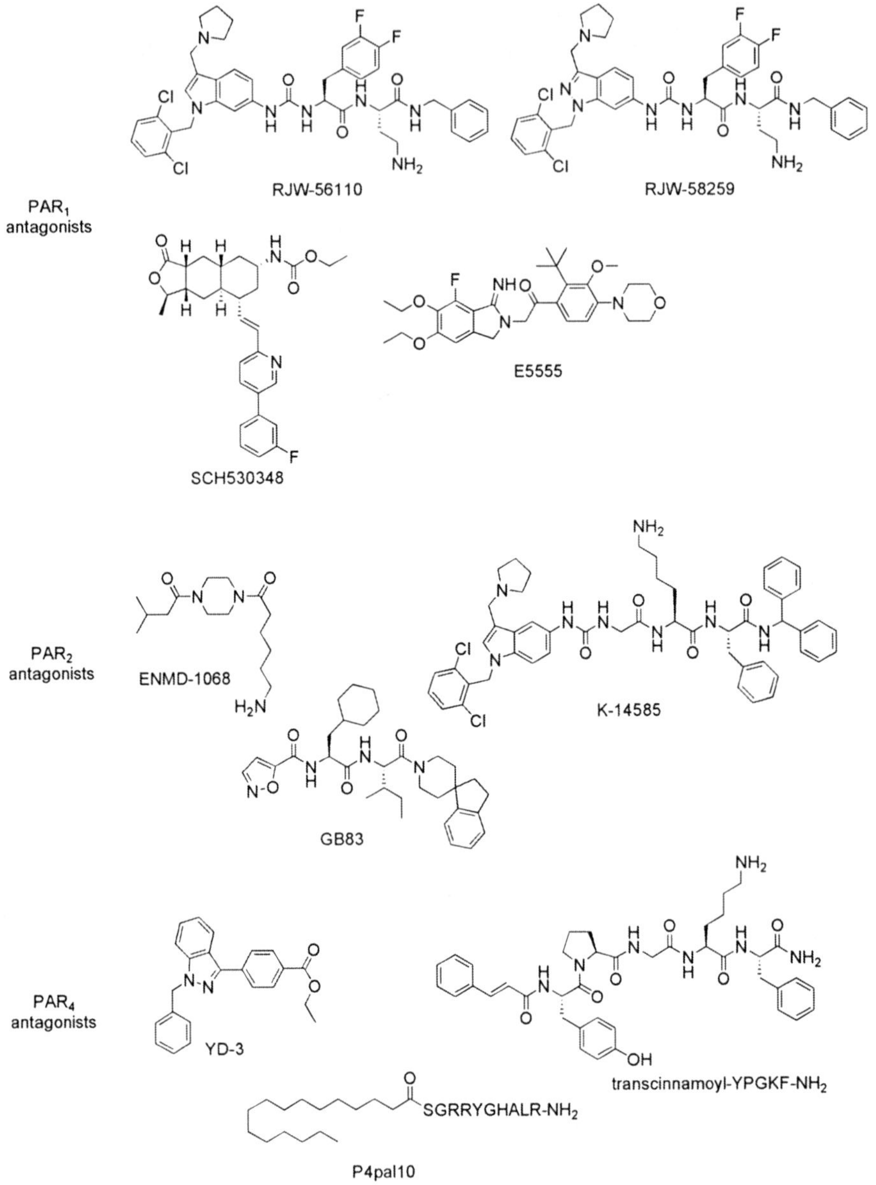

Fig. 3 PAR antagonists. The structures of nonpeptide and peptide antagonists for PARs 1, 2, and 4 are shown, as discussed in the text

led to the indazole-based compound RJW-58259 (Fig. 2), which was shown to have an improved in vivo cardiovascular safety profile [101].

Another series of PAR_1 antagonists have been based on the natural product himbacine. The development of these compounds has culminated in the compound

SCH530348 which is a synthetic tricyclic 3-phenylpyridine based molecule [100, 102]. SCH530348 is currently in late phase clinical trials where it shows efficacy in reducing major adverse coronary events with little bleeding liability in patients with acute coronary syndrome [103]. Phase III studies are ongoing (1) to assess the efficacy of this drug in improving clinical outcomes for patients with non-ST-segment elevation myocardial infarction (thrombin receptor antagonist for clinical event reduction: TRA-CER Trial) (Clinical trial registration number: NCT00527943) [104] or (2) to evaluate the drug's effect in patients with a documented history of atherosclerotic disease (TRA 2°P)-TIMI 50 [105].

Another compound, E5555, chemically identified as 1-(3-tert-Butyl-4-methoxy-5-morpholinophenyl)-2-(5,6-di-ethoxy-7-fluoro-1-imino-1,3-dihydro-2*H*-isoindol-2yl)ethanone hydrobromide, is reported to be in phase II clinical trials. This compound inhibits thrombin receptor activating peptide (TRAP) binding to PAR_1 and inhibits thrombin-stimulated platelet aggregation with an IC_{50} of 0.064 μM [106].

5.2 *Antagonists for PARs 2 and 4*

There has been less success with developing PAR_2 antagonists, despite the synthesis of a peptide-based PAR_2 antagonist that blocks trypsin, but not PAR_2-activating peptide triggering of PAR_2 [107] and a relatively low potency nonpeptide mimetic that blocks activation of the receptor by a PAR_2-activating peptide and by trypsin [108]. The peptides FSLLRY-NH_2 and LSIGRL-NH_2 are able to inhibit trypsin-stimulated calcium signaling via PAR_2 in human HEK cells and in rat PAR_2 expressing KNRK cells with reported IC_{50} values of 50 and 200 μM, respectively. These compounds, however, do not inhibit activation of the receptor by PAR_2-activating peptides. N1-3-methylbutyryl-N4-6-aminohexanoyl-piperazine (ENMD-1068) is a nonpeptide antagonist of PAR_2 that has shown promise in attenuating inflammation in a mouse model of arthritis [108], but has a low potency.

Two recent reports describe a number of peptidomimetic antagonists of PAR_2 that can block PAR_2 activating peptide-stimulated calcium signaling and NFκB reporter activity. One series of compounds compete with low micromolar K_i values for the receptor binding of a high affinity PAR_2 radio-ligand [109]. One of these, N-[1-(2,6-dichlorophenyl)methyl]-3-(1-pyrrolidinylmethyl)-1*H*-indol-5-yl]aminocarbonyl]-glycinyl-L-lysinyl-L-phenylalanyl-N-benzhydrylamide (K-14585) [110] was shown further to have some unusual agonist–antagonist properties. K1485 inhibited PAR_2 activating peptide-stimulated calcium signaling but not ERK signaling. Conversely, the compound also stimulated p38 MAPK signaling through PAR_2 independently of $G\alpha_q$ coupling. K14585 could further inhibit the phosphorylation and DNA binding of the inflammatory transcription factor, p65 NFκB, but also stimulated a NF-κB reporter signal to the same level as PAR_2 activating peptide. These interesting complex actions of K14585 that illustrate an unusual case of "biased" PAR_2 signaling merit further study. Finally, a second series of

PAR_2 antagonists have been described [111], one of which (GB83: 5-Isoxazoyl-Cha-Ile-spiroindane-1,40-piperidine) is able to inhibit PAR_2 activation by both proteases and other PAR_2 agonists with an IC_{50} of about 2 μM. It remains to be seen if these newly described PAR2 antagonists will lead to compounds of therapeutic utility.

Although the development of small molecule PAR_2 antagonists for clinical use in vivo has not yet been successful, the use of receptor-targeted antisera that block activation of the receptor by proteinases shows promise. Thus, in a murine joint inflammation model, the use of a polyclonal or monoclonal antibody that targets the PAR_2 cleavage-activation site, thereby inhibiting tryptic activation has been shown to diminish the inflammatory response [108]. In keeping with the clinical use of targeted monoclonal antibodies to treat inflammatory diseases, employing a PAR_2-targeted monoclonal that can prevent the activation of PAR_2 by multiple serine proteinases would be of considerable value.

PAR_4 antagonist development is also in its infancy. Very few compounds have been described, with the most promising one being YD-3 [1-benzyl-3(ethoxycarbonylphenyl)-indazole] which inhibits thrombin stimulated PAR_4 activation with an IC_{50} of 30 μM [112, 113]. It is possible that YD3 may affect PAR_4 in a species-specific way (e.g., targeting human but not rodent PAR_4). The inhibition of rodent platelet PAR_4 to block thrombin or PAR_4-activating peptide aggregation has also been described with the tethered ligand-derived peptide, trans-cinnamoyl-YPGKF-NH_2 [114]. The use of this antagonist to block PAR_4 on human platelets has yet to be studied. Another novel strategy for inhibition of PAR_4 is through the use of so-called pepducins, that are N-palmitoylated peptides of 7–12 amino acids with sequences matching the third intracellular loops of the PARs or other G-protein-coupled receptors. These lipid-conjugated peptides are thought to act as cell-penetrating "dominant-negative" inhibitors of the intracellular interactions between the PARs and their cognate G-protein signaling partners [115]. The PAR_4 antagonist pepducin P4pal-10 (N-palmitoyl-SGRRYGHALR-NH_2) is able to prolong bleeding time and prevent systemic platelet activation in mice [115, 116].

6 Summary and Looking to the Future

The ability of proteinases to generate hormone-like signals in cells either directly by regulating cell surface receptors or indirectly by generating peptide hormone agonists adds a novel dimension to the many biological roles that proteinases can play. Although proteinases can signal by a variety of mechanisms as outlined in Fig. 2, it is signaling via the PARs (Fig. 1) that has recently captured the imagination of those working in the area of signal transduction. Given the ability of coagulation pathway proteinases to signal via the PARs, the enzymes provide an essential link between the clotting system and the innate immune response responsible for triggering pain and inflammation. To understand this physiological role of proteinases, a pharmacological approach has proved of considerable utility.

This approach has included (1) bioassays done both in vitro and in vivo using a number of animal models ranging from intestinal to joint inflammation, (2) employing a "structure–activity" approach for evaluating the activation of PARs by both tethered and soluble peptide sequences, (3) the use of receptor antagonists, and (4) the use of PAR_1-PAR_2-PAR_4 null animals. This pharmacological approach, combined with a molecular pharmacological approach has led to the development of therapeutically targeted PAR_1 antagonists that show promise of complementing the traditional anticoagulant therapies which have been used to date in the setting of cardiovascular disease. Further, an in-depth study of the molecular pharmacology of the PARs themselves has revealed unique receptor characteristics that also have both physiological and therapeutic implications. In particular, the "biased" signaling via PAR_1, depending on its activation by either thrombin or APC, is of note. Since PAR_1 can clearly activate unique responses that are either inflammatory or anti-inflammatory, the design of agents that can trigger either of these responses selectively by PAR_1 will have therapeutic utility in the future. A similar scenario can be predicted for PAR_2 that can exhibit "biased" signaling, depending on its interactions (or not) with β-arrestins. In sum, the pharmacological approach to studying the roles of proteinases in inflammation, with a focus on the PARs, has led in directions that could not have been predicted when PAR_1 was first cloned. This area of investigation, using the pharmacological approaches outlined in this chapter shows great promise of teaching us many new lessons in the future about the roles of proteinases and their receptors in inflammation.

Acknowledgements Work in the authors' laboratory is supported by grants from the Canadian Institutes of Health Research. KKH and RR were supported by postdoctoral fellowships from the Alberta Heritage Foundation for Medical Research (currently titled: Alberta Innovates Health Solutions).

References

1. Puente XS, Sanchez LM, Gutierrez-Fernandez A, Velasco G, Lopez-Otin C (2005) A genomic view of the complexity of mammalian proteolytic systems. Biochem Soc Trans 33:331–334
2. Werle E, Erdos EG (1954) A new hypotensive, enterotropic and hysterotropic substance in human urine. Naunyn Schmiedebergs Arch Exp Pathol Pharmakol 223:234–243
3. Erdos EG (2002) Kinins, the long march–a personal view. Cardiovasc Res 54:485–491
4. Steiner DF, Cunningham D, Spigelman L, Aten B (1967) Insulin biosynthesis: evidence for a precursor. Science 157:697–700
5. Rieser P, Rieser CH (1964) Anabolic responses of diaphragm muscle to insulin and to other pancreatic proteins. Proc Soc Exp Biol Med 116:669–671
6. Rieser P (1967) The insulin-like action of pepsin and pepsinogen. Acta Endocrinol (Copenh) 54:375–379
7. Kono T, Barham FW (1971) Insulin-like effects of trypsin on fat cells. Localization of the metabolic steps and the cellular site affected by the enzyme. J Biol Chem 246:6204–6209
8. Shoelson SE, White MF, Kahn CR (1988) Tryptic activation of the insulin receptor. Proteolytic truncation of the alpha-subunit releases the beta-subunit from inhibitory control. J Biol Chem 263:4852–4860

9. Burger MM (1970) Proteolytic enzymes initiating cell division and escape from contact inhibition of growth. Nature 227:170–171
10. Sefton BM, Rubin H (1970) Release from density dependent growth inhibition by proteolytic enzymes. Nature 227:843–845
11. Chen LB, Buchanan JM (1975) Mitogenic activity of blood components. I. Thrombin and prothrombin. Proc Natl Acad Sci USA 72:131–135
12. Carney DH, Cunningham DD (1977) Initiation of check cell division by trypsin action at the cell surface. Nature 268:602–606
13. Carney DH, Cunningham DD (1978) Transmembrane action of thrombin initiates chick cell division. J Supramol Struct 9:337–350
14. Lopez-Otin C, Overall CM (2002) Protease degradomics: a new challenge for proteomics. Nat Rev Mol Cell Biol 3:509–519
15. Quesada V, Ordonez GR, Sanchez LM, Puente XS, Lopez-Otin C (2009) The Degradome database: mammalian proteases and diseases of proteolysis. Nucleic Acids Res 37: D239–D243
16. Puente XS, Lopez-Otin C (2004) A genomic analysis of rat proteases and protease inhibitors. Genome Res 14:609–622
17. Kessenbrock K, Plaks V, Werb Z (2010) Matrix metalloproteinases: regulators of the tumor microenvironment. Cell 141:52–67
18. Hawthorne S, Hamilton R, Walker BJ, Walker B (2004) Utilization of biotinylated diphenyl phosphonates for disclosure of serine proteases. Anal Biochem 326:273–275
19. Pan Z, Jeffery DA, Chehade K, Beltman J, Clark JM, Grothaus P, Bogyo M, Baruch A (2006) Development of activity-based probes for trypsin-family serine proteases. Bioorg Med Chem Lett 16:2882–2885
20. Fonovic M, Bogyo M (2008) Activity-based probes as a tool for functional proteomic analysis of proteases. Expert Rev Proteomics 5:721–730
21. Paulick MG, Bogyo M (2008) Application of activity-based probes to the study of enzymes involved in cancer progression. Curr Opin Genet Dev 18:97–106
22. Kato D, Boatright KM, Berger AB, Nazif T, Blum G, Ryan C, Chehade KA, Salvesen GS, Bogyo M (2005) Activity-based probes that target diverse cysteine protease families. Nat Chem Biol 1:33–38
23. Hansen KK, Sherman PM, Cellars L, Andrade-Gordon P, Pan Z, Baruch A, Wallace JL, Hollenberg MD, Vergnolle N (2005) A major role for proteolytic activity and proteinase-activated receptor-2 in the pathogenesis of infectious colitis. Proc Natl Acad Sci USA 102:8363–8368
24. Matthews DJ, Wells JA (1993) Substrate phage: selection of protease substrates by monovalent phage display. Science 260:1113–1117
25. Sharma N, Oikonomopoulou K, Ito K, Renaux B, Diamandis EP, Hollenberg MD, Rancourt DE (2008) Substrate specificity determination of mouse implantation serine proteinase and human kallikrein-related peptidase 6 by phage display. Biol Chem 389:1097–1105
26. Gioia M, Foster LJ, Overall CM (2009) Cell-based identification of natural substrates and cleavage sites for extracellular proteases by SILAC proteomics. Methods Mol Biol 539:131–153
27. Kawabata A, Saifeddine M, Al-Ani B, Leblond L, Hollenberg MD (1999) Evaluation of proteinase-activated receptor-1 (PAR1) agonists and antagonists using a cultured cell receptor desensitization assay: activation of PAR2 by PAR1-targeted ligands. J Pharmacol Exp Ther 288:358–370
28. Ramachandran R, Mihara K, Mathur M, Rochdi MD, Bouvier M, Defea K, Hollenberg MD (2009) Agonist-biased signaling via proteinase activated receptor-2: differential activation of calcium and mitogen-activated protein kinase pathways. Mol Pharmacol 76:791–801
29. Oikonomopoulou K, Batruch I, Smith CR, Soosaipillai A, Diamandis EP, Hollenberg MD (2010) Functional proteomics of kallikrein-related peptidases in ovarian cancer ascites fluid. Biol Chem 391:381–390

30. Oikonomopoulou K, Hansen KK, Baruch A, Hollenberg MD, Diamandis EP (2008) Immunofluorometric activity-based probe analysis of active KLK6 in biological fluids. Biol Chem 389:747–756
31. Knecht W, Cottrell GS, Amadesi S, Mohlin J, Skaregarde A, Gedda K, Peterson A, Chapman K, Hollenberg MD, Vergnolle N et al (2007) Trypsin IV or mesotrypsin and p23 cleave protease-activated receptors 1 and 2 to induce inflammation and hyperalgesia. J Biol Chem 282:26089–26100
32. Cuatrecasas P (1971) Properties of the insulin receptor of isolated fat cell membranes. J Biol Chem 246:7265–7274
33. Cuatrecasas P (1969) Interaction of insulin with the cell membrane: the primary action of insulin. Proc Natl Acad Sci USA 63:450–457
34. Magnaldo I, Pouyssegur J, Paris S (1988) Thrombin exerts a dual effect on stimulated adenylate cyclase in hamster fibroblasts, an inhibition via a GTP-binding protein and a potentiation via activation of protein kinase C. Biochem J 253:711–719
35. Rasmussen UB, Vouret-Craviari V, Jallat S, Schlesinger Y, Pages G, Pavirani A, Lecocq JP, Pouyssegur J, Van Obberghen-Schilling E (1991) cDNA cloning and expression of a hamster alpha-thrombin receptor coupled to Ca2+ mobilization. FEBS Lett 288:123–128
36. Vu TK, Hung DT, Wheaton VI, Coughlin SR (1991) Molecular cloning of a functional thrombin receptor reveals a novel proteolytic mechanism of receptor activation. Cell 64:1057–1068
37. Molino M, Blanchard N, Belmonte E, Tarver AP, Abrams C, Hoxie JA, Cerletti C, Brass LF (1995) Proteolysis of the human platelet and endothelial cell thrombin receptor by neutrophil-derived cathepsin G. J Biol Chem 270:11168–11175
38. Renesto P, Si-Tahar M, Moniatte M, Balloy V, Van Dorsselaer A, Pidard D, Chignard M (1997) Specific inhibition of thrombin-induced cell activation by the neutrophil proteinases elastase, cathepsin G, and proteinase 3: evidence for distinct cleavage sites within the aminoterminal domain of the thrombin receptor. Blood 89:1944–1953
39. Schechter NM, Brass LF, Lavker RM, Jensen PJ (1998) Reaction of mast cell proteases tryptase and chymase with protease activated receptors (PARs) on keratinocytes and fibroblasts. J Cell Physiol 176:365–373
40. Nakayama T, Hirano K, Shintani Y, Nishimura J, Nakatsuka A, Kuga H, Takahashi S, Kanaide H (2003) Unproductive cleavage and the inactivation of protease-activated receptor-1 by trypsin in vascular endothelial cells. Br J Pharmacol 138:121–130
41. Nakayama T, Hirano K, Hirano M, Nishimura J, Kuga H, Nakamura K, Takahashi S, Kanaide H (2004) Inactivation of protease-activated receptor-1 by proteolytic removal of the ligand region in vascular endothelial cells. Biochem Pharmacol 68:23–32
42. Kuliopulos A, Covic L, Seeley SK, Sheridan PJ, Helin J, Costello CE (1999) Plasmin desensitization of the PAR1 thrombin receptor: kinetics, sites of truncation, and implications for thrombolytic therapy. Biochemistry 38:4572–4585
43. Dulon S, Cande C, Bunnett NW, Hollenberg MD, Chignard M, Pidard D (2003) Proteinase-activated receptor-2 and human lung epithelial cells: disarming by neutrophil serine proteinases. Am J Respir Cell Mol Biol 28:339–346
44. Cumashi A, Ansuini H, Celli N, De Blasi A, O'Brien PJ, Brass LF, Molino M (2001) Neutrophil proteases can inactivate human PAR3 and abolish the co-receptor function of PAR3 on murine platelets. Thromb Haemost 85:533–538
45. Sambrano GR, Huang W, Faruqi T, Mahrus S, Craik C, Coughlin SR (2000) Cathepsin G activates protease-activated receptor-4 in human platelets. J Biol Chem 275:6819–6823
46. Ossovskaya VS, Bunnett NW (2004) Protease-activated receptors: contribution to physiology and disease. Physiol Rev 84:579–621
47. Compton SJ, Renaux B, Wijesuriya SJ, Hollenberg MD (2001) Glycosylation and the activation of proteinase-activated receptor 2 (PAR(2)) by human mast cell tryptase. Br J Pharmacol 134:705–718

48. Hollenberg MD, Yang SG, Laniyonu AA, Moore GJ, Saifeddine M (1992) Action of thrombin receptor polypeptide in gastric smooth muscle: identification of a core pentapeptide retaining full thrombin-mimetic intrinsic activity. Mol Pharmacol 42:186–191
49. Scarborough RM, Naughton MA, Teng W, Hung DT, Rose J, Vu TK, Wheaton VI, Turck CW, Coughlin SR (1992) Tethered ligand agonist peptides. Structural requirements for thrombin receptor activation reveal mechanism of proteolytic unmasking of agonist function. J Biol Chem 267:13146–13149
50. Vassallo RR Jr, Kieber-Emmons T, Cichowski K, Brass LF (1992) Structure-function relationships in the activation of platelet thrombin receptors by receptor-derived peptides. J Biol Chem 267:6081–6085
51. Sabo T, Gurwitz D, Motola L, Brodt P, Barak R, Elhanaty E (1992) Structure-activity studies of the thrombin receptor activating peptide. Biochem Biophys Res Commun 188:604–610
52. Ahlquist RP (1948) A study of the adrenotropic receptors. Am J Physiol 153:586–600
53. Hollenberg MD, Laniyonu AA, Saifeddine M, Moore GJ (1993) Role of the amino- and carboxyl-terminal domains of thrombin receptor-derived polypeptides in biological activity in vascular endothelium and gastric smooth muscle: evidence for receptor subtypes. Mol Pharmacol 43:921–930
54. Hollenberg MD, Saifeddine M, al-Ani B, Kawabata A (1997) Proteinase-activated receptors: structural requirements for activity, receptor cross-reactivity, and receptor selectivity of receptor-activating peptides. Can J Physiol Pharmacol 75:832–841
55. Laniyonu AA, Hollenberg MD (1995) Vascular actions of thrombin receptor-derived polypeptides: structure-activity profiles for contractile and relaxant effects in rat aorta. Br J Pharmacol 114:1680–1686
56. Kinlough-Rathbone RL, Perry DW, Guccione MA, Rand ML, Packham MA (1993) Degranulation of human platelets by the thrombin receptor peptide SFLLRN: comparison with degranulation by thrombin. Thromb Haemost 70:1019–1023
57. Coughlin SR (2005) Protease-activated receptors in hemostasis, thrombosis and vascular biology. J Thromb Haemost 3:1800–1814
58. Hollenberg MD, Compton SJ (2002) International Union of Pharmacology XXVIII. Proteinase-activated receptors. Pharmacol Rev 54:203–217
59. McLaughlin JN, Patterson MM, Malik AB (2007) Protease-activated receptor-3 (PAR3) regulates PAR1 signaling by receptor dimerization. Proc Natl Acad Sci USA 104:5662–5667
60. Ostrowska E, Reiser G (2008) The protease-activated receptor-3 (PAR-3) can signal autonomously to induce interleukin-8 release. Cell Mol Life Sci 65:970–981
61. Kahn ML, Zheng YW, Huang W, Bigornia V, Zeng D, Moff S, Farese RV Jr, Tam C, Coughlin SR (1998) A dual thrombin receptor system for platelet activation. Nature 394:690–694
62. Hansen KK, Saifeddine M, Hollenberg MD (2004) Tethered ligand-derived peptides of proteinase-activated receptor 3 (PAR3) activate PAR1 and PAR2 in Jurkat T cells. Immunology 112:183–190
63. Vergnolle N, Macnaughton WK, Al-Ani B, Saifeddine M, Wallace JL, Hollenberg MD (1998) Proteinase-activated receptor 2 (PAR2)-activating peptides: identification of a receptor distinct from PAR2 that regulates intestinal transport. Proc Natl Acad Sci USA 95:7766–7771
64. Buresi MC, Vergnolle N, Sharkey KA, Keenan CM, Andrade-Gordon P, Cirino G, Cirillo D, Hollenberg MD, MacNaughton WK (2005) Activation of proteinase-activated receptor-1 inhibits neurally evoked chloride secretion in the mouse colon in vitro. Am J Physiol Gastrointest Liver Physiol 288:G337–G345
65. Vergnolle N, Hollenberg MD, Sharkey KA, Wallace JL (1999) Characterization of the inflammatory response to proteinase-activated receptor-2 (PAR2)-activating peptides in the rat paw. Br J Pharmacol 127:1083–1090

66. Vergnolle N, Hollenberg MD, Wallace JL (1999) Pro- and anti-inflammatory actions of thrombin: a distinct role for proteinase-activated receptor-1 (PAR1). Br J Pharmacol 126:1262–1268
67. Steinhoff M, Vergnolle N, Young SH, Tognetto M, Amadesi S, Ennes HS, Trevisani M, Hollenberg MD, Wallace JL, Caughey GH et al (2000) Agonists of proteinase-activated receptor 2 induce inflammation by a neurogenic mechanism. Nat Med 6:151–158
68. Cenac N, Coelho AM, Nguyen C, Compton S, Andrade-Gordon P, MacNaughton WK, Wallace JL, Hollenberg MD, Bunnett NW, Garcia-Villar R et al (2002) Induction of intestinal inflammation in mouse by activation of proteinase-activated receptor-2. Am J Pathol 161:1903–1915
69. Hollenberg MD, Saifeddine M, Sandhu S, Houle S, Vergnolle N (2004) Proteinase-activated receptor-4: evaluation of tethered ligand-derived peptides as probes for receptor function and as inflammatory agonists in vivo. Br J Pharmacol 143:443–454
70. McGuire JJ, Dai J, Andrade-Gordon P, Triggle CR, Hollenberg MD (2002) Proteinase-activated receptor-2 (PAR2): vascular effects of a PAR2-derived activating peptide via a receptor different than PAR2. J Pharmacol Exp Ther 303:985–992
71. Kenakin T (2007) Collateral efficacy in drug discovery: taking advantage of the good (allosteric) nature of 7TM receptors. Trends Pharmacol Sci 28:407–415
72. Urban JD, Clarke WP, von Zastrow M, Nichols DE, Kobilka B, Weinstein H, Javitch JA, Roth BL, Christopoulos A, Sexton PM et al (2007) Functional selectivity and classical concepts of quantitative pharmacology. J Pharmacol Exp Ther 320:1–13
73. Vouret-Craviari V, Van Obberghen-Schilling E, Scimeca JC, Van Obberghen E, Pouyssegur J (1993) Differential activation of p44mapk (ERK1) by alpha-thrombin and thrombin-receptor peptide agonist. Biochem J 289(Pt 1):209–214
74. McLaughlin JN, Shen L, Holinstat M, Brooks JD, Dibenedetto E, Hamm HE (2005) Functional selectivity of G protein signaling by agonist peptides and thrombin for the protease-activated receptor-1. J Biol Chem 280:25048–25059
75. Riewald M, Ruf W (2005) Protease-activated receptor-1 signaling by activated protein C in cytokine-perturbed endothelial cells is distinct from thrombin signaling. J Biol Chem 280:19808–19814
76. Feistritzer C, Riewald M (2005) Endothelial barrier protection by activated protein C through PAR1-dependent sphingosine 1-phosphate receptor-1 crossactivation. Blood 105:3178–3184
77. Finigan JH, Dudek SM, Singleton PA, Chiang ET, Jacobson JR, Camp SM, Ye SQ, Garcia JG (2005) Activated protein C mediates novel lung endothelial barrier enhancement: role of sphingosine 1-phosphate receptor transactivation. J Biol Chem 280:17286–17293
78. Russo A, Soh UJ, Paing MM, Arora P, Trejo J (2009) Caveolae are required for protease-selective signaling by protease-activated receptor-1. Proc Natl Acad Sci USA 106:6393–6397
79. Bae JS, Yang L, Rezaie AR (2007) Receptors of the protein C activation and activated protein C signaling pathways are colocalized in lipid rafts of endothelial cells. Proc Natl Acad Sci USA 104:2867–2872
80. Bae JS, Yang L, Rezaie AR (2008) Lipid raft localization regulates the cleavage specificity of protease activated receptor 1 in endothelial cells. J Thromb Haemost 6:954–961
81. Kaneider NC, Leger AJ, Agarwal A, Nguyen N, Perides G, Derian C, Covic L, Kuliopulos A (2007) 'Role reversal' for the receptor PAR1 in sepsis-induced vascular damage. Nat Immunol 8:1303–1312
82. Blackburn JS, Brinckerhoff CE (2008) Matrix metalloproteinase-1 and thrombin differentially activate gene expression in endothelial cells via PAR-1 and promote angiogenesis. Am J Pathol 173:1736–1746
83. Blackburn JS, Liu I, Coon CI, Brinckerhoff CE (2009) A matrix metalloproteinase-1/protease activated receptor-1 signaling axis promotes melanoma invasion and metastasis. Oncogene 28:4237–4248

84. Trivedi V, Boire A, Tchernychev B, Kaneider NC, Leger AJ, O'Callaghan K, Covic L, Kuliopulos A (2009) Platelet matrix metalloprotease-1 mediates thrombogenesis by activating PAR1 at a cryptic ligand site. Cell 137:332–343
85. Ayoub MA, Trinquet E, Pfleger KD, Pin JP (2010) Differential association modes of the thrombin receptor PAR1 with G{alpha}i1, G{alpha}12, and {beta}-arrestin 1. FASEB J 24(9):3522–3535
86. Al-Ani B, Hansen KK, Hollenberg MD (2004) Proteinase-activated receptor-2: key role of amino-terminal dipeptide residues of the tethered ligand for receptor activation. Mol Pharmacol 65:149–156
87. Lohse MJ, Benovic JL, Codina J, Caron MG, Lefkowitz RJ (1990) Beta-Arrestin: a protein that regulates beta-adrenergic receptor function. Science 248:1547–1550
88. DeWire SM, Ahn S, Lefkowitz RJ, Shenoy SK (2007) Beta-arrestins and cell signaling. Annu Rev Physiol 69:483–510
89. DeFea KA, Zalevsky J, Thoma MS, Dery O, Mullins RD, Bunnett NW (2000) Beta-arrestin-dependent endocytosis of proteinase-activated receptor 2 is required for intracellular targeting of activated ERK1/2. J Cell Biol 148:1267–1281
90. Zoudilova M, Kumar P, Ge L, Wang P, Bokoch GM, Defea KA (2007) Beta-Arrestin-dependent regulation of the Cofilin Pathway downstream of protease-activated receptor-2. J Biol Chem 282:20634–20646
91. Zoudilova M, Min J, Richards HL, Carter D, Huang T, DeFea KA (2010) Beta-Arrestins scaffold cofilin with chronophin to direct localized actin filament severing and membrane protrusions downstream of protease-activated receptor-2. J Biol Chem 285:14318–14329
92. Awasthi V, Mandal SK, Papanna V, Rao LV, Pendurthi UR (2007) Modulation of tissue factor-factor VIIa signaling by lipid rafts and caveolae. Arterioscler Thromb Vasc Biol 27:1447–1455
93. Ruda EM, Petty A, Scrutton MC, Tuffin DP, Manley PW (1988) Identification of small peptide analogues having agonist and antagonist activity at the platelet thrombin receptor. Biochem Pharmacol 37:2417–2426
94. Ruda EM, Scrutton MC, Manley PW, Tuffin DP (1990) Thrombin receptor antagonists. Structure-activity relationships for the platelet thrombin receptor and effects on prostacyclin synthesis by human umbilical vein endothelial cells. Biochem Pharmacol 39:373–381
95. Rasmussen UB, Gachet C, Schlesinger Y, Hanau D, Ohlmann P, Van Obberghen-Schilling E, Pouyssegur J, Cazenave JP, Pavirani A (1993) A peptide ligand of the human thrombin receptor antagonizes alpha-thrombin and partially activates platelets. J Biol Chem 268:14322–14328
96. Natarajan S, Riexinger D, Peluso M, Seiler SM (1995) 'Tethered ligand' derived pentapeptide agonists of thrombin receptor: a study of side chain requirements for human platelet activation and GTPase stimulation. Int J Pept Protein Res 45:145–151
97. Bernatowicz MS, Klimas CE, Hartl KS, Peluso M, Allegretto NJ, Seiler SM (1996) Development of potent thrombin receptor antagonist peptides. J Med Chem 39:4879–4887
98. Andrade-Gordon P, Maryanoff BE, Derian CK, Zhang HC, Addo MF, Darrow AL, Eckardt AJ, Hoekstra WJ, McComsey DF, Oksenberg D et al (1999) Design, synthesis, and biological characterization of a peptide-mimetic antagonist for a tethered-ligand receptor. Proc Natl Acad Sci USA 96:12257–12262
99. Ahn HS, Arik L, Boykow G, Burnett DA, Caplen MA, Czarniecki M, Domalski MS, Foster C, Manna M, Stamford AW et al (1999) Structure-activity relationships of pyrroloquinazolines as thrombin receptor antagonists. Bioorg Med Chem Lett 9:2073–2078
100. Chackalamannil S, Wang Y, Greenlee WJ, Hu Z, Xia Y, Ahn HS, Boykow G, Hsieh Y, Palamanda J, Agans-Fantuzzi J et al (2008) Discovery of a novel, orally active himbacine-based thrombin receptor antagonist (SCH 530348) with potent antiplatelet activity. J Med Chem 51:3061–3064
101. Zhang HC, Derian CK, Andrade-Gordon P, Hoekstra WJ, McComsey DF, White KB, Poulter BL, Addo MF, Cheung WM, Damiano BP et al (2001) Discovery and optimization

of a novel series of thrombin receptor (par-1) antagonists: potent, selective peptide mimetics based on indole and indazole templates. J Med Chem 44:1021–1024
102. Chackalamannil S, Xia Y, Greenlee WJ, Clasby M, Doller D, Tsai H, Asberom T, Czarniecki M, Ahn HS, Boykow G et al (2005) Discovery of potent orally active thrombin receptor (protease activated receptor 1) antagonists as novel antithrombotic agents. J Med Chem 48:5884–5887
103. Goto S, Yamaguchi T, Ikeda Y, Kato K, Yamaguchi H, Jensen P (2010) Safety and exploratory efficacy of the novel thrombin receptor (PAR-1) antagonist SCH530348 for non-ST-segment elevation acute coronary syndrome. J Atheroscler Thromb 17:156–164
104. TRA*CER EaSC (2009) The Thrombin Receptor Antagonist for Clinical Event Reduction in Acute Coronary Syndrome (TRA*CER) trial: study design and rationale. Am Heart J 158 (327–334):e324
105. Morrow DA, Scirica BM, Fox KA, Berman G, Strony J, Veltri E, Bonaca MP, Fish P, McCabe CH, Braunwald E (2009) Evaluation of a novel antiplatelet agent for secondary prevention in patients with a history of atherosclerotic disease: design and rationale for the Thrombin-Receptor Antagonist in Secondary Prevention of Atherothrombotic Ischemic Events (TRA 2 degrees P)-TIMI 50 trial. Am Heart J 158(335–341):e333
106. Serebruany VL, Kogushi M, Dastros-Pitei D, Flather M, Bhatt DL (2009) The in-vitro effects of E5555, a protease-activated receptor (PAR)-1 antagonist, on platelet biomarkers in healthy volunteers and patients with coronary artery disease. Thromb Haemost 102:111–119
107. Al-Ani B, Saifeddine M, Wijesuriya SJ, Hollenberg MD (2002) Modified proteinase-activated receptor-1 and -2 derived peptides inhibit proteinase-activated receptor-2 activation by trypsin. J Pharmacol Exp Ther 300:702–708
108. Kelso EB, Lockhart JC, Hembrough T, Dunning L, Plevin R, Hollenberg MD, Sommerhoff CP, McLean JS, Ferrell WR (2006) Therapeutic promise of proteinase-activated receptor-2 antagonism in joint inflammation. J Pharmacol Exp Ther 316:1017–1024
109. Kanke T, Kabeya M, Kubo S, Kondo S, Yasuoka K, Tagashira J, Ishiwata H, Saka M, Furuyama T, Nishiyama T et al (2009) Novel antagonists for proteinase-activated receptor 2: inhibition of cellular and vascular responses in vitro and in vivo. Br J Pharmacol 158:361–371
110. Goh FG, Ng PY, Nilsson M, Kanke T, Plevin R (2009) Dual effect of the novel peptide antagonist K-14585 on proteinase-activated receptor-2-mediated signalling. Br J Pharmacol 158:1695–1704
111. Barry GD, Suen JY, Le GT, Cotterell A, Reid RC, Fairlie DP (2010) Novel agonists and antagonists for human protease activated receptor 2. J Med Chem 53:7428–7440
112. Wu CC, Huang SW, Hwang TL, Kuo SC, Lee FY, Teng CM (2000) YD-3, a novel inhibitor of protease-induced platelet activation. Br J Pharmacol 130:1289–1296
113. Wu CC, Hwang TL, Liao CH, Kuo SC, Lee FY, Lee CY, Teng CM (2002) Selective inhibition of protease-activated receptor 4-dependent platelet activation by YD-3. Thromb Haemost 87:1026–1033
114. Hollenberg MD, Saifeddine M (2001) Proteinase-activated receptor 4 (PAR4): activation and inhibition of rat platelet aggregation by PAR4-derived peptides. Can J Physiol Pharmacol 79:439–442
115. Covic L, Gresser AL, Talavera J, Swift S, Kuliopulos A (2002) Activation and inhibition of G protein-coupled receptors by cell-penetrating membrane-tethered peptides. Proc Natl Acad Sci USA 99:643–648
116. Covic L, Misra M, Badar J, Singh C, Kuliopulos A (2002) Pepducin-based intervention of thrombin-receptor signaling and systemic platelet activation. Nat Med 8:1161–1165

Serine and Cysteine Proteases and Their Inhibitors as Antimicrobial Agents and Immune Modulators

Bénédicte Manoury, Ali Roghanian, and Jean-Michel Sallenave

Abstract Proteases are not merely restricted to digestive purposes and remodeling of extracellular matrix and tissues, but are also key factors for the induction of physiological immune responses. This induction can be direct, through the degradation of pathogens within phagolysosomes, or indirect, through the activation of key pattern recognition receptors (PRRs), such as toll-like receptors (TLRs). Unfortunately, excess production of proteases leads to maladaptive host responses and excess tissue inflammation and damage. Although the mechanisms described here will apply to a variety of different organs, we will deal chiefly with processes occurring in the lung, in pathological conditions such as chronic obstructive pulmonary disease (COPD) and cystic fibrosis (CF). To combat these deleterious effects of proteases, the host fortunately produces antiproteases, which directly counteract the proteolytic activities of proteases. In addition to this "straightforward" effect, novel "defensin-like" activities for these molecules are clearly now emerging, as it has recently been demonstrated that protease inhibitors can themselves help in restoring tissue homeostasis by inducing innate and adaptive responses, such as through their interaction with dendritic cells (DCs).

Bénédicte Manoury and Ali Roghanian authors contributed equally.

B. Manoury
Institut Curie U932, 24 rue d'Ulm, 75005 Paris, France

A. Roghanian
Cancer Sciences Division, University of Southampton School of Medicine, Southampton General Hospital, Southampton SO16 6YD, UK

J.-M. Sallenave (✉)
Institut Pasteur, Unité de Défense Innée et Inflammation et Inserm U874, 25 rue du Dr. Roux, 75724 Paris Cedex 15, France

Inserm U874, 25 rue du Dr. Roux, 75724 Paris Cedex 15, France

Université Paris, 7-Denis Diderot, Paris, France
e-mail: jms@pasteur.fr

N. Vergnolle and M. Chignard (eds.), *Proteases and Their Receptors in Inflammation*, Progress in Inflammation Research, DOI 10.1007/978-3-0348-0157-7_2,

Keywords Adjuvant • Antiproteases • Asparagine endopeptidase (AEP) • Dendritic cell (DC) • Elafin • Immune responses • Inflammation • Macrophage • Neutrophil elastase (NE) • Proteases • Secretory leukocyte protease inhibitor (SLPI) • Toll-like receptor (TLR)

1 Introduction

Proteases are classified on the basis of catalytic mechanism, and five known distinct classes are described: metallo, aspartic, cysteine, serine, and threonine. In humans, metallopeptidases are extremely diverse as they encompass 24 families, whereas cysteine proteases are represented by 19 families, serine proteases 17, and aspartic and threonine peptidases are represented by three families. For further generic information about this "degradome," we refer the reader to recent reviews including [1] and [2].

Until recent times, the action of proteases was believed to be restricted to digestive purposes, extracellular modeling and/or remodeling of tissues, mainly through proteolytic activity on interstitial molecules, occurring throughout homeostasis and development or, in aberrant maladaptive circumstances, during disease pathogenesis. This view has clearly become untenable as proteases are clearly involved in a myriad of homeostatic as well as pathological processes. Similarly, several novel physiological functions have been attributed to endogenous antiproteases including antimicrobial and immunomodulatory activities.

We will discuss in this chapter the actions of proteases and antiproteases on physiological immune induction and inflammatory processes, as well as proteases-driven maladaptive responses. Although the mechanisms described here will apply to a variety of different organs, we will deal chiefly with processes occurring in the lung, as the protease/antiprotease balance in other tissues will be addressed by other contributors in this issue.

2 Toll-Like Receptors and Dendritic Cells in the Induction of Immune Responses

Mucosal surfaces are the first barriers against infections and their role is paramount in the prevention of systemic dissemination of pathogens. To perform this role in an unchallenged naive host, the latter uses both innate and adaptive immunity. The innate immune system is genetically programmed to detect invariant features of invading microbes. In contrast, the adaptive immune system, which is composed of T and B lymphocytes, employs antigen receptors that are not encoded in the germline but are generated de novo in each organism. Thus, adaptive immune responses are highly specific. The best-characterized microbial sensors are the so-called PRRs of the innate immune system, which detect relatively invariant molecular patterns found in most micro-organisms [3]. These structures are referred to as pathogen-associated molecular patterns (PAMPs). Microbial pathogens are recognized through multiple, distinct PRRs that can be broadly categorized into secreted, transmembrane, and cytosolic classes. The transmembrane PRRs include the TLR family and the C-type lectins. TLRs in mammals are either expressed on the plasma membrane or in endosomal/lysosomal organelles [4]. Cell-surface TLRs recognize conserved microbial patterns that are accessible on the cell surface, such as lipopolysaccharide (LPS) of gram-negative bacteria (TLR4), lipoteichoic acids of gram-positive bacteria and bacterial lipoproteins (TLR1/TLR2 and TLR2/TLR6), and flagellin (TLR5), whereas endosomal TLRs mainly detect microbial nucleic acids, such as double-stranded RNA (dsRNA) (TLR3), single-stranded RNA (ssRNA) (TLR7), and dsDNA (TLR9) [5–8].

Innate immune cells bearing TLRs include DCs, macrophages, and neutrophils, among others. DCs are crucial immune cells detecting micro-organisms and linking innate to adaptive immunity. TLR signaling is linked to MyD88- and TRIF-dependent signaling pathways that regulate the activation of different transcription factors, such as nuclear factor (NF)-κB. Specific interaction between TLRs and their ligands activates NF-κB resulting in enhanced inflammatory cytokine responses, induction of DC maturation (e.g., upregulation of CD40, CD80, CD83, and CD86) and chemokine receptors (e.g., CCR7) [9]. These features have for a long time indicated that, in particular, TLR triggering switches the immature DC phenotype to an inflammatory phenotype that is capable of inducing adaptive immune responses, instructing both antigen-specific $CD4^+$ and $CD8^+$ T-cell responses and humoral responses.

2.1 Role of TLR9 in Inflammation and Immunity

Some studies suggest a role for TLR9 in the triggering of innate immune response to protozoan parasites as well as for some bacteria and viruses. For example, TLR9 is required for the development of the Th1-type inflammatory responses that follow oral infection with *Toxoplasma gondii* in mice from some inbred strains and is also implicated in the control of parasitemia during infection with *Trypanosoma cruzi*. The hemozoin pigment of *Plasmodium* or some parasite DNA associated with the pigment results in signaling through TLR9. More recently it has been shown that the early natural killer (NK) cell response to infection with *Leishmania donovani* was dependent on the secretion of IL-12 by myeloid DCs triggered in response to TLR9 stimulation [10]. TLR9-deficient ($TLR9^{-/-}$) mice have been recently described to be more susceptible to infection with *Leishmania major*. DCs lacking TLR9 failed to be activated by *L. major* probably suggesting that the DNA of *L. major* is a TLR9 ligand. Furthermore, *L. major*-infected $TLR9^{-/-}$ DCs were unable to stimulate $CD4^+$ T cells [11]. TLR9 ligands are known to be ssDNA carrying unmethylated CpG motifs [12]. A vast array of data indicates that TLR9 plays a key role in DNA-induced immunity and links it with a role in acquired immunity through the activation of various cell types such as plasmacytoid DCs (pDCs), conventional DCs (cDCs), and B cells. Analysis of TLR9$^{-/-}$ mice revealed that TLR9 is essential not only for proinflammatory cytokines production and other inflammatory responses but it also plays a role in the induction of Th1 acquired immune response and in the proliferation of B cells. In addition, TLR9 also recognizes bacterial and viral DNA. In particular, TLR9 cooperates with TLR2 to induce innate immune response against *Mycobacterium tuberculosis*. TLR9 also plays an important role in the fight against infections with *Brucella*, *Streptococcus pneumoniae*, and could be involved in recognition and clearance of *Helicobacter*. TLR9-mediated antiviral responses are largely documented. Indeed, mouse cytomegalovirus, herpes simplex virus type 1 and 2, and adenovirus are recognized by TLR9 on pDCs which produce high amount of interferon (IFN)-α in response to this stimulation. Recently, natural DNA repetitive extragenic sequences from *Pseudomonas aeruginosa* have been shown to strongly stimulate TLR9 [13]. In addition, signaling through TLR9 appears to be important in *P. aeruginosa* keratitis, and silencing TLR9 signaling reduces inflammation but contributes to decreased bacterial killing in the cornea [14].

3 Role of Proteases in the Induction of Immunity

3.1 Cysteine Proteases

Cysteine proteases were historically shown to have an important role in antigen presentation and the induction of immunity [15]. They are constitutively expressed in most cell types, especially in macrophages and DCs. They contain a cysteine thiol as part of their catalytic site and are related to papain and belong to the C1 family. Among them, cathepsins B, C, F, L, H, K, L, S, V and W have been isolated. Some of these enzymes are endopeptidases, whereas others are either amino or carboxy exopeptidases (see Table 1). Another endopeptidase named asparagine

Table 1 Lysosomal proteases

Cathepsin	Location	Family	Cleavage pattern	Phenotype/function
B	Lysosomes	Cysteine	Carboxypeptidase	Lysosomal apoptosis pathway and tumor spreading
C	Endo/lysosomes	Cysteine	Aminopeptidase	Serine protease activation
F	Lysosomes	Cysteine	Endopeptidase	Ii processing
K	Lysosomes	Cysteine	Endopeptidase	TLR9 signaling
L(V)	Lysosomes	Cysteine	Endopeptidase	CD4 and NK T cells tymic selection
S	Endo/lysosomes	Cysteine	Endopeptidase	MHC class II pathway, Ii chain processing
X(Z)	Endo/lysosomes	Cysteine	Carboxypeptidase	T-cell migration
D,E	Lysosomes	Aspartic	Endopeptidase	Lysosomal storage, early cell death
AEP	Endo/lysosomes	Cysteine	Asparagine sites	MHC class II pathway, cathepsins maturation and TLR processing

endopeptidase (AEP) or legumain is unrelated to the papain-like cysteine protease family such as cathepsin B and L and is grouped together with the caspases, separases, and some bacterial proteases in clan CD [16–18]. Most of these enzymes are synthesized as precursors and targeted to the endocytic pathway. For example, the N- and C-terminal propeptides of AEP are auto-cleaved in the lysosomal compartments to generate a 46 kDa mature form, which can be further processed into a 36 kDa fragment [19].

Acidic pH is a prerequisite for maturation of most of these enzymes and so their greatest activity is found in lysosomal compartments. Their main function is to provide ligands for the MHC class II-restricted antigenic pathway. MHC class II molecules access the endolysosomal compartments to bind peptides and display them on the surface of DCs to trigger $CD4^{+}$ T-cell response. Indeed, the uptake of exogenous antigen into DCs is followed by protease-mediated degradation in endolysosomal compartments. These proteases also process the invariant chain (Ii), a chaperone molecule which associates with MHC class II molecules in the endoplasmic reticulum (RE). Cathepsin L and cathepsin S are the best characterized proteases to proteolyse Ii [20]. The endolysosomal proteases have probably a redundant role in the selection of the peptides which will be presented at the DCs surface. However, there are examples where some antigens require a particular protease. Indeed, AEP is unique among lysosomal cysteine proteases, in that it is insensitive to leupeptin and cleaves on the carboxyl terminal sides of asparagine residues. AEP initiates the processing of tetanus toxin in human B cells, destroys an immuno-dominant peptide of myelin basic protein (MBP – an autoantigen implicated in the autoimmune disease multiple sclerosis) and performs the early steps of degradation of the Ii chain in human B-EBV cells [21–23].

3.2 Asparagine Endopeptidase, TLR7/9 Pathway and Antigen Presentation in DCs

DCs are heterogeneous and consist of various DC subsets among which TLR expression and function differ. pDC is a DC subset which differs from cDC and can produce vast amounts of type I interferon upon bacterial and viral infection. pDCs only express TLR7 and TLR9. Thus, pDCs can be regarded as a DC subset specialized for detecting nucleic acids mainly through TLR7/9. In mice, cross-presentation has been considered a unique property of cDCs. This crucial mechanism in microbial immunity allows exogenous antigen to be delivered into the MHC class I pathway to initiate cytotoxic T-cell response. However, recently, it has been shown that stimulation by TLR 7/9 also licences pDCs to cross-present [24].

Little is known about how endosomal TLRs and their ligands are targeted to the endocytic pathway. TLRs are sensitive to chloroquine, a lysomotropic agent that neutralizes acidic compartments indicating a role for endo/lysosomal proteases for their signaling. Indeed, recent findings have described the importance of proteolysis for TLR9 function [25, 26]. It has recently been shown that mouse TLR9 is non-functional until it is subjected to proteolytic cleavage in the endosomes. Upon stimulation, full-length TLR9 is cleaved into a C-terminal fragment which is highly dependent on AEP in DCs. A recruitment of TLR9 and a boost in AEP activity, which was induced shortly after TLR9 stimulation, was shown to promote TLR9 cleavage and correlated with an increased acidification in endosomes and lysosomes. Moreover, mutating a putative AEP cleavage site in TLR9 strongly decreases its signaling in DCs suggesting perhaps that a direct cleavage of TLR9 by AEP is required for this process. These results demonstrated that TLR9 requires a proteolytic cleavage for its signaling and identified a key endocytic protease playing a critical role in this process in DCs [27]. Interestingly, in contrast, TLR9 processing does not rely on AEP in macrophages probably because of the already highly acidic milieu found in the endocytic pathway of macrophages in comparison to DCs, thus allowing many proteases (and not only AEP) such as cathepsins B, L, K and S to perform TLR9 degradation [25, 27] and thus, TLR9 proteolysis has been proposed to restrict receptor activation to endosomal/lysosomal compartments and to prevent TLRs from responding to selfnucleic acids. Other endosomal TLRs, and in particular TLR7, are also probably subjected to a similar proteolytic maturation but this remains to be fully investigated (unpublished data).

Several studies have suggested that intracellular TLRs can be targeted directly from the ER, where they reside, to endosomes in which they signal. Relatedly, mouse and human genomic studies have identified UNC93B1, which encodes for a 12-membrane spanning molecule highly conserved in the ER, as a key regulator in the transport of endosomal TLRs. The third mutation (UNC93B mutation) results in a phenotype where no signaling occurs via the intracellular TLRs 3, 7 and 9 and also diminishes presentation of exogenous antigen [28, 29]. However, the exact role played by UNC93B1 in these processes remains to be fully elucidated.

4 Proteases and Maladaptive Inflammation

Proteases produced by inflammatory cells such as neutrophils and macrophages play a crucial role in the first line of defense against invading bacteria, fungi and protozoa, either by directly killing pathogens or by inducing immune recognition, e.g., via TLRs. Individuals with cyclic neutropenia, a disease characterized by mutations in the gene encoding neutrophil elastase (NE), commonly experience recurrent bacterial infections, highlighting their critical importance in this respect. Neutrophils contain at least four types of granules: azurophil granules, specific granules, gelatinase granules, and secretory granules [30, 31]. In addition to proteases, these granules are an important reservoir of other antimicrobial proteins, such as defensins, and components of the respiratory burst oxidase [32]. It has also been suggested that these granules contain a wide range of membrane-bound receptors (e.g., CD11b/CD18 [33] and *N*-formyl-methionyl-leucyl-phenylalanine

[fMLP] receptor) for endothelial adhesion molecules, extracellular matrix proteins, bacterial products, and soluble mediators of inflammation [30, 32]. In addition to these molecules, a novel antimicrobial mechanism for neutrophils has recently been described, with the demonstration that neutrophils form neutrophil extracellular traps (NET) that could potentially bind, disarm and kill pathogens extracellularly [34–37]. DNA is the major structural component of NETs and it provides the backbone on which the proteinaceous effectors such as proteases are anchored to [34].

Although all of the effects described above are beneficial to the host, chronic and persistent presence of neutrophils is a hallmark of lung pathologies such as COPD and CF. There is certainly an excess of neutrophil chemoattractants such as IL-8 and leukotriene B_4 (LTB_4) recovered in bronchoalveolar lavage (BAL) fluid of these patients [38, 39]. Bacteria present in high concentrations in these pathologies also provide additional chemoattractants for neutrophils. Furthermore, neutrophils may survive longer in the airways of CF/COPD patients because of the production of excess concentrations of granulocyte macrophage-colony stimulating factor (GM-CSF) and the relative lack of IL-10, which, when present, promotes neutrophil apoptosis [38–41]. Moreover, cleavage of the phosphatidylserine receptor (PSR) and CD14 by NE could specifically disrupt phagocytosis of apoptotic neutrophils by macrophages [42, 43]. On the other hand, the decreased mucociliary clearance in CF/COPD leads to longer retention of apoptotic neutrophils causing them to necrose, hence releasing their toxic agents, e.g., NE, into the affected airways. In turn, NE contributes to the vicious circle of chronic inflammatory airway disease by inducing mucin production in airway epithelial cells [44–46]. Mucins, normally beneficial in microbial infections, by binding and removing bacteria via the mucociliary ladder, can be detrimental in chronic pathologies, by clogging the airways and providing an appropriate milieu for bacterial growth and colonization [47]. NE also reduces ciliary beat frequency resulting in marked disruption of epithelial cells [48], and induces goblet cell metaplasia which is dependent on its proteolytic activity [49–52].

In addition to the direct deleterious effect of proteases (such as NE) on innate immune effectors, these mediators also have a negative effect on immune cells such as DCs. For years, the nature of the elusive lung DCs was poorly understood, but with increasing interest in the role of adaptive immunity in the pathophysiology of human CF, COPD and emphysema, interest in further characterization of specific DC subsets in normal and diseased lungs arose [53–55]. In that context, we and others have shown that NE could be instrumental in the elicitation of this breach in host defense, through its action on DCs. Indeed, we demonstrated that NE is able to disable mature DC function by reducing the level of DC surface costimulatory molecules (CSMs), interfering both with the ability of immature DCs to mature in response to bacterial LPS and by reducing the allostimulatory activity of these cells, resulting in reduced Th1 cytokine production [56]. Similarly, neutrophils and culture supernatants of unprimed/primed neutrophils are able to downregulate human monocyte-derived DCs allostimulatory function in vitro [57]. This effect was associated with the amount of NE released by neutrophils, which in turn converted immature myeloid DCs into transforming growth factor (TGF)-β1-secreting cells [57]. These in vitro observations are further supported by an earlier report showing that APCs isolated from BAL fluid of CF patients were unable to present antigen and stimulate T-cell responses [58], despite appropriate responses from systemic APCs (monocytic cells). However, although the characteristics and functional properties of lung DCs can be easily studied in animal models, very few and in most cases contradictory data from their human counterparts are currently available [55].

4.1 Neutrophil Elastase

Human NE is a serine protease found in the azurophil granules of the neutrophil. The highly cationic glycoprotein product contains 218 amino acids and four disulfide bridges, and is a member of the serine protease family [59]. The catalytic site of the NE molecule is composed of the triad His41-Asp99-Ser173, in which the γ-oxygen of serine becomes a powerful nucleophile, able to attack a suitably located carbonyl group on the target substrate [60]. Neutrophils release NE upon exposure to various cytokines and chemoattractants, including tumor necrosis factor (TNF)-α, interleukin (IL)-8, C5a, LPS, and a tripeptide derived from bacterial wall fMLP [61]. The concentration of NE in neutrophils exceeds 5 mM [62], and each neutrophil contains approximately 400 NE-positive granules. Although NE is most abundant in neutrophils, small amounts are expressed by monocytes and T cells [63, 64]. NE has broad substrate specificity and is capable of degrading a wide range of extracellular matrix proteins, including elastin, collagen (types I–IV), fibronectin, laminin, and proteoglycans. Additionally, many biological molecules like cytokines and their receptors contain putative cleavage sites for neutrophil serine proteases. Indeed, as expected, many receptors, cytokines and other molecules have been found to be natural substrates for NE (Table 2).

Like the cysteine protease family described above, NE possesses potent microbicidal activity and is speculated to assist with phagocytosis of pathogens by activated neutrophils [65]. To determine the contribution of NE in combating bacterial infections, NE-deficient ($NE^{-/-}$) mice were generated [62] and shown to be more susceptible to sepsis and death following intraperitoneal infection with gram-negative (*Klebsiella pneumoniae*, *P. aeroginosa*, and *Escherichia coli*) but not gram-positive (*Staphylococcus aureus*) bacteria. NE is required for maximal intracellular killing of *P. aeruginosa* by neutrophils, as it degrades the major outer membrane protein F, a protein with important functions, including porin activity, maintenance of structural integrity, and sensing of host immune system activation [66]. In addition, in vitro incubation of NE with *E. coli* leads to a loss of bacterial integrity and lysis of bacteria [62]. Indeed, the primary sequence of outer membrane protein A (OmpA) amino acid has multiple NE-preferred cleavage sites and NE was shown to directly degrade purified OmpA of *E. coli* in vitro [62]. Furthermore, NE degrades virulence factors of enterobacteria such as *Salmonella enterica* serovar

Table 2 Summary of the expanding list of natural neutrophil elastase (NE) substrates

Target	Hypothetical biological function	References
Receptors		
Proteinase-activated receptor-1 (PAR-1)	Inactivation, modulation of response	[76, 150]
PAR-2	Inactivation, modulation of response	[150–152]
PAR-3	Inactivation, modulation of response	[153]
IL-2Rα	Inhibiting cellular response and prolongation of cytokine half-life time	[154]
TNF-RII	Inhibiting cellular response and prolongation of cytokine half-life time	[155]
C5aR (CD88)	Inhibition of chemotaxis, feedback mechanism	[156]
CR1 (CD35)	Inhibition of complement signaling	[157]
Urokinase R (CD87)	Regulation of cell migration	[158]
Granulocyte-colony stimulating factor receptor (G-CSF-R)	Growth inhibition	[159, 160]
CD43 (sialophorin)	Regulation of adhesion	[161, 162]
CD14	Inhibition of LPS-mediated cell activation/apoptotic cell recognition	[163]
CD2, CD4, and CD8	Impairment of T lymphocytes	[164]
CD40, CD80, and CD86	Impairment of DCs	[56]
Soluble IL-6 receptor	Regulation of inflammation	[165]
CXC chemokine receptor 1 (CXCR1)	Regulation of cell migration	[166]
Cytokines/chemokines		
TNF-α	Regulation of inflammation	[167]
IL-2	Regulation of inflammation	[63]
IL-6	Regulation of inflammation	[168]
IL-8	Regulation of inflammation	[169]
IL-12p40	Regulation of inflammation	(unpublished)
G-CSF	Growth inhibition	[159]
Integrins/others		
Intercellular adhesion molecule-1 (ICAM-1)	Regulation of adhesion	[170, 171]
Vascular endothelium cadherin	Regulation of adhesion	[172]
Proepithelin	Regulation of wound healing	[173]
Tissue factor pathway inhibitor (TFPI)	Regulation of coagulation and intravascular thrombus growth	[174]
Matrix metalloprotease-9 (MMP-9)	Regulation of proteolysis	[175]
Tissue inhibitor of metalloprotease-1 (TIMP-1)	Regulation of proteolysis	[175]
Basic fibroblast growth factor (bFGF)	Regulation of angiogenesis	[176]
Vascular endothelial growth factor (VEGF)	Regulation of angiogenesis	[176, 177]
Laminin-332 (laminin-5)	Regulation of cell migration	[178]
Surfactant protein D (SP-D)		[179]

(continued)

Table 2 (continued)

Target	Hypothetical biological function	References
	Regulation of inflammation/innate immunity	
Insulin receptor substrate-1 (IRS-1)	Regulation of cell growth	[180]
von Willebrand factor (VWF)	Regulation of cell hemostasis	[181]
Cut homeobox 1 (CUX1)	Regulation of gene expression	[182]
Plasma factor XIII (FXIII)	Regulation of coagulation	[183]
AlphaIIb β3	Regulation of adhesion	[184]

Typhimurium, *Shigella flexneri*, *Yersinia enterocolitica* and *Streptococcus pneumoniae* [67]. Thus, in the absence of NE these bacteria escape from the phagolysosome leading to their increased survival in the cytoplasm of infected neutrophils [68]. Finally, NE is able to suppress flagellin transcription in *P. aeruginosa.* Flagellin suppression by NE could elucidate how and why CF patients undergo cyclical exacerbations of the inflammatory lung disease caused by *P. aeruginosa.* When neutrophil numbers and thus NE concentrations are low, *P. aeruginosa* may proliferate, assemble a flagellum, and release flagellin, stimulating a robust inflammatory response in the patient's airways [69].

4.1.1 NE Signaling Activity

It has been suggested that NE signals via the cell surface membrane-bound TLR4 [70], by activating the NF-κB signaling pathway [71–73]. A more recent study, however, proposed that IL-1R1/MyD88 signaling and inflammasome activation, but not TLRs, are critical for NE-induced lung inflammation and emphysema in murine models [74]. Additionally, NE has been reported to induce apoptosis, thus contributing to the pathogenesis of inflammatory injury in the respiratory tract. NE-induced apoptosis of lung epithelial cells is mediated by a proteinase-activated receptor-1 (PAR1)-triggered pathway involving activation of NF-κB and p53, and a PUMA- and Bax-dependent increase in mitochondrial permeability leading to activation of distal caspases [75, 76].

4.2 Endogenous Protease Inhibitors

To modulate the multiple activities of proteases (including NE), either beneficial, but also potentially deleterious (see above), the body synthesizes antiproteases. We will concentrate our discussion on NE inhibitors, as other inhibitors will be described in this issue by other contributors. These NE inhibitors can be broadly classified into two groups, the "alarm" and the "systemic" antiproteases. Systemic antiproteases, such as α1-protease inhibitor (α1-PI), are produced mainly by hepatocytes. However, during infection, the activity of locally produced mucosal alarm antiproteases such as SLPI and elafin may add an extra edge to the host defense armamentarium, as will be discussed below (reviewed in [77]and [78]).

4.2.1 Alarm Antiproteases

SLPI and elafin alarm *antiproteases* have been isolated and characterized under a variety of names in adult and fetal tissues [78]. They belong to the family of whey-acidic protein (WAP) proteins and are produced by epithelial cells and cells of the immune system. Importantly, alarm antiproteases are generated locally in areas of infection or neutrophil infiltration and are upregulated by pathogen- and inflammation-associated factors, including cytokines and NE itself [79]. In addition to their antiprotease properties, and because of their biochemical characteristics (heavily disulphide-bonded, low molecular mass cationic peptides, present at mucosal sites), elafin and SLPI have recently been proposed to possess "defensin/cathelicidin-like" properties [77, 78, 80].

Elafin

Elafin was simultaneously isolated from the skin of psoriatic patients [81, 82] and from the sputum of COPD subjects [83, 84]. Elafin gene was cloned and sequenced by Saheki and colleagues in 1992 [85] and by Sallenave and Silva in 1993 [86], and shown to code for a 117-amino acids protein, of which the first 22 amino acids represent a hydrophobic signal peptide. Elafin is produced as a 9.9-kDa full-length non-glycosylated cationic protein composed of an N-terminal "cementoin" domain which facilitates transglutaminase-mediated cross-linkage on to polymers or extracellular matrix components and a globular C-terminus, containing the protease inhibitor moiety [87]. The elafin molecule shares ~40% homology with SLPI and has been shown to be a more specific inhibitor of proteases than SLPI, since it inhibits NE, porcine pancreatic enzyme, and proteinase 3 [83, 88, 89], but does not inhibit cathepsin G, trypsin, or chymotrypsin [83, 88].

The regulation of elafin expression during inflammation has been well studied. In vitro, bronchial and alveolar epithelial cells produce little elafin protein, but the quantity of elafin recovered from the supernatant can be greatly enhanced by addition of the inflammatory cytokines IL-1 and TNF-α [79]. These cytokines induce similar increases in expression of elafin from keratinocytes in vitro [90]. The c-jun, p38 mitogen-activated protein (MAP) kinase, and NF-κB pathways are thought to be implicated in the elafin response to inflammatory cytokines [91–93]. Of note, the cytokine-mediated increase in elafin production by epithelial cells is greater than the increase in SLPI production [79]. Hence, whereas SLPI has been described as providing a baseline antiprotease shield and can be isolated from bronchial lavage samples from healthy individuals [94–96], elafin might be of greater significance during an inflammatory challenge to the lungs. In keeping with this notion, elafin mRNA expression in bronchial epithelial cells is increased by free NE, which is found in abundance at times of inflammation [97, 98].

Although inhibition of NE activity has historically been considered to be the primary role of elafin, recent work has highlighted further properties of this cationic molecule. Simpson and colleagues [99] demonstrated that elafin has antibacterial activity against gram-negative *P. aeruginosa* and gram-positive *S. aureus*, and further established that, while antiprotease activity resides exclusively in the C-terminus, the majority of antimicrobial activity of elafin resides in its N-terminal domain [99]. In support of these findings, supernatants of *P. aeruginosa* could induce elafin production in human keratinocytes, and elafin inhibits growth of *P. aeruginosa* in vitro, but not *E. coli* [100, 101]. Further, adenovirus (Ad)-mediated augmentation of human elafin in murine lungs was shown to protect the lungs against *P. aeruginosa*-mediated injury, and also reduced bacterial numbers. Similarly, overexpression of elafin using the Ad-strategy dramatically improved the clearance of *S. aureus* in vitro and in vivo [102]. In these studies, concomitant anti-inflammatory activities have been demonstrated, which can probably be explained by an inhibition of the AP-1 and NF-κB pathways [103, 104]. More recently, using wild-type and CD14 knockout mice, Wilkinson and co-workers demonstrated the opsonic activity for elafin against *P. aeruginosa*, both in vitro and in vivo [105]. In an extension of these data, there is evidence that elafin binds both smooth and rough forms of LPS in vitro and could potentially modulate immune responses depending on the microenvironment [106].

We have also shown that elafin exhibits chemotactic activity for leukocytes locally in the lung [107, 108], while, conversely, downregulating inflammation systemically [108]. In keeping with this immunomodulatory activity, we demonstrated that overexpression of elafin in murine lungs results in a higher number of $CD11c^{+}/MHCII^{+}$ DCs with an activated phenotype, as evidenced by expression of higher levels of co-stimulatory molecules CSMs (CD80 and CD86), and higher levels of Th1-biased cytokines IL-12p40, TNF-α, and IFN-γ in their broncholaveolar (BAL) fluids [109].

Secretory Leukocyte Protease Inhibitor

Secretory leukocyte protease inhibitor (SLPI) is an 11.7-kDa protein that was first isolated from human parotid gland secretions [110]. SLPI orthologs have also been demonstrated in mice, rats, pigs, and sheep [111–113]. It is a non-glycosylated, highly basic, acid-stable, cysteine-rich, 107-amino acid, single-chain polypeptide [110]. The tertiary structure of the SLPI molecule resembles a boomerang, with each arm carrying one domain [114]. The four-in-each-domain disulfide bridges formed between the cysteine residues, as well as the two-domain interaction, contribute to the conformation and efficacy of the molecule [115]. SLPI provides a significant component of the human antiprotease shield within the lung. Through its C-terminal domain, SLPI gives significant protection against proteases, such as NE and the serine protease cathepsin G [116]. SLPI is produced by various inflammatory cells, such as neutrophils [117], mast cells [118], and macrophages [119]. It is estimated that SLPI is present at concentrations of 0.1–2 μg/ml in BAL fluid [120, 121] and 2.5 μg/ml in nasal secretions [122].

It is believed that SLPI also shields the tissues against inflammatory products by downregulating the macrophage responses against bacterial LPS. Patients with sepsis have elevated circulating SLPI levels and LPS is the key mediator in bacterial endotoxic shock [96, 123, 124]. LPS seems to induce SLPI production by macrophages directly or by way of IL-1β, TNF-α, IL-6, and IL-10 [125, 126]. SLPI, like elafin, in turn inhibits the downstream components of the NF-κB pathway by protecting the inhibitor of NF-κ (I-κB) from degradation by the ubiquitin-proteosome pathway [103]. SLPI is believed to enter cells, becoming rapidly localized to the cytoplasm and nucleus where it affects NF-κB activation by binding directly to NF-κB binding sites in a site-specific manner [127]. Thus, SLPI renders macrophages unable to release pro-inflammatory cytokines and nitric oxide [125]. These data have been confirmed by in vivo studies demonstrating that SLPI knockout mice show increased susceptibility to endotoxic shock, and macrophages and B lymphocytes from the same mice show increased activation after administration of LPS [128].

In addition to its NE inhibitory and immunomodulatory activities, SLPI, like elafin, possesses broad-spectrum antibactericidal, antiviral, and antifungal properties [115, 129–134].

The Systemic Antiprotease α1-Protease Inhibitor

The systemic antiprotease α1-PI (also called α1-antitrypsin) is a 52-kDa secreted glycoprotein and is the prototypic member of the serine protease inhibitor (serpin) superfamily of proteins, which has a major role in inactivating NE and other proteases, such as cathepsin G and proteinase 3. Although some epithelial surfaces and cells of the immune system may produce small quantities of systemic antiproteases, such as α1-PI [135, 136], these inhibitors are produced primarily by hepatocytes [137, 138]. The production of α1-PI by alveolar macrophages is upregulated by pro-inflammatory cytokines and bacterial LPS [139]. Also, the cytokine oncostatin M is a major inducer of α1-PI in bronchial epithelial cells [135, 140].

The importance of α1-PI in the lung has historically been inferred from genetic studies: α1-PI deficiency is a genetic disorder that affects about 1 in 2,000–5,000 individuals. α1-PI deficiency is characterized by a decrease in levels of secreted α1-PI, which results in early-onset of emphysema in affected individuals. Although it was originally believed that genetic emphysema was caused by this decreased secretion of α1-PI in the respiratory tract, leading to unopposed and prolonged NE activity [141], recent evidence suggests that the mutated Z variant of α1-PI, when polymerized, may be pro-inflammatory when secreted, acting as an important chemoattractant for neutrophils in the α1-PI-deficient lung and adding to the excessive neutrophil and NE burden [137, 142].

In addition to its role as an antiprotease, like elafin and SLPI, α1-PI possesses important pleiotropic anti- or pro-inflammatory properties, depending upon the conditions. These effects include blocking of the pro-inflammatory effects of human NE [143, 144], and regulating expression of pro-inflammatory cytokines such as TNF-α, IL-6, IL-8, IL-1β, and monocyte chemotactic protein (MCP)-1 by monocytes [145, 146]. Both the native and polymerized forms of α1-PI have been shown to possess similar effects as monocyte stimulators, with pro-inflammatory effects at low doses, and anti-inflammatory activities at physiologically normal doses [145]. This strengthens the concept that some of the apparently contradictory effects of these inhibitors reported in the literature may be due to differences in dosage between experimental protocols.

Lastly, α1-PI could also inhibit alveolar cell apoptosis in vivo [147]. Thus direct inhibition of active NE [75] and caspase-3 [148] by α1-PI may represent a novel anti-apoptotic mechanism relevant to disease processes characterized by excessive structural cell apoptosis, oxidative stress, and inflammation in the airways [149].

5 Conclusions

Here, we have described the important role of proteases in immune functions, not only in the direct degradation of micro-organisms and antigen presentation, but also in the induction of inflammatory responses. We have also discussed the importance of protease inhibitors in the modulation of maladaptive responses caused by extracellularly released proteases. Finally, we described novel bioactivities of elastase inhibitors, such as antimicrobial and adjuvant-like functions. These latter functions are likely to be exploited further for the treatment of individuals prone to developing CF and COPD, especially to combat frequent episodes of lung infections, either in a therapeutic (antimicrobial activity) or prophylactic (vaccination) fashion.

References

1. Rawlings ND, Morton FR, Barrett AJ (2006) MEROPS: the peptidase database. Nucleic Acids Res 34:D270–D272
2. Pardo A, Selman M, Kaminski N (2008) Approaching the degradome in idiopathic pulmonary fibrosis. Int J Biochem Cell Biol 40:1141–1155
3. Anderson KV (2000) Toll signaling pathways in the innate immune response. Curr Opin Immunol 12:13–19
4. Takeda K, Akira S (2005) Toll-like receptors in innate immunity. Int Immunol 17:1–14
5. Takeuchi O, Hoshino K, Kawai T, Sanjo H, Takada H, Ogawa T, Takeda K, Akira S (1999) Differential roles of TLR2 and TLR4 in recognition of gram-negative and gram-positive bacterial cell wall components. Immunity 11:443–451

6. Hayashi F, Smith KD, Ozinsky A, Hawn TR, Yi EC, Goodlett DR, Eng JK, Akira S, Underhill DM, Aderem A (2001) The innate immune response to bacterial flagellin is mediated by Toll-like receptor 5. Nature 410:1099–1103
7. Hacker H, Vabulas RM, Takeuchi O, Hoshino K, Akira S, Wagner H (2000) Immune cell activation by bacterial CpG-DNA through myeloid differentiation marker 88 and tumor necrosis factor receptor-associated factor (TRAF)6. J Exp Med 192:595–600
8. Diebold SS, Kaisho T, Hemmi H, Akira S, Reis e Sousa C (2004) Innate antiviral responses by means of TLR7-mediated recognition of single-stranded RNA. Science 303:1529–1531
9. Janeway CA Jr (2002) A trip through my life with an immunological theme. Annu Rev Immunol 20:1–28
10. Kumagai Y, Takeuchi O, Akira S (2008) TLR9 as a key receptor for the recognition of DNA. Adv Drug Deliv Rev 60:795–804
11. Abou Fakher FH, Rachinel N, Klimczak M, Louis J, Doyen N (2009) TLR9-dependent activation of dendritic cells by DNA from Leishmania major favors Th1 cell development and the resolution of lesions. J Immunol 182:1386–1396
12. Hemmi H, Takeuchi O, Kawai T, Kaisho T, Sato S, Sanjo H, Matsumoto M, Hoshino K, Wagner H, Takeda K et al (2000) A Toll-like receptor recognizes bacterial DNA. Nature 408:740–745
13. Magnusson M, Tobes R, Sancho J, Pareja E (2007) Cutting edge: natural DNA repetitive extragenic sequences from gram-negative pathogens strongly stimulate TLR9. J Immunol 179:31–35
14. Huang X, Barrett RP, McClellan SA, Hazlett LD (2005) Silencing Toll-like receptor-9 in *Pseudomonas aeruginosa* keratitis. Invest Ophthalmol Vis Sci 46:4209–4216
15. Blum JS, Cresswell P (1988) Role for intracellular proteases in the processing and transport of class II HLA antigens. Proc Natl Acad Sci USA 85:3975–3979
16. Chen JM, Dando PM, Rawlings ND, Brown MA, Young NE, Stevens RA, Hewitt E, Watts C, Barrett AJ (1997) Cloning, isolation, and characterization of mammalian legumain, an asparaginyl endopeptidase. J Biol Chem 272:8090–8098
17. Chen JM, Dando PM, Stevens RA, Fortunato M, Barrett AJ (1998) Cloning and expression of mouse legumain, a lysosomal endopeptidase. Biochem J 335(Pt 1):111–117
18. Uhlmann F, Wernic D, Poupart MA, Koonin EV, Nasmyth K (2000) Cleavage of cohesin by the CD clan protease separin triggers anaphase in yeast. Cell 103:375–386
19. Li DN, Matthews SP, Antoniou AN, Mazzeo D, Watts C (2003) Multistep autoactivation of asparaginyl endopeptidase in vitro and in vivo. J Biol Chem 278:38980–38990
20. Chapman HA (2006) Endosomal proteases in antigen presentation. Curr Opin Immunol 18:78–84
21. Manoury B, Hewitt EW, Morrice N, Dando PM, Barrett AJ, Watts C (1998) An asparaginyl endopeptidase processes a microbial antigen for class II MHC presentation. Nature 396:695–699
22. Manoury B, Mazzeo D, Fugger L, Viner N, Ponsford M, Streeter H, Mazza G, Wraith DC, Watts C (2002) Destructive processing by asparagine endopeptidase limits presentation of a dominant T cell epitope in MBP. Nat Immunol 3:169–174
23. Manoury B, Mazzeo D, Li DN, Billson J, Loak K, Benaroch P, Watts C (2003) Asparagine endopeptidase can initiate the removal of the MHC class II invariant chain chaperone. Immunity 18:489–498
24. Mouries J, Moron G, Schlecht G, Escriou N, Dadaglio G, Leclerc C (2008) Plasmacytoid dendritic cells efficiently cross-prime naive T cells in vivo after TLR activation. Blood 112:3713–3722
25. Park B, Brinkmann MM, Spooner E, Lee CC, Kim YM, Ploegh HL (2008) Proteolytic cleavage in an endolysosomal compartment is required for activation of Toll-like receptor 9. Nat Immunol 9:1407–1414
26. Ewald SE, Lee BL, Lau L, Wickliffe KE, Shi GP, Chapman HA, Barton GM (2008) The ectodomain of Toll-like receptor 9 is cleaved to generate a functional receptor. Nature 456:658–662

27. Sepulveda FE, Maschalidi S, Colisson R, Heslop L, Ghirelli C, Sakka E, Lennon-Dumenil AM, Amigorena S, Cabanie L, Manoury B (2009) Critical role for asparagine endopeptidase in endocytic Toll-like receptor signaling in dendritic cells. Immunity 31:737–748
28. Tabeta K, Hoebe K, Janssen EM, Du X, Georgel P, Crozat K, Mudd S, Mann N, Sovath S, Goode J et al (2006) The Unc93b1 mutation 3d disrupts exogenous antigen presentation and signaling via Toll-like receptors 3, 7 and 9. Nat Immunol 7:156–164
29. Kim YM, Brinkmann MM, Paquet ME, Ploegh HL (2008) UNC93B1 delivers nucleotide-sensing toll-like receptors to endolysosomes. Nature 452:234–238
30. Borregaard N, Cowland JB (1997) Granules of the human neutrophilic polymorphonuclear leukocyte. Blood 89:3503–3521
31. Faurschou M, Borregaard N (2003) Neutrophil granules and secretory vesicles in inflammation. Microbes Infect 5:1317–1327
32. Hager M, Cowland JB, Borregaard N (2010) Neutrophil granules in health and disease. J Intern Med 268:25–34
33. Calafat J, Kuijpers TW, Janssen H, Borregaard N, Verhoeven AJ, Roos D (1993) Evidence for small intracellular vesicles in human blood phagocytes containing cytochrome b558 and the adhesion molecule CD11b/CD18. Blood 81:3122–3129
34. Balloy V, Sallenave JM, Crestani B, Dehoux M, Chignard M (2003) Neutrophil DNA contributes to the antielastase barrier during acute lung inflammation. Am J Respir Cell Mol Biol 28:746–753
35. Brinkmann V, Reichard U, Goosmann C, Fauler B, Uhlemann Y, Weiss DS, Weinrauch Y, Zychlinsky A (2004) Neutrophil extracellular traps kill bacteria. Science 303:1532–1535
36. Medina E (2009) Neutrophil extracellular traps: a strategic tactic to defeat pathogens with potential consequences for the host. J Innate Immun 1:176–180
37. Papayannopoulos V, Zychlinsky A (2009) NETs: a new strategy for using old weapons. Trends Immunol 30:513–521
38. Bonfield TL, Panuska JR, Konstan MW, Hilliard KA, Hilliard JB, Ghnaim H, Berger M (1995) Inflammatory cytokines in cystic fibrosis lungs. Am J Respir Crit Care Med 152:2111–2118
39. Barnes PJ, Shapiro SD, Pauwels RA (2003) Chronic obstructive pulmonary disease: molecular and cellular mechanisms. Eur Respir J 22:672–688
40. Armstrong DS, Grimwood K, Carlin JB, Carzino R, Gutierrez JP, Hull J, Olinsky A, Phelan EM, Robertson CF, Phelan PD (1997) Lower airway inflammation in infants and young children with cystic fibrosis. Am J Respir Crit Care Med 156:1197–1204
41. Bonfield TL, Konstan MW, Berger M (1999) Altered respiratory epithelial cell cytokine production in cystic fibrosis. J Allerg Clin Immunol 104:72–78
42. Vandivier RW, Fadok VA, Hoffmann PR, Bratton DL, Penvari C, Brown KK, Brain JD, Accurso FJ, Henson PM (2002) Elastase-mediated phosphatidylserine receptor cleavage impairs apoptotic cell clearance in cystic fibrosis and bronchiectasis. J Clin Invest 109:661–670
43. Henriksen PA, Devitt A, Kotelevtsev Y, Sallenave JM (2004) Gene delivery of the elastase inhibitor elafin protects macrophages from neutrophil elastase-mediated impairment of apoptotic cell recognition. FEBS Lett 574:80–84
44. Voynow JA, Gendler SJ, Rose MC (2006) Regulation of mucin genes in chronic inflammatory airway diseases. Am J Respir Cell Mol Biol 34:661–665
45. Voynow JA, Young LR, Wang Y, Horger T, Rose MC, Fischer BM (1999) Neutrophil elastase increases MUC5AC mRNA and protein expression in respiratory epithelial cells. Am J Physiol 276:L835–843
46. Kohri K, Ueki IF, Nadel JA (2002) Neutrophil elastase induces mucin production by ligand-dependent epidermal growth factor receptor activation. Am J Physiol Lung Cell Mol Physiol 283:L531–540
47. Rose MC, Voynow JA (2006) Respiratory tract mucin genes and mucin glycoproteins in health and disease. Physiol Rev 86:245–278

48. Amitani R, Wilson R, Rutman A, Read R, Ward C, Burnett D, Stockley RA, Cole PJ (1991) Effects of human neutrophil elastase and *Pseudomonas aeruginosa* proteinases on human respiratory epithelium. Am J Respir Cell Mol Biol 4:26–32
49. Lucey EC, Stone PJ, Breuer R, Christensen TG, Calore JD, Catanese A, Franzblau C, Snider GL (1985) Effect of combined human neutrophil cathepsin G and elastase on induction of secretory cell metaplasia and emphysema in hamsters, with in vitro observations on elastolysis by these enzymes. Am Rev Respir Dis 132:362–366
50. Breuer R, Christensen TG, Lucey EC, Stone PJ, Snider GL (1985) Quantitative study of secretory cell metaplasia induced by human neutrophil elastase in the large bronchi of hamsters. J Lab Clin Med 105:635–640
51. Voynow JA, Fischer BM, Malarkey DE, Burch LH, Wong T, Longphre M, Ho SB, Foster WM (2004) Neutrophil elastase induces mucus cell metaplasia in mouse lung. Am J Physiol Lung Cell Mol Physiol 287:L1293–1302
52. Voynow JA, Rubin BK (2009) Mucins, mucus, and sputum. Chest 135:505–512
53. von Garnier C, Filgueira L, Wikstrom M, Smith M, Thomas JA, Strickland DH, Holt PG, Stumbles PA (2005) Anatomical location determines the distribution and function of dendritic cells and other APCs in the respiratory tract. J Immunol 175:1609–1618
54. Vermaelen K, Pauwels R (2005) Pulmonary dendritic cells. Am J Respir Crit Care Med 172:530–551
55. Tsoumakidou M, Demedts IK, Brusselle GG, Jeffery PK (2008) Dendritic cells in chronic obstructive pulmonary disease: new players in an old game. Am J Respir Crit Care Med 177:1180–1186
56. Roghanian A, Drost EM, MacNee W, Howie SE, Sallenave JM (2006) Inflammatory lung secretions inhibit dendritic cell maturation and function via neutrophil elastase. Am J Respir Crit Care Med 174:1189–1198
57. Maffia PC, Zittermann SE, Scimone ML, Tateosian N, Amiano N, Guerrieri D, Lutzky V, Rosso D, Romeo HE, Garcia VE et al (2007) Neutrophil elastase converts human immature dendritic cells into transforming growth factor-beta1-secreting cells and reduces allostimulatory ability. Am J Pathol 171:928–937
58. Knight AM, Lucocq JM, Prescott AR, Ponnambalam S, Watts C (1997) Antigen endocytosis and presentation mediated by human membrane IgG1 in the absence of the Ig(alpha)/Ig(beta) dimer. EMBO J 16:3842–3850
59. Bode W, Meyer E Jr, Powers JC (1989) Human leukocyte and porcine pancreatic elastase: X-ray crystal structures, mechanism, substrate specificity, and mechanism-based inhibitors. Biochemistry 28:1951–1963
60. Kelly E, Greene CM, McElvaney NG (2008) Targeting neutrophil elastase in cystic fibrosis. Expert Opin Ther Targets 12:145–157
61. Lee WL, Downey GP (2001) Leukocyte elastase: physiological functions and role in acute lung injury. Am J Respir Crit Care Med 164:896–904
62. Belaaouaj A, McCarthy R, Baumann M, Gao Z, Ley TJ, Abraham SN, Shapiro SD (1998) Mice lacking neutrophil elastase reveal impaired host defense against gram negative bacterial sepsis. Nat Med 4:615–618
63. Ariel A, Yavin EJ, Hershkoviz R, Avron A, Franitza S, Hardan I, Cahalon L, Fridkin M, Lider O (1998) IL-2 induces T cell adherence to extracellular matrix: inhibition of adherence and migration by IL-2 peptides generated by leukocyte elastase. J Immunol 161:2465–2472
64. Shapiro SD (1994) Elastolytic metalloproteinases produced by human mononuclear phagocytes. Potential roles in destructive lung disease. Am J Respir Crit Care Med 150: S160–S164
65. Reeves EP, Lu H, Jacobs HL, Messina CG, Bolsover S, Gabella G, Potma EO, Warley A, Roes J, Segal AW (2002) Killing activity of neutrophils is mediated through activation of proteases by K + flux. Nature 416:291–297

66. Hirche TO, Benabid R, Deslee G, Gangloff S, Achilefu S, Guenounou M, Lebargy F, Hancock RE, Belaaouaj A (2008) Neutrophil elastase mediates innate host protection against Pseudomonas aeruginosa. J Immunol 181:4945–4954
67. Standish AJ, Weiser JN (2009) Human neutrophils kill Streptococcus pneumoniae via serine proteases. J Immunol 183:2602–2609
68. Weinrauch Y, Drujan D, Shapiro SD, Weiss J, Zychlinsky A (2002) Neutrophil elastase targets virulence factors of enterobacteria. Nature 417:91–94
69. Sonawane A, Jyot J, During R, Ramphal R (2006) Neutrophil elastase, an innate immunity effector molecule, represses flagellin transcription in *Pseudomonas aeruginosa*. Infect Immun 74:6682–6689
70. Devaney JM, Greene CM, Taggart CC, Carroll TP, O'Neill SJ, McElvaney NG (2003) Neutrophil elastase up-regulates interleukin-8 via toll-like receptor 4. FEBS Lett 544:129–132
71. Geraghty P, Rogan MP, Greene CM, Boxio RM, Poiriert T, O'Mahony M, Belaaouaj A, O'Neill SJ, Taggart CC, McElvaney NG (2007) Neutrophil elastase up-regulates cathepsin B and matrix metalloprotease-2 expression. J Immunol 178:5871–5878
72. Lee KY, Ho SC, Lin HC, Lin SM, Liu CY, Huang CD, Wang CH, Chung KF, Kuo HP (2006) Neutrophil-derived elastase induces TGF-beta1 secretion in human airway smooth muscle via NF-kappaB pathway. Am J Respir Cell Mol Biol 35:407–414
73. Walsh DE, Greene CM, Carroll TP, Taggart CC, Gallagher PM, O'Neill SJ, McElvaney NG (2001) Interleukin-8 up-regulation by neutrophil elastase is mediated by MyD88/IRAK/TRAF-6 in human bronchial epithelium. J Biol Chem 276:35494–35499
74. Couillin I, Vasseur V, Charron S, Gasse P, Tavernier M, Guillet J, Lagente V, Fick L, Jacobs M, Coelho FR et al (2009) IL-1R1/MyD88 signaling is critical for elastase-induced lung inflammation and emphysema. J Immunol 183:8195–8202
75. Suzuki T, Yamashita C, Zemans RL, Briones N, Van Linden A, Downey GP (2009) Leukocyte elastase induces lung epithelial apoptosis via a PAR-1-, NF-kappaB-, and p53-dependent pathway. Am J Respir Cell Mol Biol 41:742–755
76. Renesto P, Si-Tahar M, Moniatte M, Balloy V, Van Dorsselaer A, Pidard D, Chignard M (1997) Specific inhibition of thrombin-induced cell activation by the neutrophil proteinases elastase, cathepsin G, and proteinase 3: evidence for distinct cleavage sites within the aminoterminal domain of the thrombin receptor. Blood 89:1944–1953
77. Roghanian A, Sallenave JM (2008) Neutrophil elastase (NE) and NE inhibitors: canonical and noncanonical functions in lung chronic inflammatory diseases (cystic fibrosis and chronic obstructive pulmonary disease). J Aerosol Med Pulm Drug Deliv 21:125–144
78. Sallenave JM (2010) Secretory leukocyte protease inhibitor and elafin/trappin-2: versatile mucosal antimicrobials and regulators of immunity. Am J Respir Cell Mol Biol 42:635–643
79. Sallenave JM, Shulmann J, Crossley J, Jordana M, Gauldie J (1994) Regulation of secretory leukocyte proteinase inhibitor (SLPI) and elastase-specific inhibitor (ESI/elafin) in human airway epithelial cells by cytokines and neutrophilic enzymes. Am J Respir Cell Mol Biol 11:733–741
80. Williams SE, Brown TI, Roghanian A, Sallenave JM (2006) SLPI and elafin: one glove, many fingers. Clin Sci (Lond) 110:21–35
81. Wiedow O, Luademann J, Utecht B (1991) Elafin is a potent inhibitor of proteinase 3. Biochem Biophys Res Commun 174:6–10
82. Schalkwijk J, van Vlijmen IM, Alkemade JA, de Jongh GJ (1993) Immunohistochemical localization of SKALP/elafin in psoriatic epidermis. J Invest Dermatol 100:390–393
83. Sallenave JM, Ryle AP (1991) Purification and characterization of elastase-specific inhibitor. Sequence homology with mucus proteinase inhibitor. Biol Chem Hoppe Seyler 372:13–21
84. Sallenave JM, Marsden MD, Ryle AP (1992) Isolation of elafin and elastase-specific inhibitor (ESI) from bronchial secretions. Evidence of sequence homology and immunological cross-reactivity. Biol Chem Hoppe Seyler 373:27–33

85. Saheki T, Ito F, Hagiwara H, Saito Y, Kuroki J, Tachibana S, Hirose S (1992) Primary structure of the human elafin precursor preproelafin deduced from the nucleotide sequence of its gene and the presence of unique repetitive sequences in the prosegment. Biochem Biophys Res Commun 185:240–245
86. Sallenave JM, Silva A (1993) Characterization and gene sequence of the precursor of elafin, an elastase-specific inhibitor in bronchial secretions. Am J Respir Cell Mol Biol 8:439–445
87. Nara K, Ito S, Ito T, Suzuki Y, Ghoneim MA, Tachibana S, Hirose S (1994) Elastase inhibitor elafin is a new type of proteinase inhibitor which has a transglutaminase-mediated anchoring sequence termed "cementoin". J Biochem 115:441–448
88. Wiedow O, Schroder JM, Gregory H, Young JA, Christophers E (1990) Elafin: an elastase-specific inhibitor of human skin. Purification, characterization, and complete amino acid sequence. J Biol Chem 265:14791–14795
89. Ying QL, Simon SR (2001) Kinetics of the inhibition of proteinase 3 by elafin. Am J Respir Cell Mol Biol 24:83–89
90. Tanaka N, Fujioka A, Tajima S, Ishibashi A, Hirose S (2000) Elafin is induced in epidermis in skin disorders with dermal neutrophilic infiltration: interleukin-1 beta and tumour necrosis factor-alpha stimulate its secretion in vitro. Br J Dermatol 143:728–732
91. Pfundt R, van Vlijmen-Willems I, Bergers M, Wingens M, Cloin W, Schalkwijk J (2001) In situ demonstration of phosphorylated c-jun and p38 MAP kinase in epidermal keratinocytes following ultraviolet B irradiation of human skin. J Pathol 193:248–255
92. Pfundt R, Wingens M, Bergers M, Zweers M, Frenken M, Schalkwijk J (2000) TNF-alpha and serum induce SKALP/elafin gene expression in human keratinocytes by a p38 MAP kinase-dependent pathway. Arch Dermatol Res 292:180–187
93. Bingle L, Tetley TD, Bingle CD (2001) Cytokine-mediated induction of the human elafin gene in pulmonary epithelial cells is regulated by nuclear factor-kappaB. Am J Respir Cell Mol Biol 25:84–91
94. Van Seuningen I, Audie JP, Gosselin B, Lafitte JJ, Davril M (1995) Expression of human mucous proteinase inhibitor in respiratory tract: a study by in situ hybridization. J Histochem Cytochem 43:645–648
95. Tremblay GM, Sallenave JM, Israel-Assayag E, Cormier Y, Gauldie J (1996) Elafin/elastase-specific inhibitor in bronchoalveolar lavage of normal subjects and farmer's lung. Am J Respir Crit Care Med 154:1092–1098
96. Sallenave JM, Donnelly SC, Grant IS, Robertson C, Gauldie J, Haslett C (1999) Secretory leukocyte proteinase inhibitor is preferentially increased in patients with acute respiratory distress syndrome. Eur Respir J 13:1029–1036
97. Reid PT, Marsden ME, Cunningham GA, Haslett C, Sallenave JM (1999) Human neutrophil elastase regulates the expression and secretion of elafin (elastase-specific inhibitor) in type II alveolar epithelial cells. FEBS Lett 457:33–37
98. van Wetering S, van der Linden AC, van Sterkenburg MA, de Boer WI, Kuijpers AL, Schalkwijk J, Hiemstra PS (2000) Regulation of SLPI and elafin release from bronchial epithelial cells by neutrophil defensins. Am J Physiol Lung Cell Mol Physiol 278:L51–58
99. Simpson AJ, Maxwell AI, Govan JR, Haslett C, Sallenave JM (1999) Elafin (elastase-specific inhibitor) has anti-microbial activity against gram-positive and gram-negative respiratory pathogens. FEBS Lett 452:309–313
100. Meyer-Hoffert U, Wichmann N, Schwichtenberg L, White PC, Wiedow O (2003) Supernatants of Pseudomonas aeruginosa induce the Pseudomonas-specific antibiotic elafin in human keratinocytes. Exp Dermatol 12:418–425
101. Bellemare A, Vernoux N, Morisset D, Bourbonnais Y (2008) Human pre-elafin inhibits a Pseudomonas aeruginosa-secreted peptidase and prevents its proliferation in complex media. Antimicrob Agents Chemother 52:483–490
102. McMichael JW, Maxwell AI, Hayashi K, Taylor K, Wallace WA, Govan JR, Dorin JR, Sallenave JM (2005) Antimicrobial activity of murine lung cells against Staphylococcus aureus is increased in vitro and in vivo after elafin gene transfer. Infect Immun 73:3609–3617

103. Henriksen PA, Hitt M, Xing Z, Wang J, Haslett C, Riemersma RA, Webb DJ, Kotelevtsev YV, Sallenave JM (2004) Adenoviral gene delivery of elafin and secretory leukocyte protease inhibitor attenuates NF-kappa B-dependent inflammatory responses of human endothelial cells and macrophages to atherogenic stimuli. J Immunol 172:4535–4544
104. Butler MW, Robertson I, Greene CM, O'Neill SJ, Taggart CC, McElvaney NG (2006) Elafin prevents lipopolysaccharide-induced AP-1 and NF-kappaB activation via an effect on the ubiquitin-proteasome pathway. J Biol Chem 281:34730–34735
105. Wilkinson TS, Dhaliwal K, Hamilton TW, Lipka AF, Farrell L, Davidson DJ, Duffin R, Morris AC, Haslett C, Govan JR et al (2009) Trappin-2 promotes early clearance of Pseudomonas aeruginosa through CD14-dependent macrophage activation and neutrophil recruitment. Am J Pathol 174:1338–1346
106. McMichael JW, Roghanian A, Jiang L, Ramage R, Sallenave JM (2005) The antimicrobial antiproteinase elafin binds to lipopolysaccharide and modulates macrophage responses. Am J Respir Cell Mol Biol 32:443–452
107. Simpson AJ, Cunningham GA, Porteous DJ, Haslett C, Sallenave JM (2001) Regulation of adenovirus-mediated elafin transgene expression by bacterial lipopolysaccharide. Hum Gene Ther 12:1395–1406
108. Sallenave JM, Cunningham GA, James RM, McLachlan G, Haslett C (2003) Regulation of pulmonary and systemic bacterial lipopolysaccharide responses in transgenic mice expressing human elafin. Infect Immun 71:3766–3774
109. Roghanian A, Williams SE, Sheldrake TA, Brown TI, Oberheim K, Xing Z, Howie SE, Sallenave JM (2006) The antimicrobial/elastase inhibitor elafin regulates lung dendritic cells and adaptive immunity. Am J Respir Cell Mol Biol 34:634–642
110. Thompson RC, Ohlsson K (1986) Isolation, properties, and complete amino acid sequence of human secretory leukocyte protease inhibitor, a potent inhibitor of leukocyte elastase. Proc Natl Acad Sci USA 83:6692–6696
111. Zitnik RJ, Zhang J, Kashem MA, Kohno T, Lyons DE, Wright CD, Rosen E, Goldberg I, Hayday AC (1997) The cloning and characterization of a murine secretory leukocyte protease inhibitor cDNA. Biochem Biophys Res Commun 232:687–697
112. Song X, Zeng L, Jin W, Thompson J, Mizel DE, Lei K, Billinghurst RC, Poole AR, Wahl SM (1999) Secretory leukocyte protease inhibitor suppresses the inflammation and joint damage of bacterial cell wall-induced arthritis. J Exp Med 190:535–542
113. Brown TI, Mistry R, Gray R, Imrie M, Collie DD, Sallenave JM (2005) Characterization of the ovine ortholog of secretory leukoprotease inhibitor. Mamm Genome 16:621–630
114. Grutter MG, Fendrich G, Huber R, Bode W (1988) The 2.5 A X-ray crystal structure of the acid-stable proteinase inhibitor from human mucous secretions analysed in its complex with bovine alpha-chymotrypsin. EMBO J 7:345–351
115. Hiemstra PS, Maassen RJ, Stolk J, Heinzel-Wieland R, Steffens GJ, Dijkman JH (1996) Antibacterial activity of antileukoprotease. Infect Immun 64:4520–4524
116. Gauthier F, Fryksmark U, Ohlsson K, Bieth JG (1982) Kinetics of the inhibition of leukocyte elastase by the bronchial inhibitor. Biochim Biophys Acta 700:178–183
117. Sallenave JM, Si Tahar M, Cox G, Chignard M, Gauldie J (1997) Secretory leukocyte proteinase inhibitor is a major leukocyte elastase inhibitor in human neutrophils. J Leukoc Biol 61:695–702
118. Westin U, Lundberg E, Wihl JA, Ohlsson K (1999) The effect of immediate-hypersensitivity reactions on the level of SLPI, granulocyte elastase, alpha1-antitrypsin, and albumin in nasal secretions, by the method of unilateral antigen challenge. Allergy 54:857–864
119. Mihaila A, Tremblay GM (2001) Human alveolar macrophages express elafin and secretory leukocyte protease inhibitor. Z Naturforsch C 56:291–297
120. Vogelmeier C, Biedermann T, Maier K, Mazur G, Behr J, Krombach F, Buhl R (1997) Comparative loss of activity of recombinant secretory leukoprotease inhibitor and alpha 1-protease inhibitor caused by different forms of oxidative stress. Eur Respir J 10:2114–2119

121. Kouchi I, Yasuoka S, Ueda Y, Ogura T (1993) Analysis of secretory leukocyte protease inhibitor (SLPI) in bronchial secretions from patients with hypersecretory respiratory diseases. Tokushima J Exp Med 40:95–107
122. Lee CH, Igarashi Y, Hohman RJ, Kaulbach H, White MV, Kaliner MA (1993) Distribution of secretory leukoprotease inhibitor in the human nasal airway. Am Rev Respir Dis 147:710–716
123. Grobmyer SR, Barie PS, Nathan CF, Fuortes M, Lin E, Lowry SF, Wright CD, Weyant MJ, Hydo L, Reeves F et al (2000) Secretory leukocyte protease inhibitor, an inhibitor of neutrophil activation, is elevated in serum in human sepsis and experimental endotoxemia. Crit Care Med 28:1276–1282
124. Duits LA, Tjabringa GS, Aarts NJ, Hiemstra PS, Nibbering PH, van Dissel JT, Van't Wout JW (2003) Plasma secretory leukocyte protease inhibitor in febrile patients. Clin Microbiol Infect 9:605–613
125. Jin FY, Nathan C, Radzioch D, Ding A (1997) Secretory leukocyte protease inhibitor: a macrophage product induced by and antagonistic to bacterial lipopolysaccharide. Cell 88:417–426
126. Gipson TS, Bless NM, Shanley TP, Crouch LD, Bleavins MR, Younkin EM, Sarma V, Gibbs DF, Tefera W, McConnell PC et al (1999) Regulatory effects of endogenous protease inhibitors in acute lung inflammatory injury. J Immunol 162:3653–3662
127. Taggart CC, Cryan SA, Weldon S, Gibbons A, Greene CM, Kelly E, Low TB, O'Neill SJ, McElvaney NG (2005) Secretory leucoprotease inhibitor binds to NF-kappaB binding sites in monocytes and inhibits p65 binding. J Exp Med 202:1659–1668
128. Nakamura A, Mori Y, Hagiwara K, Suzuki T, Sakakibara T, Kikuchi T, Igarashi T, Ebina M, Abe T, Miyazaki J et al (2003) Increased susceptibility to LPS-induced endotoxin shock in secretory leukoprotease inhibitor (SLPI)-deficient mice. J Exp Med 197:669–674
129. McNeely TB, Dealy M, Dripps DJ, Orenstein JM, Eisenberg SP, Wahl SM (1995) Secretory leukocyte protease inhibitor: a human saliva protein exhibiting anti-human immunodeficiency virus 1 activity in vitro. J Clin Invest 96:456–464
130. McNeely TB, Shugars DC, Rosendahl M, Tucker C, Eisenberg SP, Wahl SM (1997) Inhibition of human immunodeficiency virus type 1 infectivity by secretory leukocyte protease inhibitor occurs prior to viral reverse transcription. Blood 90:1141–1149
131. Tomee JF, Hiemstra PS, Heinzel-Wieland R, Kauffman HF (1997) Antileukoprotease: an endogenous protein in the innate mucosal defense against fungi. J Infect Dis 176:740–747
132. Hocini H, Becquart P, Bouhlal H, Adle-Biassette H, Kazatchkine MD, Belec L (2000) Secretory leukocyte protease inhibitor inhibits infection of monocytes and lymphocytes with human immunodeficiency virus type 1 but does not interfere with transcytosis of cell-associated virus across tight epithelial barriers. Clin Diagn Lab Immunol 7:515–518
133. King AE, Critchley HO, Kelly RW (2003) Innate immune defences in the human endometrium. Reprod Biol Endocrinol 1:116
134. Chattopadhyay A, Gray LR, Patton LL, Caplan DJ, Slade GD, Tien HC, Shugars DC (2004) Salivary secretory leukocyte protease inhibitor and oral candidiasis in human immunodeficiency virus type 1-infected persons. Infect Immun 72:1956–1963
135. Sallenave JM, Tremblay GM, Gauldie J, Richards CD (1997) Oncostatin M, but not interleukin-6 or leukemia inhibitory factor, stimulates expression of alpha1-proteinase inhibitor in A549 human alveolar epithelial cells. J Interferon Cytokine Res 17:337–346
136. Paakko P, Kirby M, du Bois RM, Gillissen A, Ferrans VJ, Crystal RG (1996) Activated neutrophils secrete stored alpha 1-antitrypsin. Am J Respir Crit Care Med 154:1829–1833
137. Lomas DA, Mahadeva R (2002) Alpha1-antitrypsin polymerization and the serpinopathies: pathobiology and prospects for therapy. J Clin Invest 110:1585–1590
138. Stecenko AA, Brigham KL (2003) Gene therapy progress and prospects: alpha-1 antitrypsin. Gene Ther 10:95–99
139. Chughtai B, O'Riordan TG (2004) Potential role of inhibitors of neutrophil elastase in treating diseases of the airway. J Aerosol Med 17:289–298

140. Boutten A, Venembre P, Seta N, Hamelin J, Aubier M, Durand G, Dehoux MS (1998) Oncostatin M is a potent stimulator of alpha1-antitrypsin secretion in lung epithelial cells: modulation by transforming growth factor-beta and interferon-gamma. Am J Respir Cell Mol Biol 18:511–520
141. Campbell EJ, Campbell MA, Boukedes SS, Owen CA (1999) Quantum proteolysis by neutrophils: implications for pulmonary emphysema in alpha 1-antitrypsin deficiency. J Clin Invest 104:337–344
142. Parmar JS, Mahadeva R, Reed BJ, Farahi N, Cadwallader KA, Keogan MT, Bilton D, Chilvers ER, Lomas DA (2002) Polymers of alpha(1)-antitrypsin are chemotactic for human neutrophils: a new paradigm for the pathogenesis of emphysema. Am J Respir Cell Mol Biol 26:723–730
143. Stockley RA, Bayley DL, Unsal I, Dowson LJ (2002) The effect of augmentation therapy on bronchial inflammation in alpha1-antitrypsin deficiency. Am J Respir Crit Care Med 165:1494–1498
144. Spencer LT, Paone G, Krein PM, Rouhani FN, Rivera-Nieves J, Brantly ML (2004) Role of human neutrophil peptides in lung inflammation associated with alpha1-antitrypsin deficiency. Am J Physiol Lung Cell Mol Physiol 286:L514–520
145. Aldonyte R, Jansson L, Janciauskiene S (2004) Concentration-dependent effects of native and polymerised alpha1-antitrypsin on primary human monocytes, in vitro. BMC Cell Biol 5:11
146. Janciauskiene S, Larsson S, Larsson P, Virtala R, Jansson L, Stevens T (2004) Inhibition of lipopolysaccharide-mediated human monocyte activation, in vitro, by alpha1-antitrypsin. Biochem Biophys Res Commun 321:592–600
147. Petrache I, Fijalkowska I, Medler TR, Skirball J, Cruz P, Zhen L, Petrache HI, Flotte TR, Tuder RM (2006) Alpha-1 antitrypsin inhibits caspase-3 activity, preventing lung endothelial cell apoptosis. Am J Pathol 169:1155–1166
148. Petrache I, Fijalkowska I, Zhen L, Medler TR, Brown E, Cruz P, Choe KH, Taraseviciene-Stewart L, Scerbavicius R, Shapiro L et al (2006) A novel antiapoptotic role for alpha1-antitrypsin in the prevention of pulmonary emphysema. Am J Respir Crit Care Med 173:1222–1228
149. Modrykamien A, Stoller JK (2009) Alpha-1 antitrypsin (AAT) deficiency – what are the treatment options? Expert Opin Pharmacother 10:2653–2661
150. Loew D, Perrault C, Morales M, Moog S, Ravanat C, Schuhler S, Arcone R, Pietropaolo C, Cazenave JP, van Dorsselaer A et al (2000) Proteolysis of the exodomain of recombinant protease-activated receptors: prediction of receptor activation or inactivation by MALDI mass spectrometry. Biochemistry 39:10812–10822
151. Dulon S, Cande C, Bunnett NW, Hollenberg MD, Chignard M, Pidard D (2003) Proteinase-activated receptor-2 and human lung epithelial cells: disarming by neutrophil serine proteinases. Am J Respir Cell Mol Biol 28:339–346
152. Dulon S, Leduc D, Cottrell GS, D'Alayer J, Hansen KK, Bunnett NW, Hollenberg MD, Pidard D, Chignard M (2005) Pseudomonas aeruginosa elastase disables proteinase-activated receptor 2 in respiratory epithelial cells. Am J Respir Cell Mol Biol 32:411–419
153. Cumashi A, Ansuini H, Celli N, De Blasi A, O'Brien PJ, Brass LF, Molino M (2001) Neutrophil proteases can inactivate human PAR3 and abolish the co-receptor function of PAR3 on murine platelets. Thromb Haemost 85:533–538
154. Bank U, Reinhold D, Schneemilch C, Kunz D, Synowitz HJ, Ansorge S (1999) Selective proteolytic cleavage of IL-2 receptor and IL-6 receptor ligand binding chains by neutrophil-derived serine proteases at foci of inflammation. J Interferon Cytokine Res 19:1277–1287
155. Porteu F, Brockhaus M, Wallach D, Engelmann H, Nathan CF (1991) Human neutrophil elastase releases a ligand-binding fragment from the 75-kDa tumor necrosis factor (TNF) receptor. Comparison with the proteolytic activity responsible for shedding of TNF receptors from stimulated neutrophils. J Biol Chem 266:18846–18853

156. Tralau T, Meyer-Hoffert U, Schroder JM, Wiedow O (2004) Human leukocyte elastase and cathepsin G are specific inhibitors of C5a-dependent neutrophil enzyme release and chemotaxis. Exp Dermatol 13:316–325
157. Sadallah S, Hess C, Miot S, Spertini O, Lutz H, Schifferli JA (1999) Elastase and metalloproteinase activities regulate soluble complement receptor 1 release. Eur J Immunol 29:3754–3761
158. Beaufort N, Leduc D, Rousselle JC, Magdolen V, Luther T, Namane A, Chignard M, Pidard D (2004) Proteolytic regulation of the urokinase receptor/CD87 on monocytic cells by neutrophil elastase and cathepsin G. J Immunol 172:540–549
159. Hunter MG, Druhan LJ, Massullo PR, Avalos BR (2003) Proteolytic cleavage of granulocyte colony-stimulating factor and its receptor by neutrophil elastase induces growth inhibition and decreased cell surface expression of the granulocyte colony-stimulating factor receptor. Am J Hematol 74:149–155
160. Piper MG, Massullo PR, Loveland M, Druhan LJ, Kindwall-Keller TL, Ai J, Copelan A, Avalos BR (2010) Neutrophil elastase downmodulates native G-CSFR expression and granulocyte-macrophage colony formation. J Inflamm (Lond) 7:5
161. Halbwachs-Mecarelli L, Bessou G, Lesavre P, Renesto P, Chignard M (1996) Neutrophil serine proteases are most probably involved in the release of CD43 (leukosialin, sialophorin) from the neutrophil membrane during cell activation. Blood 87:1200–1202
162. Weber S, Babina M, Hermann B, Henz BM (1997) Leukosialin (CD43) is proteolytically cleaved from stimulated HMC-1 cells. Immunobiology 197:82–96
163. Le-Barillec K, Si-Tahar M, Balloy V, Chignard M (1999) Proteolysis of monocyte CD14 by human leukocyte elastase inhibits lipopolysaccharide-mediated cell activation. J Clin Invest 103:1039–1046
164. Doring G, Frank F, Boudier C, Herbert S, Fleischer B, Bellon G (1995) Cleavage of lymphocyte surface antigens CD2, CD4, and CD8 by polymorphonuclear leukocyte elastase and cathepsin G in patients with cystic fibrosis. J Immunol 154:4842–4850
165. Duvoix A, Mackay RM, Henderson N, McGreal E, Postle A, Reid K, Clark H (2010) Physiological concentration of calcium inhibits elastase-induced cleavage of a functional recombinant fragment of surfactant protein D. Immunobiology 216:72–9
166. Hartl D, Latzin P, Hordijk P, Marcos V, Rudolph C, Woischnik M, Krauss-Etschmann S, Koller B, Reinhardt D, Roscher AA et al (2007) Cleavage of CXCR1 on neutrophils disables bacterial killing in cystic fibrosis lung disease. Nat Med 13:1423–1430
167. van Kessel KP, van Strijp JA, Verhoef J (1991) Inactivation of recombinant human tumor necrosis factor-alpha by proteolytic enzymes released from stimulated human neutrophils. J Immunol 147:3862–3868
168. Bank U, Kupper B, Reinhold D, Hoffmann T, Ansorge S (1999) Evidence for a crucial role of neutrophil-derived serine proteases in the inactivation of interleukin-6 at sites of inflammation. FEBS Lett 461:235–240
169. Leavell KJ, Peterson MW, Gross TJ (1997) Human neutrophil elastase abolishes interleukin-8 chemotactic activity. J Leukoc Biol 61:361–366
170. Champagne B, Tremblay P, Cantin A, St Pierre Y (1998) Proteolytic cleavage of ICAM-1 by human neutrophil elastase. J Immunol 161:6398–6405
171. Robledo O, Papaioannou A, Ochietti B, Beauchemin C, Legault D, Cantin A, King PD, Daniel C, Alakhov VY, Potworowski EF et al (2003) ICAM-1 isoforms: specific activity and sensitivity to cleavage by leukocyte elastase and cathepsin G. Eur J Immunol 33:1351–1360
172. Hermant B, Bibert S, Concord E, Dublet B, Weidenhaupt M, Vernet T, Gulino-Debrac D (2003) Identification of proteases involved in the proteolysis of vascular endothelium cadherin during neutrophil transmigration. J Biol Chem 278:14002–14012
173. Zhu J, Nathan C, Jin W, Sim D, Ashcroft GS, Wahl SM, Lacomis L, Erdjument-Bromage H, Tempst P, Wright CD et al (2002) Conversion of proepithelin to epithelins: roles of SLPI and elastase in host defense and wound repair. Cell 111:867–878

174. Massberg S, Grahl L, von Bruehl ML, Manukyan D, Pfeiler S, Goosmann C, Brinkmann V, Lorenz M, Bidzhekov K, Khandagale AB et al (2010) Reciprocal coupling of coagulation and innate immunity via neutrophil serine proteases. Nat Med 16:887–896
175. Jackson PL, Xu X, Wilson L, Weathington NM, Clancy JP, Blalock JE, Gaggar A (2010) Human neutrophil elastase-mediated cleavage sites of MMP-9 and TIMP-1: implications to cystic fibrosis proteolytic dysfunction. Mol Med 16:159–166
176. Ai S, Cheng XW, Inoue A, Nakamura K, Okumura K, Iguchi A, Murohara T, Kuzuya M (2007) Angiogenic activity of bFGF and VEGF suppressed by proteolytic cleavage by neutrophil elastase. Biochem Biophys Res Commun 364:395–401
177. Kurtagic E, Jedrychowski MP, Nugent MA (2009) Neutrophil elastase cleaves VEGF to generate a VEGF fragment with altered activity. Am J Physiol Lung Cell Mol Physiol 296: L534–L546
178. Mydel P, Shipley JM, Adair-Kirk TL, Kelley DG, Broekelmann TJ, Mecham RP, Senior RM (2008) Neutrophil elastase cleaves laminin-332 (laminin-5) generating peptides that are chemotactic for neutrophils. J Biol Chem 283:9513–9522
179. Cooley J, McDonald B, Accurso FJ, Crouch EC, Remold-O'Donnell E (2008) Patterns of neutrophil serine protease-dependent cleavage of surfactant protein D in inflammatory lung disease. J Leukoc Biol 83:946–955
180. Houghton AM, Rzymkiewicz DM, Ji H, Gregory AD, Egea EE, Metz HE, Stolz DB, Land SR, Marconcini LA, Kliment CR et al (2010) Neutrophil elastase-mediated degradation of IRS-1 accelerates lung tumor growth. Nat Med 16:219–223
181. Raife TJ, Cao W, Atkinson BS, Bedell B, Montgomery RR, Lentz SR, Johnson GF, Zheng XL (2009) Leukocyte proteases cleave von Willebrand factor at or near the ADAMTS13 cleavage site. Blood 114:1666–1674
182. Goulet B, Markovic Y, Leduy L, Nepveu A (2008) Proteolytic processing of cut homeobox 1 by neutrophil elastase in the MV4;11 myeloid leukemia cell line. Mol Cancer Res 6:644–653
183. Bagoly Z, Fazakas F, Komaromi I, Haramura G, Toth E, Muszbek L (2008) Cleavage of factor XIII by human neutrophil elastase results in a novel active truncated form of factor XIII A subunit. Thromb Haemost 99:668–674
184. Si-Tahar M, Pidard D, Balloy V, Moniatte M, Kieffer N, Van Dorsselaer A, Chignard M (1997) Human neutrophil elastase proteolytically activates the platelet integrin alphaIIbbeta3 through cleavage of the carboxyl terminus of the alphaIIb subunit heavy chain. Involvement in the potentiation of platelet aggregation. J Biol Chem 272:11636–11647

Kallikrein Protease Involvement in Skin Pathologies Supports a New View of the Origin of Inflamed Itchy Skin

Azza Eissa and Eleftherios P. Diamandis

Abstract Skin barrier defects in common dermatological diseases, such as atopic dermatitis and psoriasis, are mostly attributed to anomalies in T-cell immunity. A new viewpoint of inflammatory dermatoses onset was recently suggested, in which barrier defects trigger secretion of pro-inflammatory mediators by stressed keratinocyte cells, which activate the T-cell immune system and further deteriorate the barrier. Herein, we review epidermal keratinocytes as active immune cells. In particular, we focus on recent groundbreaking evidence on the role of keratinocyte-secreted kallikreins as inflammation and allergy mediators. Kallikreins are skin surface proteases known for their role in digesting adhesion proteins and maintaining barrier integrity and function. Kallikrein hyperactivity in skin pathologies was recently shown to mediate inflammation secondary to inherited and acquired barrier defects, in support of the epidermal roots of inflamed and itchy skin. Hence, future therapy design should be directed toward ameliorating keratinocyte-induced barrier defects and inflammation, alone or in combination with dampening T-cell immune responses.

A. Eissa
Department of Laboratory Medicine and Pathobiology, University of Toronto, Toronto, ON M5S 1A8, Canada

Department of Pathology and Laboratory Medicine, Mount Sinai Hospital, Joseph and Wolf Lebovic Centre, 60 Murray Street, Toronto, ON M5T 3L9, Canada

E.P. Diamandis (✉)
Department of Laboratory Medicine and Pathobiology, University of Toronto, Toronto, ON M5S 1A8, Canada

Department of Pathology and Laboratory Medicine, Mount Sinai Hospital, Joseph and Wolf Lebovic Centre, 60 Murray Street, Toronto, ON M5T 3L9, Canada

Department of Clinical Biochemistry, University Health Network, Toronto, ON M5G 1X5, Canada
e-mail: ediamandis@mtsinai.on.ca

N. Vergnolle and M. Chignard (eds.), *Proteases and Their Receptors in Inflammation*, Progress in Inflammation Research, DOI 10.1007/978-3-0348-0157-7_3,

Keywords Allergy • Antimicrobial barrier • Atopic dermatitis • Cathelicidin • Desquamation • Inflammation • Kallikrein serine proteases • Netherton syndrome • Permeability barrier • Proteinase-activated receptors • Psoriasis • Rosacea • Stratum corneum

1 Introduction

Impaired skin barrier is a hallmark of many common inflammatory skin diseases, including psoriasis vulgaris and atopic dermatitis. Two opposing paradigms have been proposed to explain concurrent barrier defects and inflammatory symptoms in skin disorders. The "inside-out" theory postulates that skin barrier breakdown is a secondary response to the inflammation process that occurs due to activation of immune cells by allergen and/or irritants. This premise is supported by the fact that neutrophil serine proteases, such as leukocyte elastase and cathepsin G, released during inflammation, dissociate keratinocytes from epidermal sheets [1] and Th2-generated cytokines, such as IL-4, inhibit proper skin barrier formation [2, 3]. On the other hand, the "outside-in" theory postulates that skin barrier defects drive the inflammatory response. Harsh environmental stimuli, UV radiation, chemical, physical or pathogen penetration through the skin surface causes stressed keratinocyte cells in the epidermis to secrete pro-inflammatory cytokines such as TNF-α, IFN-γ, IL-1, and GM-CSF [4, 5]. These cytokines act in an autocrine fashion to induce keratinocyte proliferation to repair the wounded area, and in paracrine and endocrine fashion to stimulate local and systemic inflammatory responses. The current theories remain under intense debate, even though aberrant lymphocyte activation is often considered the root cause of inflammatory skin diseases, in line with the "inside-out" theory.

Very recently, researchers have begun to challenge the prevailing dogma that inflammatory skin diseases, such as atopic dermatitis (eczema), are mainly allergic diseases that lead to barrier defects such as dryness and rashes. Instead, a growing number of experts currently consider a structural barrier defect to be the primary cause, instigating immunologic problems seen in AD patients. For instance, while 80% of nonatopic dermatitis patients progress to develop "atopy" characterized by elevated IgE levels and immune hyperactivity, 20% of patients continue as "nonatopic" having barrier defects without developing elevated IgE [6]. The extent of barrier abnormality in these patients parallels disease severity [6, 7], further suggesting that a defective barrier drives the majority of immune hyperactivity in AD by allowing penetration of allergens through the skin, which stimulates immune cells. Furthermore, changes in at least three groups of barrier genes encoding epidermal structural proteins, proteases, and protease inhibitors were found to increase one's susceptibility to develop AD or psoriasis inflammatory skin diseases [6, 8]. For example, several loss-of-function mutations in the gene encoding *filament aggregating protein* (filaggrin or FLG) have been identified in a considerable number of AD patients [9–11]. Loss of FLG results in barrier deformation and loss of the skin's natural moisturizing factor (NMF) and hydration. More than half AD patients do not have FLG mutations and even those who have FLG mutations can improve, suggesting that defects in other barrier proteins can contribute to the impaired barrier in this disease and that compensatory mechanisms are operating to restore normal barrier function [12]. FLG mutations are known to underlie ichthyosis vulgaris (IV), a disorder of skin barrier formation and one of the most prevalent single-gene disorders in humans [13]. Interestingly, atopic dermatitis, allergic rhinitis, and asthma are common in approximately two-thirds of ichthyosis vulgaris patients [10, 14]. Hence, skin barrier defects, inflammation, and allergy appear to be largely interlinked. Moreover, the discovery that experimentally induced barrier injuries initiate keratinocyte hyperproliferation and cytokine release to activate lymphocytes [15] suggested the existence of a positive feedback loop, where barrier breach contributes to immune cell activation (Fig. 1).

Untangling the complex interaction of environmental and genetic challenges causing barrier defects to epidermal keratinocyte and immune cell response systems is a major avenue of future research. Although the skin's immune role is mainly attributed to dendritic Langerhans cells' recruitment of T cells, stressed keratinocytes are now known to have a major immune function as a "cytokine factory" which instigates inflammation upon barrier breach. Intriguingly, nonstressed epidermal keratinocytes are also immunologically active. Recent studies have shown that keratinocytes secrete innate factors that are capable of triggering skin inflammation in a cell autonomous manner and independent of skin barrier breach [16, 17]. Among these newly discovered epidermal pro-inflammatory factors is an important family of secreted serine proteases known as *kallikrein-related peptidases* (KLKs).

Serine proteases are proteolytic enzymes that hydrolyze peptide bonds by a serine-directed nucleophilic attack mechanism, which ultimately results in the target protein's irreversible activation, inactivation, and/or degradation. In this chapter, we will review proteolysis as a key regulatory mechanism of skin barrier integrity and inflammation. In particular, we will focus on the role of epidermal kallikrein serine protease and serine protease inhibitor balance in maintaining a healthy skin barrier and how offsets to this balance affect normal barrier function and trigger inflammation. We will begin by providing a short overview of inherent mechanisms governing normal skin barrier structure and function. Then, we will discuss how kallikreins modulate normal skin barrier functions and trigger inflammation upon barrier breach. We will also highlight how kallikrein proteases were recently shown to mediate inflammation independently of environmental stimuli and barrier defects. Specific skin pathologies will be discussed to illuminate how environmental, genetic, and innate pathways modulate barrier integrity and inflammation via kallikrein protease activity. The increased knowledge of kallikrein-related peptidase "epidermal" and "immune" functions is central to our understanding of the interlinked "inside-out" and "outside-in" theories of inflammatory skin diseases characterized by severe barrier dysfunctions.

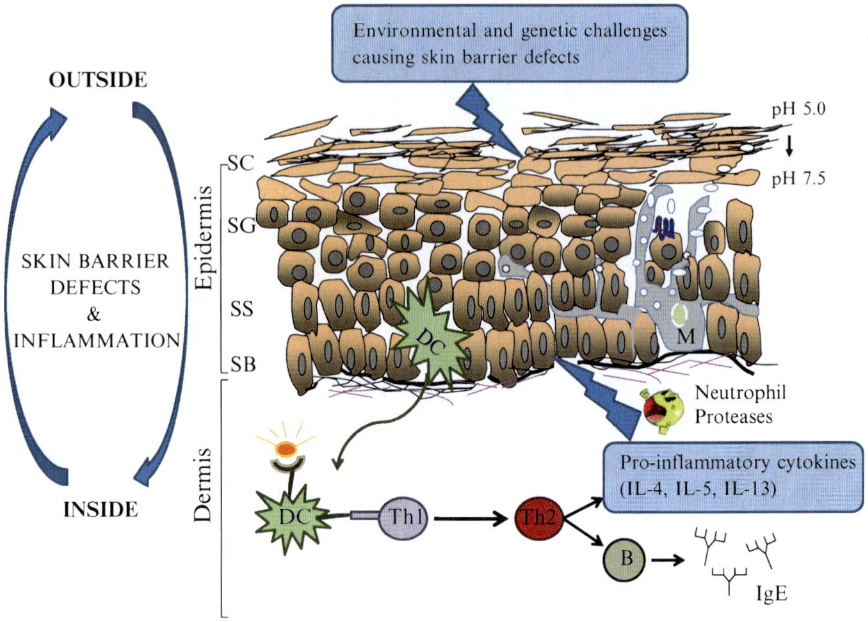

Fig. 1 The "outside-in" and "inside-out" vicious cycle of barrier defects and inflammation in inflammatory skin diseases. Environmental challenges (i.e., mechanical trauma, chemical detergents, pathogens, allergens like Der P 1 protease produced by house dust mites, etc.) and genetic abnormalities in barrier-related proteins affect one or more components of the stratum corneum (SC) barrier causing its breakdown. A defective barrier with abnormally thin stratum corneum (SC) permits increased transepidermal water loss (TWEL) and entry of pathogens and allergens. As a result, stressed keratinocytes in the stratum granulosum (SG) secrete cytokines that activate inflammation in response to barrier defects, hence the term "outside-in." Pathogen access through a defective barrier presents antigens to dendritic Langerhans cells, which stimulates a Th1 → Th2 shift in T-cell immunophenotype, which in turn further deteriorates the barrier, forming an "outside-inside-outside" feedback loop. The "inside-out" theory is based largely on genetic defects leading to overproduction of Th2 cells causing allergy via IgE overproduction, inflammation via cytokine release and skin barrier defects via neutrophil proteases, all of which are reminiscent of symptoms of inflammatory skin diseases such as atopic dermatitis. *SC* Stratum corneum, *SG* Stratum granulosum, *SS* Stratum spinosum, *SB* Stratum basale, *DC* Dendritic cells, *M* Melanocytes

2 Overview of Normal Human Skin Barrier Structure and Function

The skin's outer layer, the epidermis, is the first line of defense against harsh environment and insults by pathogens and allergens. Mature human epidermis consists of four layers: the stratum basale (SB), stratum spinosum (SS), stratum granulosum (SG) and stratum corneum (SC), in the order of increased differentiation from the lowest layer to the outer skin surface. The majority of skin barrier functions are attributed to the uppermost epidermal layer, the stratum corneum [18].

The stratum corneum (SC) layer gets renewed every 2–4 weeks via an elegant differentiation program of keratinocyte cells, which represent the major cell constituents of skin epidermis. Keratinocytes at the basal layer (SB) withdraw from the cell cycle, detach from the basement membrane, and proliferate upwards to differentiate into intermediate spinous (SS) and granular layers (SG). Basal keratinocyte proliferation and differentiation culminates with the processes of terminal differentiation and cornification at the stratum granulosum/stratum corneum (SG/SC) interface, where granular keratinocytes:

- Transport and secrete cargos via their lamellar granules (LG) into the SG/SC extracellular space. These cargos include structural proteins, adhesion proteins, lipids, lipid-processing enzymes, antimicrobial peptides, and a cocktail of proteases and protease inhibitors
- Replace their plasma membrane with a tough insoluble protein and lipid envelope known as the cornified envelope (CE)
- Aggregate their keratin intermediate filaments via filaggrin (FLG), causing collapse of their cytoskeleton into flattened squames
- Lose their nuclei and sub-cellular organelles to get terminally differentiated into nonviable anucleated cells, known as "corneocytes"

The last step of skin barrier formation, which does not involve any further cell differentiation, is known as "desquamation," which refers to corneocyte shedding off the skin surface as a result of regulated degradation of adhesion proteins linking uppermost corneocytes by endogenous proteases [19]. Inherent terminal keratinocyte differentiation and corneocyte desquamation ensue in parallel to maintain the SC barrier thickness relatively constant.

Up until the last two decades, the SC was viewed as a dead layer of nonviable corneocyte cells embedded in a lamellar lipid sea, which was often represented by the skin barrier "brick and mortar" model of vertically stacked corneocyte "bricks" held together by an extracellular lipid "mortar" [20]. With further studies, it became apparent that the SC layer is full of exciting metabolic activity and is more dynamic than previously thought, as several terminal keratinocyte differentiation products and metabolic processes take place in its extracellular milieu to regulate barrier functions (Table 1).

Skin barrier functions is an umbrella term that encompasses the skin's structural and tensile strength, lipid permeability barrier limiting water and electrolyte loss, antimicrobial barrier preventing entry of pathogens and microbes, and other functions such as antioxidant barrier and protection from ultraviolet irradiation. Although these co-localized SC barrier functions are highly inter-dependent, their molecular and biochemical basis tend to differ. For example, in the SC, filaggrin serves as a template for the assembly of corneocytes' cornified envelope forming flat and tough corneocyte "bricks" in the outer barrier. Filaggrin ultimately dissociates to free amino acids that form the skin's NMF, which creates a hydrated and acidic skin surface. SC acidifying factors include fatty acids from sebum, lactic and amino acids from sweat, and free fatty acids processed from epidermal phospholipids [21, 22]. Moreover, lipid precursor processing by β-glucocerebrosidase,

Table 1 Skin barrier functions are regulated by several persistent metabolic activities in stratum corneum interstices[a]

SC interstitial molecules	Examples	Specific role	Barrier function
Structural proteins	Corneodesmosomes: desmoglein 1 (DSG1), desmoglein 4 (DSG4), desmocollin 1 (DSC1), corneodesmosin (CDSN) Cornified envelope proteins such as Filaggrin	Act as anchoring molecules that attach neighboring corneocytes Build dense matrix of cytoskeleton fibrils that create the tough corneocyte exterior	Integrity and cohesion Mechanical resistance
Natural moisturizing factor	Amino acid end-degradation products of the corneocyte protein Filaggrin	Regulate skin hydration Maintain acidic skin surface	Chemical integrity Antimicrobial barrier
Proteases	Serine proteases: kallikreins Cysteine proteases: stratum corneum thiol protease (SCTP) Aspartic proteases: cathepsin D	Hydrolyze proteins into peptides (Corneodesmosomes are common protease targets) Lipid processing, Filaggrin processing	Physical integrity and cohesion Antimicrobial barrier
Protease inhibitors	Lympho-epithelial Kazal-type inhibitors of serine proteases Cystatin A SLPI Elafin Alpha-2 macroglobulin	Inhibit and regulate protease activity in human epidermis Inhibit exogenous proteases such as dust mite proteases (Der P1)	Integrity and cohesion Antimicrobial barrier
Antimicrobial peptides	Cathelicidin (hCAP18) Defensins	Kill gram-positive and gram-negative bacteria Some have antiviral activity Some have pro-inflammatory activity	Antimicrobial barrier
Sulfatases	Steroid sulfatase	Converts cholesterol sulfate into cholesterol	Permeability[b]
Lipases	Acid lipase Triglycerol lipase Phospholipase A_2	Convert phospholipids and triglycerides into free fatty acids	Permeability[b]
Glycosidases	β-glucocerbrosidase	Converts glycosylceramides into ceramides	Permeability[b]
Ceramidases	Acid sphingomyleinase	Converts sphingomyelin into ceramides	Permeability[b]

[a]Modified from [80]

[b]Permeability is regulated by having a proper ratio of ceramides, cholesterol, nonessential fatty acids forming lipid lamellar layers in the SC milieu

acidic sphingomyelinase and secretory phospholipase A2 enzymes into ceramides and free fatty acids generates mature lipid lamellar membranes that hamper transepidermal water loss (TEWL) and form the SC permeability barrier [23, 24].

In addition to its hydrophobic content and acidic pH, the skin surface contains antimicrobial peptides (AP), and certain keratinocyte, eccrine and sebaceous gland-derived proteases and protease inhibitors which comprise its outermost antimicrobial shield. Additionally, specialized SC extracellular adhesion proteins, known as corneodesmosomes, are incorporated into the corneocyte envelope to adhere adjacent corneocytes together and maintain the SC structural barrier.

It is important to note that SC integrity and the majority, if not all, of its barrier functions are governed by its inherent calcium and pH gradients. Human SC extracellular calcium levels increase and pH levels decrease from the lower SG/SC border to the uppermost skin surface [25]. The innate increase in epidermal calcium concentrations regulates terminal keratinocyte differentiation, lamellar granule (LG) secretion and cornified envelope formation, while the innate decrease in SC pH levels, from pH 7.0 to pH 5.0 at the skin surface, regulates desquamation, lipid permeability, and antimicrobial barrier integrity.

3 Kallikrein-Related Peptidases as Regulators of Skin Barrier Functions

KLKs are a family of 15 serine proteases (KLK1 to KLK15) encoded by a multigene cluster on chromosome 19q13.4 and translated as pre-pro-peptide chains [26]. All KLK proteins are secreted as latent pro-KLK enzymes, which require trypsin-like cleavage after their pro-peptide sequence for activation, except for pro-KLK4. Once active, the majority of KLKs function as extracellular trypsin-like serine proteases, which cleave substrates after arginine or lysine. KLK3, KLK7, and KLK9 are chymotrypsin-like serine proteases that cleave substrates after phenylalanine or tyrosine. Some KLKs, such as KLK14, display dual trypsin and chymotrypsin-like activities.

Of the 15 KLK-related peptidases present in the human body, eight KLKs co-localize in human epidermis in addition to the parent tissue KLK1 [27]. These KLKs are expressed in the SC, upper SG, sebaceous glands, eccrine sweat glands, hair follicles, and nerves [28, 29]. They are considered markers of terminal keratinocyte differentiation based on their enhanced expression in the upper stratum granulosum and stratum corneum. KLK5 and KLK7 are the most extensively studied epidermal kallikreins as they were originally isolated from normal human stratum corneum tissue in both active and inactive forms [30, 31]. These proteases were previously dubbed stratum corneum trypsin-like enzyme (SCTE) and chymotrypsin-like enzyme (SCCE), respectively. Both enzymes have optimum activity at alkaline to neutral pH, and retain considerable activity at acidic pH of 5, within the normal stratum corneum pH gradient [32].

Trypsin-like KLK5 and chymotrypsin-like KLK7 are often referred to as "desquamatory enzymes" due to their ability to cleave corneodesmosomes, such as desmocollin 1 (DSC1), desmoglein 1 (DSG1) and the glycoprotein corneodesmosin

(CDSN), resulting in regulated corneocyte shedding during normal skin cell turnover [30, 33, 34]. Desquamation upon degradation of corneodesmosomes by KLKs must be tightly regulated to maintain a fine balance between adequate barrier breakdown for normal cell renewal and preserving the barrier's integrity to prevent pathogens and irritants from penetrating through. Given KLK irreversible proteolytic processing of substrate targets such as corneodesmosomes, several regulatory mechanisms are operative to maintain KLK protease activity in check. Among these regulatory modes is an epidermal proteolytic cascade that regulates their activity and function via zymogen activation. Similar to serine proteases of the coagulation cascade, latent pro-KLKs in human SC are activated extracellularly, in a step-wise manner, by trypsin-like cleavage of their pro-peptide. KLK5 or SCTE autoactivates and activates pro-KLK7, pro-KLK14, and pro-KLK8, all of which are catalytically active serine proteases in human skin surface [32, 35, 36]. KLK5 has been proposed to act as an initiator of the pro-KLK activation cascade in upper SC [32] and has also been shown to activate other epidermal enzymes such as pro-Elastase-2 (pro-ELA2), which is also implicated in skin barrier function [37]. Furthermore, KLK8 has been proposed to control desquamation via its activation of co-localized pro-KLKs [38]. The activity of KLKs in this proteolytic cascade, and thereby the rate of desquamation, is controlled by inherent pH and ions of the SC and by a cocktail of co-localized serine protease inhibitors. For instance, chymotrypsin-like serine protease KLK7 is inhibited by skin-derived antileukoproteinase SKALP/elafin, secretory leukocyte protease inhibitor (SLPI), cystatin protease inhibitor A and M/E, and lymphoepithelial Kazal-type inhibitor (LEKTI) domains [39–41], while trypsin-like KLK5 is inhibited by epidermal LEKTI proteins encoded by the *SPINK5*, *SPINK6*, and *SPINK9* genes [42–45]. Many of these protease inhibitors, namely cystatin A, are secreted in the sweat to cover the skin surface and inhibit exogenous proteases such as house dust mites Der P1 proteases [46].

In addition to regulating endogenous KLK protease optimal activity, the SC pH also regulates KLK binding efficiency to epidermal inhibitors. The current model of KLK-regulation of desquamation is based largely on pH-dependent epidermal inhibition of KLK activity. Although KLK5 and KLK7 are expected to be optimally active in the neutral pH of lower SC, they bind inhibitory LEKTI domains efficiently at this pH [42]. As the pH becomes progressively more acidic in upper SC layers, KLK5 and KLK7 dissociate from LEKTI domains to degrade corneodesmosomes resulting in regulated corneocyte shedding off the skin surface.

Although not optimal for KLK activity, the acidity of uppermost SC is optimal for lipid processing enzymes which regulate permeability barrier formation [23, 24, 47, 48]. Disruptions to the permeability barrier by chemicals, detergents, and mechanical force reduce intracellular calcium ions and elevate SC pH, which trigger different homeostatic repair responses in the underlying epidermis. A sudden injury-induced decline in calcium ions triggers amplified secretion of lipids by lamellar granules (LG) to accelerate lamellar membrane formation and barrier recovery [24]. On the other hand, trauma-induced elevation in SC pH activates kallikrein serine proteases, which play an opposite role in lipid permeability barrier formation [24, 47, 49]. Hyperactive trypsin-like KLK5 and/or KLK14 activate cell surface proteinase-activated receptor-2 (PAR2) in upper granular keratinocytes resulting in suppression of LG secretion [24, 50]. Proteinase-activated receptors (PARs) are activated by serine protease proteolysis at the N-terminal, exposing a "tethered" receptor-triggering ligand. Although paradoxical to the goal of barrier recovery, KLK-PAR2-mediated suppression of LG secretion is thought to coincide with cornification, where granular keratinocytes begin to lose their organelles and secretory competence [6]. Moreover, kallikrein hyperactivity results in increased degradation of corneodesmosomes and overdesquamation. Thus, from a holistic viewpoint, kallikrein hyperactivity can be considered a trauma-induced epidermal stress signal that aims to restore the barrier by inducing desquamation and terminal keratinocyte differentiation to enhance cell turnover and barrier recovery. Consistently, KLK8 knock-out mice skin suffers from delayed recovery after chemical, physical, and UV-induced barrier impairment [38, 51, 52], while KLK7-overexpressing mice have thick skin as a result of keratinocyte hyperproliferation [53]. Studies have also shown that following barrier disruption serine protease activity increases in the SC. However, inhibition of these serine proteases, but no other protease types, accelerated permeability barrier recovery [24]. This improvement in barrier function is thought to be due to inhibition of KLK-mediated overdesquamation and KLK-PAR2-mediated suppression of LG-lipid secretion. Hence, kallikreins seem to play opposing protective and damaging roles during barrier breaches. It is likely that initial increases in KLK activity upon acute barrier breaches are beneficial as a stress response or repair mechanism. However, sustained KLK hyperactivity is likely to cause the skin to enter a pathological state as alluded to by recent studies of their signaling pathways in inflammatory skin diseases.

3.1 Kallikreins as Mediators of Inflammation in Skin Diseases

As mentioned above, environmental insults to the stratum corneum result in perturbation of the barrier integrity and generate an array of positive and negative downstream alarm signals that initiate both homeostatic barrier and pro-inflammatory responses. Inflammation is the homeostatic response of the body to injury or irritation. When sustained or excessive, inflammation can damage the skin tissue. Similarly, failure to restore the homeostatic balance between desquamation and differentiation results in inability to restore skin barrier function after trauma. The defective barrier makes the skin vulnerable to pathogen entry, which causes the skin to enter an inflamed, and often pathologic, state. Thus, skin barrier defects and inflammation are common manifestations of a variety of skin diseases. Intriguingly, KLK protease overexpression was detected in many inflammatory skin diseases, where they were found to expand lower in SG and SS layers. Kallikrein hyperactivity in these diseases was largely attributed to their role in regulating skin desquamation and barrier function. In support of the "outside-in" theory, KLK hyperactivity due to environmental stimuli or genetic abrogation of an epidermal KLK-inhibitor was recently shown to initiate an inflammatory skin immune response via specific signaling pathways [16, 17, 54]. Three plausible mechanisms of KLK protease-mediated initiation of skin inflammation are described below:

3.1.1 Direct Activation of Pro-Inflammatory Cytokines in SC Interstices

KLK7 overexpression and hyperactivity is implicated in the pathogenesis of inflammatory skin diseases such as AD and psoriasis [54–56] through activation of pro-inflammatory cytokines such as pro-interleukin-1β (proIL-1β) [57]. Among other cytokines, IL-1β has a role in promoting inflammation and migration of Langerhans cells to lymph nodes to present antigens to naïve T cells [58]. In atopic dermatitis and psoriasis, the elevated skin surface pH or other stimuli increase serine protease activity, which is likely to result in activation of the 31-kDa pro-form of IL-1β by proteolytic cleavage [59]. Elias et al. proposed that activation of IL-1β by KLK serine proteases is a likely first step in an epidermal cytokine cascade that triggers inflammation [7]. Interestingly, transgenic mice overexpressing human KLK7 in the skin develop increased epidermal thickness, hyperkeratosis, and dermal inflammation, similar to inflammatory skin diseases in humans [54, 60].

3.1.2 Activation of Keratinocyte Cell Surface Proteinase-Activated Receptor-2 (PAR-2)-Mediated Production of Pro-Inflammatory and Pro-Allergy Cascades

KLK epidermal functions have been largely unraveled in a *SPINK5*$^{-/-}$ mouse model of Netherton syndrome [61]. Netherton syndrome (NS) is a rare autosomal recessive skin disorder characterized by the absence of the epidermal lymphoepithelial Kazal-type inhibitor (LEKTI), resulting in severe barrier dysfunction, hair shaft defects, and atopic manifestations [62, 63]. The skin of NS patients suffers from KLK serine protease hyperactivity and excessive proteolytic degradation of corneodesmosomes, leading to premature SC separation at the boundary between the SG and upper SC layers [33, 61]. As mentioned above, LEKTI domains inhibit different epidermal KLKs with different potencies, with the highest inhibitory potency toward KLK5. The SC inherent pH gradient allows for progressive release of active KLKs from LEKTI complexes, leading to regulated corneodesmosome degradation from the uppermost SC layers [42]. This fine tuning of desquamation is absent in the skin of NS patients, due to the lack of LEKTI's spatial regulation of KLKs. This results in the presence of active KLKs in the deep layers of the SC, which leads to premature detachment of the SC layer from the SG. As predicted by the "outside-in" theory of inflammatory skin disease pathogenesis, SC barrier loss causes dehydration, inflammation and susceptibility to bacterial infections and systemic allergy manifestations such as atopic dermatitis in NS patients.

By employing the Netherton syndrome (NS) mouse model and confirming results in human NS skin, Briot et al. demonstrated that KLK5 indeed induces inflammation and atopic-like lesions in NS skin via a PAR2-NFκB-mediated cytokine burst that creates a pro-Th2 inflammatory microenvironment in the underlying dermis [16]. In LEKTI-deficient epidermis, hyperactive KLK5 activates PAR2 by proteolytic cleavage, and induces NFκB-mediated ICAM, IL-8, TNF-α, and TSLP cytokine overexpression [16]. Overexpression of the proallergic cytokine TSLP implicates KLK5 hyperactivity in an innate allergy regulatory pathway, which may explain the susceptibility of the majority of NS patients to develop AD. TSLP activates dendritic Langerhans cells and enhances their migration to skin draining lymph nodes, where they trigger the differentiation of naïve CD4+ T cells into pro-allergic CD4+ Th2 cells [64–66]. The overexpression of TSLP in keratinocytes of AD patients and barrier-deficient mouse skin further supports that keratinocytes signal to the immune cells to aid in combating the flux of invasive pathogens. Hence, it is possible that the "run-on" KLK5-PAR2-NFκB pathway activated in LEKTI-deficient skin leading to inflammation and allergy is also triggered by environmentally induced barrier breach via mechanical force or harsh detergents that elevate SC pH and activate KLKs. Given that multiple trypsin-like KLKs are co-localized in human epidermis, future studies need to examine the involvement of individual kallikreins in regulating inflammatory and allergy cascades in response to various environmental stimuli.

In addition to activating inflammatory and allergy pathways, it is important to note that hyperactive KLK5 also degrades desmosomes at the SG/SC interface, causing premature SC detachment and IL-1β, IL8, and TNF-α secretion by barrier-stressed LEKTI-deficient keratinocytes. Therefore, KLK5 hyperactivity in NS skin results in an epidermal and dermal cytokine burst by independent pathways that activate "epidermal" and "immune" cell systems. The pro-inflammatory signals induced by KLK5 hyperactivity ultimately lead to the recruitment of eosinophils and mast cells, independent of environmental stimuli, barrier defects, and the adaptive immune system [16]. Hence, recent compelling data demonstrate that the innate epidermal KLK-serine protease/serine protease inhibitor balance, which regulates normal skin barrier integrity, is also integral for the regulation of PAR2-NFκB-mediated inflammatory and allergy cascades.

3.1.3 Proteolytic Processing of Cathelicidin Antimicrobial Peptides

A third mechanism whereby KLKs are now known to mediate inflammation is their inherent ability to regulate epidermal antimicrobial peptide processing, activity, and function. Cathelicidins are important effectors of the innate immune system, known for their role as "alarmins" which protect the body against infection by gram-positive, gram-negative bacteria, and some viruses [67]. Human cathelicidin is secreted by epithelial cells, including epidermal keratinocytes, as a proprotein named 18-kDa cationic antimicrobial protein or hCAP18. Cathelicidin (hCAP18) is biologically inactive. It is activated in human epidermis by KLK5 trypsin-like proteolytic processing near the C-terminal to release a 37-amino-acid-long antimicrobial peptide, named LL-37. LL-37 is a broad spectrum active antimicrobial peptide against *Escherichia coli*, *Staphylococcus aureus* and group A *Streptococcus* [68, 69] and it has antiviral activity [70]. Trypsin-like KLK5 and chymotrypsin-like KLK7 also target LL-37 antimicrobial peptide processing in human epidermis to generate shorter antimicrobial peptides that are active against *S. aureus* [71]. These shorter peptides are further degraded by KLKs resulting in inactivation after resolving the microbial challenge. For instance, KLK7 cleavage of LL-37 gives rise to short-lived antimicrobial peptides RK-31 and KR-20 and hence it may act as an inactivator of LL-37. We recently demonstrated new active epidermal KLK8 and KLK14 in vitro trypsin-like processing of LL-37 synthetic peptide to shorter KS-30, LL-29, and LL-23 active antimicrobial peptides [36], suggesting that a large web of skin surface KLK proteases may govern antimicrobial barrier activity.

Individuals with the inflammatory skin disease rosacea show facial inflammation in response to stimuli and have an exacerbated response to irritants and allergens as a result of barrier dysfunction. The facial skin of rosacea patients displays increased serine protease activity, LL-37 antimicrobial peptide overexpression and increased inflammation, compared to nonlesional areas [17]. Furthermore, rosacea skin has unique cathelicidin peptides and abundant KLK5 protease expression compared to normal skin, suggesting the involvement of KLK5 activity in rosacea pathogenesis. Subcutaneous injection of active KLK5 in mice, in amounts mimicking those observed in rosacea, was found to increase cathelicidin processing and to induce leukocyte infiltration and inflammation, confirming a pathological link between KLK5 abnormal proteolytic processing of cathelicidin and inflammation [17]. In addition to its antimicrobial and antiviral activity, LL-37 peptide is a known pro-inflammatory mediator and a chemoattractant for eosinophils and neutrophils [72, 73]. Interestingly, the skin of SPINK5$^{-/-}$ mice has an altered expression profile of cathelicidin peptides, similar to that seen in rosacea, further supporting the importance of KLK serine protease/serine protease inhibitor balance in regulating inflammation via cathelicidin processing to LL-37 and smaller peptides [17].

In atopic dermatitis, LL-37 is down-regulated in a Th2-dependent fashion. Given that the SC pH of AD patients is elevated and contains abundant expression of KLKs, compared to normal SC, it is likely that KLK hyperactivity degrades cathelicidin, particularly LL-37, to smaller peptide fragments that lack antimicrobial activity but have pro-inflammatory function. Further understanding of the regulation, timing, and extent of individual KLK processing of cathelicidins in AD skin is required to contribute to our understanding of the higher incidence of bacterial infections in inflamed AD skin.

Although much remains to be done, several ex vivo and in vivo studies on skin from patients with different inflammatory skin diseases and skin disease mouse models have demonstrated that "run-on" KLK hyperactivity results in structural skin barrier defects and inflammation via independent pathways (Table 2).

Table 2 Summary of KLK involvement in inflammatory skin disease pathologies

Skin disease	Disease pathogenesis	KLK Levels	Pathological pathways involving KLK activity	References
Atopic dermatitis (AD) (OMIM 603165)	A common chronic inflammatory, dry, itchy, allergic skin disease involving immune, endocrine, metabolic, and infectious factors	Higher trypsin and chymotrypsin-like KLK levels	Higher SC pH (due to FLG mutations causing a decrease in SC acidity and NMF or due to barrier disruptions by environmental challenges) leads to increased kallikrein serine protease activity KLK hyperactivity leads to increased desquamation, lipid permeability barrier dysfunction, pain, inflammation, and allergy onset	[6, 14, 16, 55, 60]
Psoriasis Vulgaris (PV) (OMIM 177900)	A common chronic inflammatory dermatosis characterized by erythematous plaques and hyperproliferative keratinocyte activity	Higher trypsin and chymotrypsin-like KLKs and expanded lower in the epidermis	Higher SC pH leads to increased kallikrein serine protease activity KLK hyperactivity leads to increased desquamation, IL-1β activation, inflammation and allergy onset	[54, 56]
Netherton Syndrome (NS) (OMIM 256500)	A rare autosomal recessive mutation in SPINK5 gene on chromosome 5q32 causing truncation and/or loss of the epidermal serine protease inhibitor LEKTI	Higher KLK expression, where they expand lower to the periphery of keratinocytes in the SS and SG	KLK hyperactivity (due to *SPINK5* mutations causing LEKTI inhibitor dysfunction) results in overdesquamation, inflammation, and allergy onset Matriptase activation of pro-KLK zymogens contributes to the overwhelming KLK hyperactivity in this disease Aberrant cathelicidin expression as a result of KLK hyperactivity	[16, 61, 77]
Rosacea	An inflammatory skin disorder characterized by facial lesions with erythema, papulopustules and telangiectasia. The etiology is unknown, but symptoms are exacerbated by factors that trigger innate immune responses	Higher KLK expression, where they expand lower in the epidermis, as well as aberrant expression of cathelicidin peptides, compared to normal skin	Abnormal processing of hCAP18 and LL-37 cathelicidin peptides by hyperactive KLKs, leading to inflammation, barrier dysfunction, and itching	[17, 71]

3.2 KLK Activity Is Central to Interlinked Skin Barrier Defects, Inflammation, Allergy, and Pain Pathways

In addition to its role as a protective barrier, human epidermis has a sensory function. Skin surface kallikreins may also play a role in mediating itch and pain signals. Epidermal overexpression of human KLK7 also gives rise to a chronic itchy dermatitis, where KLK7 hyperactivity in a mouse model was shown to cause epidermal hyperproliferation, decreased barrier function, and severe pruritus (or itch) [53, 74]. The pruritus observed may be due to barrier defects giving KLK proteases access to pruritic nerves. Epidermal KLKs and trypsin-like serine proteases have been suggested to trigger itch via activation of proteinase-activated receptor-2 (PAR2), expressed on afferent nerve fibers [75, 76]. LL-37 was shown to induce secretion of the pruritogenic cytokine IL-31 by human mast cells [72], which is another possible pathway whereby KLKs may trigger skin itch. The ability of trypsin-like kallikreins to initiate proteinase-activated receptor-2 (PAR2) or LL-37-mediated inflammation was recently demonstrated, but their ability to induce pain and itch via PAR2 or LL-37 remain to be further investigated. Given that PAR2 is shown to mediate pruritus associated with epidermal lesions [76], this emphasizes the relevance of epidermal KLK serine protease/serine protease inhibitor balance in triggering the pain associated with skin inflammation and allergy.

Kallikreins are now considered important epithelial mediators of interlinked barrier structure and immune functions in normal and diseased epidermis. The importance of KLK serine protease/serine protease inhibitor balance has become apparent via multiple elegant studies focusing on KLK signaling networks in LEKTI-deficient epidermis. In addition to inhibition, KLK serine protease activity is also regulated by a pro-KLK zymogen activation proteolytic cascade in upper epidermis. Recently, a granular keratinocyte membrane-bound serine protease, known as matriptase, was shown to activate KLK5 and KLK7 in vitro and in vivo [77]. Matriptase autoactivation is more efficient than KLK5 autoactivation, which makes it a better activator of KLK5 and a more likely "initiator" of a KLK activation cascade in lower epidermis. Matriptase activation of KLKs occurs in lower SC as matriptase and LEKTI co-localize at the SG/SC interface [77], the site of excessive proteolysis and epidermal separation in NS. Therefore, matriptase activation of KLKs seems to be more significant in pathological inflamed skin, where KLKs expand lower in the epidermis. Matriptase activation of KLKs is unlikely to affect normal skin desquamation, as matriptase-deficient skin retains proteolytic activity in its upper SC, similar to normal skin [77].

Given that KLKs are normally bound to LEKTI at the SG/SC interface, matriptase activation of KLKs may be important in LEKTI-deficient epidermis. In situ protease activity is predominantly confined to the upper SC of normal skin, but in LEKTI-deficient epidermis, caseinolytic protease activity is present throughout the epidermis and the underlying dermis [77]. The epidermal protease activity is largely attributed to the expanded expression of hyperactive KLKs in lower layers, while the dermal protease activity is likely due to the influx of dermal inflammatory cell proteases. It remains to be shown whether inflammatory cells secrete any of the known epidermal kallikreins, other than KLK1, which is known to activate the kininogen-kinin system in inflammation [78, 79].

Genetic abrogation of matriptase in LEKTI-deficient epidermis completely abolishes aberrant protease activity in low epidermal and dermal layers of NS skin [77]. The sites of protease activity in normal skin and LEKTI/matriptase double-deficient epidermis are identical, confirming matriptase ability to activate free KLKs in LEKTI-deficient lower epidermis. Furthermore, matriptase ablation in SPINK5$^{-/-}$ mice improves SC function, restores corneodesmosome integrity at the SG/SC interface, and prevents KLK-mediated inflammation.

In other words, due to LEKTI's absence in Netherton syndrome, KLK5 gets activated by matriptase in the lower SC to further activate itself and other pro-KLKs. This results in excessive degradation of corneodesmosomes at the SG/SC interface and in PAR2-NFκB-mediated activation of pro-inflammatory and pro-allergic cytokine burst, which promotes atopic-like and inflamed skin lesions in NS patients. The seminal findings of Sales and colleagues demonstrate the importance of maintaining balance of KLK serine proteases, protease activators such as matriptase, and protease inhibitors such as LEKTI for normal skin barrier function. Hence, our early suggestion of epidermal kallikrein roles as promiscuous mediators of various skin barrier functions [80] including inflammation [50] seems to hold true as kallikrein activity is situated at the center of many physiological, compensatory and pathological barrier functions, as shown in Fig. 2.

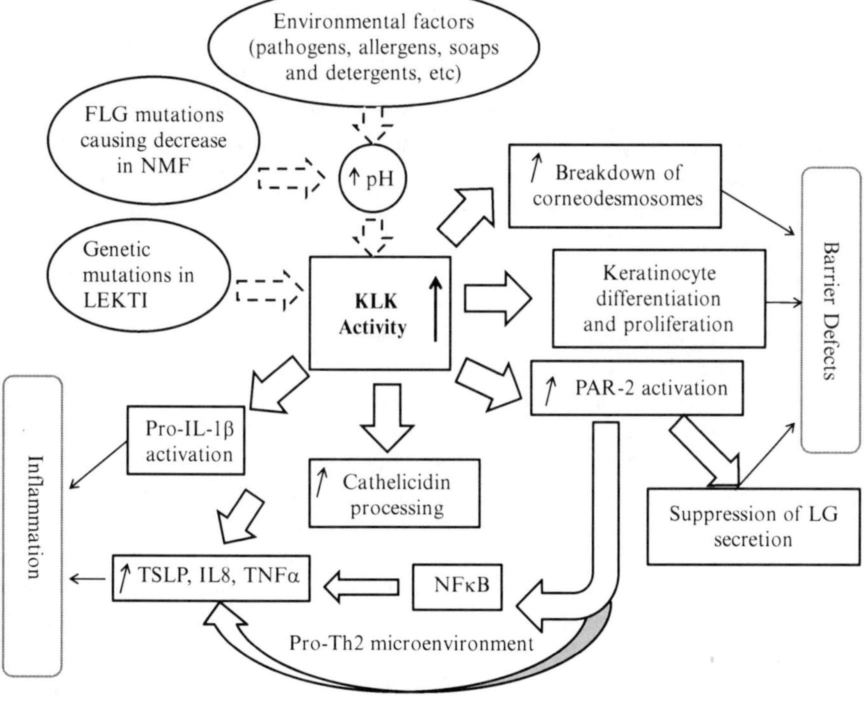

Fig. 2 Kallikrein activity is central to the interlinked "inside-out" and "outside-in" views of inflammatory skin disease pathology. Kallikrein hyperactivity (due to genetic or environmental factors) triggers independent pathways resulting in barrier defects and inflammation in skin diseases

4 Concluding Remarks and Future Perspectives

Our understanding of the "outside-in" and "inside-out" theories of inflammatory skin disease pathology and etiology has been reshaped by an explosion of recent studies focusing on innate SC biological regulatory mechanisms. Researchers and dermatology experts now attest to the primary role of epidermal factors in mediating immune responses and support the "outside-inside-outside" theory proposed by Elias and others [7], wherein barrier defects and inflammation contribute to the onset of inflammatory skin diseases in a positive feedback cycle akin to the "chicken-versus-egg" analogy. The physiological communication between epidermal keratinocytes and immune cells is important in triggering inflammation for host defense and barrier restoration after brief breaches. However, deregulated or excessive inflammation due to exacerbated stress signals by keratinocytes can cause further barrier damage and chronic disease. As discussed in this chapter, the immune role of keratinocytes is not restricted to acting as a "cytokine factory" upon barrier stress, but it also encompasses their innate secretion of a wide range of epidermal factors that maintain structural, permeability, antimicrobial and immune barrier functions in both healthy and diseased skin. At the forefront of these epidermal factors is an army of skin surface kallikrein-serine proteases, along with their activators and inhibitors.

Kallikreins mediate inflammatory and immunological reactions in skin diseases through activation of pro-inflammatory cytokines, proteolytic stimulation of keratinocyte-expressed proteinase activated receptor-2 (PAR2)-NFκB pathway, and processing of cathelicidin antimicrobial peptides. This is in addition to their regulation of normal skin desquamation and permeability barrier function. The sustained increase in kallikrein activity in inflammatory skin diseases as a result of genetic or environmental insults to the barrier could result in an array of negative downstream consequences, including: (1) barrier disruption via excessive corneodesmosome degradation, (2) PAR2-mediated inhibition of LG lipid secretion, (3) PAR2-mediated itch, (4) inflammation and immune responses. Hence, these proteases contribute to the current "outside-inside-outside" inflammatory skin disease dogma by confirming the importance of barrier proteins in instigating immune reactions, as well as barrier dysfunction. Kallikreins appear as excellent targets for inflammatory skin disease therapy development, as shown in Fig. 3. We may be better served if we dampen the body's inherent heightened response to barrier disruptions and if we enhance barrier recovery, instead of using immune suppressors, to treat diseases such as atopic dermatitis or rosacea. Small molecule inhibitors of KLK activity can be applied topically to break the vicious cycle of barrier dysfunction and inflammation in these diseases, which are commonly perceived as inflammatory diseases. In the case of Netherton syndrome, blocking matriptase may be another strategy for treatment as it may dampen KLK activation in lower SC, without affecting their regular activity in uppermost epidermis. It may be possible to create creams that will ameliorate dry and damaged atopic dermatitis or rosacea skin and keep out irritants and allergens, by dampening kallikrein

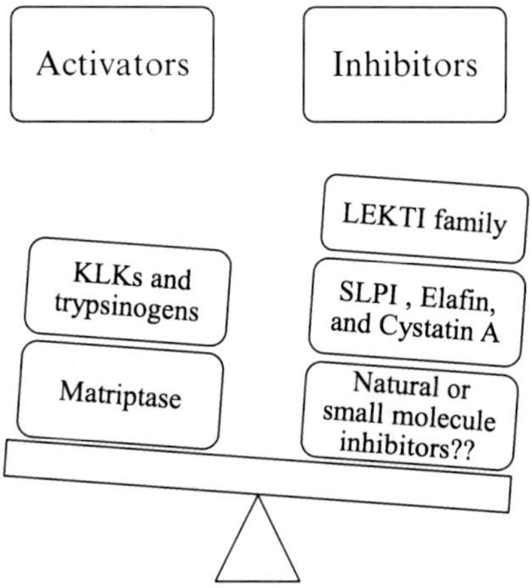

Fig. 3 The innate balance of KLK activity is integral in maintaining a healthy skin barrier. Future inflammatory skin disease therapy development may include targeting of KLK activity in the epidermis

activity. This means that along with using anti-inflammatory and/or immune suppressor medications, it may be crucial for atopic dermatitis or rosacea patients to use moisturizers and creams that prevent flare-ups and help keep the skin barrier hydrated and healthy. The hunt for specific small molecule inhibitors of kallikrein serine proteases is an ongoing endeavor in the kallikrein field with unlimited potential for dermatological applications.

Acknowledgements We would like to acknowledge the Natural Sciences and Engineering Council (NSERC) of Canada as our generous funding agency.

References

1. Ludolph-Hauser D, Schubert C, Wiedow O (1999) Structural changes of human epidermis induced by human leukocyte-derived proteases. Exp Dermatol 8:46–52
2. Hatano Y, Terashi H, Arakawa S, Katagiri K (2005) Interleukin-4 suppresses the enhancement of ceramide synthesis and cutaneous permeability barrier functions induced by tumor necrosis factor-alpha and interferon-gamma in human epidermis. J Invest Dermatol 124:786–792
3. Kurahashi R, Hatano Y, Katagiri K (2008) IL-4 suppresses the recovery of cutaneous permeability barrier functions in vivo. J Invest Dermatol 128:1329–1331
4. Homey B, Steinhoff M, Ruzicka T, Leung DY (2006) Cytokines and chemokines orchestrate atopic skin inflammation. J Allergy Clin Immunol 118:178–189
5. Segre JA (2006) Epidermal barrier formation and recovery in skin disorders. J Clin Invest 116:1150–1158
6. Cork MJ, Danby SG, Vasilopoulos Y, Hadgraft J, Lane ME, Moustafa M, Guy RH, Macgowan AL, Tazi-Ahnini R, Ward SJ (2009) Epidermal barrier dysfunction in atopic dermatitis. J Invest Dermatol 129:1892–1908
7. Elias PM, Steinhoff M (2008) "Outside-to-inside" (and now back to "outside") pathogenic mechanisms in atopic dermatitis. J Invest Dermatol 128:1067–1070
8. Guttman-Yassky E, Suarez-Farinas M, Chiricozzi A, Nograles KE, Shemer A, Fuentes-Duculan J, Cardinale I, Lin P, Bergman R, Bowcock AM et al (2009) Broad defects

in epidermal cornification in atopic dermatitis identified through genomic analysis. J Allergy Clin Immunol 124:1235–1244.e1258

9. Palmer CN, Irvine AD, Terron-Kwiatkowski A, Zhao Y, Liao H, Lee SP, Goudie DR, Sandilands A, Campbell LE, Smith FJ et al (2006) Common loss-of-function variants of the epidermal barrier protein filaggrin are a major predisposing factor for atopic dermatitis. Nat Genet 38:441–446
10. Sandilands A, Terron-Kwiatkowski A, Hull PR, O'Regan GM, Clayton TH, Watson RM, Carrick T, Evans AT, Liao H, Zhao Y et al (2007) Comprehensive analysis of the gene encoding filaggrin uncovers prevalent and rare mutations in ichthyosis vulgaris and atopic eczema. Nat Genet 39:650–654
11. Hudson TJ (2006) Skin barrier function and allergic risk. Nat Genet 38:399–400
12. Jung T, Stingl G (2008) Atopic dermatitis: therapeutic concepts evolving from new pathophysiologic insights. J Allergy Clin Immunol 122:1074–1081
13. Smith FJ, Irvine AD, Terron-Kwiatkowski A, Sandilands A, Campbell LE, Zhao Y, Liao H, Evans AT, Goudie DR, Lewis-Jones S et al (2006) Loss-of-function mutations in the gene encoding filaggrin cause ichthyosis vulgaris. Nat Genet 38:337–342
14. Elias PM, Schmuth M (2009) Abnormal skin barrier in the etiopathogenesis of atopic dermatitis. Curr Opin Allergy Clin Immunol 9:437–446
15. Swamy M, Jamora C, Havran W, Hayday A (2010) Epithelial decision makers: in search of the 'epimmunome'. Nat Immunol 11:656–665
16. Briot A, Deraison C, Lacroix M, Bonnart C, Robin A, Besson C, Dubus P, Hovnanian A (2009) Kallikrein 5 induces atopic dermatitis-like lesions through PAR2-mediated thymic stromal lymphopoietin expression in Netherton syndrome. J Exp Med 206:1135–1147
17. Yamasaki K, Di Nardo A, Bardan A, Murakami M, Ohtake T, Coda A, Dorschner RA, Bonnart C, Descargues P, Hovnanian A et al (2007) Increased serine protease activity and cathelicidin promotes skin inflammation in rosacea. Nat Med 13:975–980
18. Candi E, Schmidt R, Melino G (2005) The cornified envelope: a model of cell death in the skin. Nat Rev Mol Cell Biol 6:328–340
19. Milstone LM (2004) Epidermal desquamation. J Dermatol Sci 36:131–140
20. Elias PM (1983) Epidermal lipids, barrier function, and desquamation. J Invest Dermatol 80 (Suppl):44s–49s
21. Ohman H, Vahlquist A (1994) In vivo studies concerning a pH gradient in human stratum corneum and upper epidermis. Acta Derm Venereol 74:375–379
22. Rippke F, Schreiner V, Doering T, Maibach HI (2004) Stratum corneum pH in atopic dermatitis: impact on skin barrier function and colonization with *Staphylococcus aureus*. Am J Clin Dermatol 5:217–223
23. Hachem JP, Crumrine D, Fluhr J, Brown BE, Feingold KR, Elias PM (2003) pH directly regulates epidermal permeability barrier homeostasis, and stratum corneum integrity/cohesion. J Invest Dermatol 121:345–353
24. Hachem JP, Houben E, Crumrine D, Man MQ, Schurer N, Roelandt T, Choi EH, Uchida Y, Brown BE, Feingold KR et al (2006) Serine protease signaling of epidermal permeability barrier homeostasis. J Invest Dermatol 126:2074–2086
25. Menon GK (2002) New insights into skin structure: scratching the surface. Adv Drug Deliv Rev 54(Suppl 1):S3–S17
26. Borgono CA, Diamandis EP (2004) The emerging roles of human tissue kallikreins in cancer. Nat Rev Cancer 4:876–890
27. Komatsu N, Tsai B, Sidiropoulos M, Saijoh K, Levesque MA, Takehara K, Diamandis EP (2006) Quantification of eight tissue kallikreins in the stratum corneum and sweat. J Invest Dermatol 126:925–929
28. Komatsu N, Saijoh K, Sidiropoulos M, Tsai B, Levesque MA, Elliott MB, Takehara K, Diamandis EP (2005) Quantification of human tissue kallikreins in the stratum corneum: dependence on age and gender. J Invest Dermatol 125:1182–1189

29. Komatsu N, Saijoh K, Toyama T, Ohka R, Otsuki N, Hussack G, Takehara K, Diamandis EP (2005) Multiple tissue kallikrein mRNA and protein expression in normal skin and skin diseases. Br J Dermatol 153:274–281
30. Brattsand M, Egelrud T (1999) Purification, molecular cloning, and expression of a human stratum corneum trypsin-like serine protease with possible function in desquamation. J Biol Chem 274:30033–30040
31. Egelrud T, Lundstrom A (1991) A chymotrypsin-like proteinase that may be involved in desquamation in plantar stratum corneum. Arch Dermatol Res 283:108–112
32. Brattsand M, Stefansson K, Lundh C, Haasum Y, Egelrud T (2005) A proteolytic cascade of kallikreins in the stratum corneum. J Invest Dermatol 124:198–203
33. Descargues P, Deraison C, Prost C, Fraitag S, Mazereeuw-Hautier J, D'Alessio M, Ishida-Yamamoto A, Bodemer C, Zambruno G, Hovnanian A (2006) Corneodesmosomal cadherins are preferential targets of stratum corneum trypsin- and chymotrypsin-like hyperactivity in Netherton syndrome. J Invest Dermatol 126:1622–1632
34. Egelrud T, Lundstrom A (1990) The dependence of detergent-induced cell dissociation in non-palmo-plantar stratum corneum on endogenous proteolysis. J Invest Dermatol 95:456–459
35. Yoon H, Laxmikanthan G, Lee J, Blaber SI, Rodriguez A, Kogot JM, Scarisbrick IA, Blaber M (2007) Activation profiles and regulatory cascades of the human kallikrein-related peptidases. J Biol Chem 282:31852–31864
36. Eissa A, Amodeo V, Smith CR, Diamandis EP (2011) Kallikrein-related peptidase-8 (KLK8) is an active serine protease in human epidermis and sweat and is involved in a skin barrier proteolytic cascade. J Biol Chem 286:687–706
37. Bonnart C, Deraison C, Lacroix M, Uchida Y, Besson C, Robin A, Briot A, Gonthier M, Lamant L, Dubus P et al (2010) Elastase 2 is expressed in human and mouse epidermis and impairs skin barrier function in Netherton syndrome through filaggrin and lipid misprocessing. J Clin Invest 120:871–882
38. Kishibe M, Bando Y, Terayama R, Namikawa K, Takahashi H, Hashimoto Y, Ishida-Yamamoto A, Jiang YP, Mitrovic B, Perez D et al (2007) Kallikrein 8 is involved in skin desquamation in cooperation with other kallikreins. J Biol Chem 282:5834–5841
39. Wingens M, van Bergen BH, Hiemstra PS, Meis JF, van Vlijmen-Willems IM, Zeeuwen PL, Mulder J, Kramps HA, van Ruissen F, Schalkwijk J (1998) Induction of SLPI (ALP/HUSI-I) in epidermal keratinocytes. J Invest Dermatol 111:996–1002
40. Caubet C, Jonca N, Brattsand M, Guerrin M, Bernard D, Schmidt R, Egelrud T, Simon M, Serre G (2004) Degradation of corneodesmosome proteins by two serine proteases of the kallikrein family, SCTE/KLK5/hK5 and SCCE/KLK7/hK7. J Invest Dermatol 122:1235–1244
41. Shigemasa K, Tanimoto H, Underwood LJ, Parmley TH, Arihiro K, Ohama K, O'Brien TJ (2001) Expression of the protease inhibitor antileukoprotease and the serine protease stratum corneum chymotryptic enzyme (SCCE) is coordinated in ovarian tumors. Int J Gynecol Cancer 11:454–461
42. Deraison C, Bonnart C, Lopez F, Besson C, Robinson R, Jayakumar A, Wagberg F, Brattsand M, Hachem JP, Leonardsson G et al (2007) LEKTI fragments specifically inhibit KLK5, KLK7, and KLK14 and control desquamation through a pH-dependent interaction. Mol Biol Cell 18:3607–3619
43. Brattsand M, Stefansson K, Hubiche T, Nilsson SK, Egelrud T (2009) SPINK9: a selective, skin-specific Kazal-type serine protease inhibitor. J Invest Dermatol 129:1656–1665
44. Meyer-Hoffert U, Wu Z, Schroder JM (2009) Identification of lympho-epithelial Kazal-type inhibitor 2 in human skin as a kallikrein-related peptidase 5-specific protease inhibitor. PLoS One 4:e4372
45. Meyer-Hoffert U, Wu Z, Kantyka T, Fischer J, Latendorf T, Hansmann B, Bartels J, He Y, Glaeser R, Schroeder JM (2010) Isolation of Spink6 in human skin: a selective inhibitor of kallikrein-related peptidases. J Biol Chem 285:32174–32181

46. Kato T, Takai T, Mitsuishi K, Okumura K, Ogawa H (2005) Cystatin A inhibits IL-8 production by keratinocytes stimulated with Der p 1 and Der f 1: biochemical skin barrier against mite cysteine proteases. J Allergy Clin Immunol 116:169–176
47. Hachem JP, Man MQ, Crumrine D, Uchida Y, Brown BE, Rogiers V, Roseeuw D, Feingold KR, Elias PM (2005) Sustained serine proteases activity by prolonged increase in pH leads to degradation of lipid processing enzymes and profound alterations of barrier function and stratum corneum integrity. J Invest Dermatol 125:510–520
48. Houben E, Hachem JP, De Paepe K, Rogiers V (2008) Epidermal ceramidase activity regulates epidermal desquamation via stratum corneum acidification. Skin Pharmacol Physiol 21:111–118
49. Hachem JP, Roelandt T, Schurer N, Pu X, Fluhr J, Giddelo C, Man MQ, Crumrine D, Roseeuw D, Feingold KR et al (2010) Acute acidification of stratum corneum membrane domains using polyhydroxyl acids improves lipid processing and inhibits degradation of corneodesmosomes. J Invest Dermatol 130:500–510
50. Oikonomopoulou K, Hansen KK, Saifeddine M, Tea I, Blaber M, Blaber SI, Scarisbrick I, Andrade-Gordon P, Cottrell GS, Bunnett NW et al (2006) Proteinase-activated receptors, targets for kallikrein signaling. J Biol Chem 281:32095–32112
51. Kirihara T, Matsumoto-Miyai K, Nakamura Y, Sadayama T, Yoshida S, Shiosaka S (2003) Prolonged recovery of ultraviolet B-irradiated skin in neuropsin (KLK8)-deficient mice. Br J Dermatol 149:700–706
52. Kitayoshi H, Inoue N, Kuwae K, Chen ZL, Sato H, Ohta T, Hosokawa K, Itami S, Yoshikawa K, Yoshida S et al (1999) Effect of 12-O-tetradecanoyl-phorbol ester and incisional wounding on neuropsin mRNA and its protein expression in murine skin. Arch Dermatol Res 291:333–338
53. Ny A, Egelrud T (2004) Epidermal hyperproliferation and decreased skin barrier function in mice overexpressing stratum corneum chymotryptic enzyme. Acta Derm Venereol 84:18–22
54. Ekholm E, Egelrud T (1999) Stratum corneum chymotryptic enzyme in psoriasis. Arch Dermatol Res 291:195–200
55. Komatsu N, Saijoh K, Kuk C, Liu AC, Khan S, Shirasaki F, Takehara K, Diamandis EP (2007) Human tissue kallikrein expression in the stratum corneum and serum of atopic dermatitis patients. Exp Dermatol 16:513–519
56. Komatsu N, Saijoh K, Kuk C, Shirasaki F, Takehara K, Diamandis EP (2007) Aberrant human tissue kallikrein levels in the stratum corneum and serum of patients with psoriasis: dependence on phenotype, severity and therapy. Br J Dermatol 156:875–883
57. Nylander-Lundqvist E, Egelrud T (1997) Formation of active IL-1 beta from pro-IL-1 beta catalyzed by stratum corneum chymotryptic enzyme in vitro. Acta Derm Venereol 77:203–206
58. Wang B, Amerio P, Sauder DN (1999) Role of cytokines in epidermal Langerhans cell migration. J Leukoc Biol 66:33–39
59. Nylander-Lundqvist E, Back O, Egelrud T (1996) IL-1 beta activation in human epidermis. J Immunol 157:1699–1704
60. Hansson L, Backman A, Ny A, Edlund M, Ekholm E, Ekstrand Hammarstrom B, Tornell J, Wallbrandt P, Wennbo H, Egelrud T (2002) Epidermal overexpression of stratum corneum chymotryptic enzyme in mice: a model for chronic itchy dermatitis. J Invest Dermatol 118:444–449
61. Descargues P, Deraison C, Bonnart C, Kreft M, Kishibe M, Ishida-Yamamoto A, Elias P, Barrandon Y, Zambruno G, Sonnenberg A et al (2005) Spink5-deficient mice mimic Netherton syndrome through degradation of desmoglein 1 by epidermal protease hyperactivity. Nat Genet 37:56–65
62. Chavanas S, Bodemer C, Rochat A, Hamel-Teillac D, Ali M, Irvine AD, Bonafe JL, Wilkinson J, Taieb A, Barrandon Y et al (2000) Mutations in SPINK5, encoding a serine protease inhibitor, cause Netherton syndrome. Nat Genet 25:141–142

63. Hachem JP, Wagberg F, Schmuth M, Crumrine D, Lissens W, Jayakumar A, Houben E, Mauro TM, Leonardsson G, Brattsand M et al (2006) Serine protease activity and residual LEKTI expression determine phenotype in Netherton syndrome. J Invest Dermatol 126: 1609–1621
64. Soumelis V, Reche PA, Kanzler H, Yuan W, Edward G, Homey B, Gilliet M, Ho S, Antonenko S, Lauerma A et al (2002) Human epithelial cells trigger dendritic cell mediated allergic inflammation by producing TSLP. Nat Immunol 3:673–680
65. Ziegler SF, Artis D (2010) Sensing the outside world: TSLP regulates barrier immunity. Nat Immunol 11:289–293
66. He R, Oyoshi MK, Garibyan L, Kumar L, Ziegler SF, Geha RS (2008) TSLP acts on infiltrating effector T cells to drive allergic skin inflammation. Proc Natl Acad Sci USA 105:11875–11880
67. Yamasaki K, Gallo RL (2008) Antimicrobial peptides in human skin disease. Eur J Dermatol 18:11–21
68. Zaiou M, Nizet V, Gallo RL (2003) Antimicrobial and protease inhibitory functions of the human cathelicidin (hCAP18/LL-37) prosequence. J Invest Dermatol 120:810–816
69. Dorschner RA, Pestonjamasp VK, Tamakuwala S, Ohtake T, Rudisill J, Nizet V, Agerberth B, Gudmundsson GH, Gallo RL (2001) Cutaneous injury induces the release of cathelicidin anti-microbial peptides active against group A Streptococcus. J Invest Dermatol 117:91–97
70. Howell MD, Gallo RL, Boguniewicz M, Jones JF, Wong C, Streib JE, Leung DY (2006) Cytokine milieu of atopic dermatitis skin subverts the innate immune response to vaccinia virus. Immunity 24:341–348
71. Yamasaki K, Schauber J, Coda A, Lin H, Dorschner RA, Schechter NM, Bonnart C, Descargues P, Hovnanian A, Gallo RL (2006) Kallikrein-mediated proteolysis regulates the antimicrobial effects of cathelicidins in skin. FASEB J 20:2068–2080
72. Niyonsaba F, Ushio H, Hara M, Yokoi H, Tominaga M, Takamori K, Kajiwara N, Saito H, Nagaoka I, Ogawa H et al (2010) Antimicrobial peptides human beta-defensins and cathelicidin LL-37 induce the secretion of a pruritogenic cytokine IL-31 by human mast cells. J Immunol 184:3526–3534
73. Tjabringa GS, Ninaber DK, Drijfhout JW, Rabe KF, Hiemstra PS (2006) Human cathelicidin LL-37 is a chemoattractant for eosinophils and neutrophils that acts via formyl-peptide receptors. Int Arch Allergy Immunol 140:103–112
74. Ny A, Egelrud T (2003) Transgenic mice over-expressing a serine protease in the skin: evidence of interferon gamma-independent MHC II expression by epidermal keratinocytes. Acta Derm Venereol 83:322–327
75. Steinhoff M, Corvera CU, Thoma MS, Kong W, McAlpine BE, Caughey GH, Ansel JC, Bunnett NW (1999) Proteinase-activated receptor-2 in human skin: tissue distribution and activation of keratinocytes by mast cell tryptase. Exp Dermatol 8:282–294
76. Steinhoff M, Neisius U, Ikoma A, Fartasch M, Heyer G, Skov PS, Luger TA, Schmelz M (2003) Proteinase-activated receptor-2 mediates itch: a novel pathway for pruritus in human skin. J Neurosci 23:6176–6180
77. Sales KU, Masedunskas A, Bey AL, Rasmussen AL, Weigert R, List K, Szabo R, Overbeek PA, Bugge TH (2010) Matriptase initiates activation of epidermal pro-kallikrein and disease onset in a mouse model of Netherton syndrome. Nat Genet 42:676–683
78. Kaplan AP, Silverberg M, Ghebrehiwet B, Atkins P, Zweiman B (1989) The kallikrein-kinin system in inflammation. Adv Exp Med Biol 247A:125–136
79. Ponticelli C, Meroni PL (2009) Kallikreins and lupus nephritis. J Clin Invest 119:768–771
80. Eissa A, Diamandis EP (2008) Human tissue kallikreins as promiscuous modulators of homeostatic skin barrier functions. Biol Chem 389:669–680

Proteases from Inflammatory Cells: Regulation of Inflammatory Response

Magali Pederzoli-Ribeil, Julie Gabillet, and Véronique Witko-Sarsat

Abstract In this review, we summarize the current data pertaining to proteases mainly from polymorphonuclear neutrophil (PMN) and monocytes in the regulation of the inflammatory response. However, tryptase and chymase stored in mast cell granules, or granzymes from lymphocytes are other examples of proteases, which greatly influence several biological processes including extracellular matrix degradation, vasoconstriction, pathogen clearance, and cell death. A specific emphasis will be given to proteases from PMN, which are the first cells to be recruited to the inflammatory site. Proteases clearly modulate inflammation through cleavage of adhesion molecules, receptor implicated in pathogen recognition, phagocytosis, and production of cytokines. These cleavages can have pro or anti-inflammatory effect. In addition PMN-derived proteases can modulate the apoptosis of PMN and their uptake by macrophage, two pivotal steps in the resolution of inflammation. Deciphering the molecular mechanism governing the protease-based immune regulation should lead to novel and timely therapeutic strategies.

Keywords Apoptosis • Chemokine • Cytokine • Inflammation • Macrophage • Neutrophil • Phagocytosis • Protease

Abbreviations

ANCA	Antineutrophil cytoplasmic antibody
ADAM	A disintegrin and a metalloprotease
BAI1	Brain-specific angiogenesis inhibitor 1
CCL	CC chemokine ligand

M. Pederzoli-Ribeil • J. Gabillet • V. Witko-Sarsat (✉)
Immunology-Hematology Department, INSERM U1016, Institut Cochin, Université Paris Descartes Paris, 27, rue du faubourg Saint Jacques, 75014 Paris, France
e-mail: veronique.witko@inserm.fr

N. Vergnolle and M. Chignard (eds.), *Proteases and Their Receptors in Inflammation*, Progress in Inflammation Research, DOI 10.1007/978-3-0348-0157-7_4,

CCR	CC chemokine receptor
CR	Complement receptor
CRP	C reactive protein
ECM	Extracellular matrix
EGFR	Epithelial growth factor receptor
FcγR	Fcγ receptor
G-CSF	Granulocyte colony-stimulating factor
G-CSFR	G-CSF receptor
GM-CSF	Granulocyte macrophage colony stimulating factor
ICAM	Intercellular adhesion molecule
LFA-1	Lymphocyte function-associated antigen 1
LPC	Lysophosphatidylcholine
LPS	Lipopolysaccharide
IL	Interleukin
MAC1	Macrophage receptor 1
MADCAM1	Mucosal vascular addressin cell-adhesion molecule 1
MCP-1	Monocyte chemotactic protein-1
MFG-E8	Milk fat globule-EGF factor 8 protein
MIP	Macrophage inflammatory protein
MMP	Matrix metalloproteinase
MT1-MMP	Membrane type-1 MMP
NADPH	Nicotinamide adenine dinucleotide phosphate
NET	Neutrophils extracellular trap
PAF	Platelet-activating factor
PAR	Protease-activated receptor
PGE2	Prostaglandin E2
PLA2	Phospholipase A2
PMN	Polymorphonuclear neutrophils
PS	Phosphatidylserine
PSGL1	P-selectin glycoprotein ligand 1
PSR	PS receptor
RANTES	Regulated on activation normal T expressed and secreted
ROS	Reactive oxygen species
SAA	Serum amyloid A
SDF-1	Stromal cell-derived factor-1
SHP2	Src homology region 2 domain-containing phosphatase-1
SR-A	Scavenger A
TIM-4	T cell immunoglobulin mucin-4
TNFα	Tumor necrosis factor
TNFR	TNFα receptor
VCAM	Vascular cell-adhesion molecule 1
VLA4	Very late activation antigen-4
XIAP	X-linked Inhibitor of apoptosis protein

1 Introduction

Inflammation is a manifestation of the body's response to tissue damage and infection that may be beneficial for the defense against agents deranging its homeostasis, but also harmful for the surrounding tissues [1]. It is a coordinated response involving different cell types, which are mobilized with their own kinetics to ensure the accomplishment of the biologic program leading ultimately to tissue repair and successful resolution. Inflammatory cells include circulating cells such as polymorphonuclear neutrophils (PMN), monocytes, lymphocytes, platelets, and mast cells because they are recruited upon an inflammatory stimulus. However, resident cells such as endothelial cells play a crucial role in the course of an inflammatory process and are integral part of it.

This chapter will focus more specifically on the role of different proteases from PMN or monocytes in the modulation of the inflammatory process [2] and show that these proteases are involved in virtually all steps of the inflammatory process. The role of both cell-specific and ubiquitous proteinases and their relevant protein substrates in key processes such as adhesion, diapedesis, migration, phagocytosis, killing, and apoptosis will be highlighted (Fig. 1). This review will thus emphasize the importance of PMN-protease-dependent regulatory mechanisms in the control of the inflammatory and immune response [3]. It thus become apparent that understanding the molecular basis of the regulatory functions of proteases from inflammatory cells might be the future challenge for the next years in phagocyte research and this will presumably open new avenues of research in the modulation of inflammation [4, 5].

The initial steps of inflammation involve increased vascular permeability, release of kinins, and histamine by resident mast cells. Although these latter are mainly known for their role in allergic disorders, they have recently been shown to exert many other immune functions. Mast cells constitute a heterogeneous population residing in mucosal and connective tissues and are found in close association with endothelial cells of blood vessels and nerves. Mast cell subsets are functionally distinct and their profile of expressed mediators is altered in disease states. Upon stimulation, mast cells can release preformed granule-associated mediators, such as histamine and proteases like tryptase and chymase which are extremely important in the context of allergy (for review see [6]). PMN and monocytes, considered as "professional phagocytes" have the dual functions of immune surveillance and in situ elimination of microorganisms or cellular debris [2]. These phagocytic cells express both ubiquitous (calpain, cathepsins, caspases) and cell-specific proteases and especially PMN-derived serine-proteinases (cathepsin G, PMN elastase, and proteinase 3) [7, 8] and PMN- and macrophage-derived matrix metalloproteinases (MMP) [9] (see the dedicated chapter on Matrix Metalloproteinases in inflammatory processes), which are now recognized as instrumental in the regulation of the immune process. PMN-derived serine proteases can be released as soluble enzymes upon degranulation or associated with the plasma membrane [8, 10] thus modifying their sensitivity to inhibitors [4]. Indeed, the relative importance of protease-dependent regulation

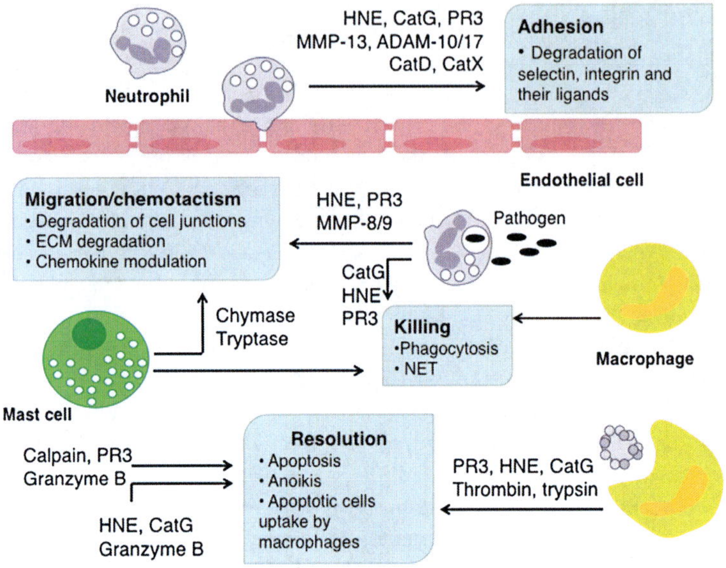

Fig. 1 Multiple roles of inflammatory cell-derived proteases in the regulation of inflammation. Proteases interfere with leukocytes recruitment by shedding selectin, integrin, and by cleaving their ligands, thus preventing leukocytes accumulation around the vessel walls. Proteases can cleave cell–cell junction proteins or ECM components thus favoring leukocytes migration. Proteases participate in the killing mechanisms and in the regulation of inflammation by cleaving chemokines and cytokines. Proteases can modulate the resolution of inflammation by interfering with neutrophil apoptosis and their phagocytosis by macrophages. *ADAM* A disintegrin and a metalloprotease, *catD* Cathepsin D, *catG* cathepsin G, *catX* cathepsin X, *ECM* extracellular matrix, *HNE* Human neutrophil elastase, *MMP* metalloprotease, *NET* neutrophils extracellular trap, *PR3* proteinase 3

depends both on the availability of key substrates and on the presence of protease inhibitors [11] (see the dedicated chapter on Proteases and Antiproteases in Inflammation).

2 Effect of Proteases on Initial Stages of Inflammation: Diapedesis and Migration of Phagocytic Cells

One initial event is the vasodilatation induced by vasoactive peptides such as kinins released by damaged tissues that result in local extravasation of leukocytes. Proteinase 3 can cleave the high-molecular weight kininogen and liberate a novel tridecapeptide, thereby initiating kallikrein-independent activation of kinin pathway that may operate during systemic inflammation [12]. To complete their functions during an inflammatory response, PMN and macrophages require a rapid transition from a circulating nonadherent state to an adherent state allowing them to migrate into tissues. These two events require adhesion and transmigration through blood vessel walls by the traditional and sequential three steps of rolling, activation, and firm adhesion [13]. Both PMN and monocytes use similar adhesion mechanisms to migrate from the blood to the tissues and various proteases have been shown to modulate these phenomena.

2.1 *The Selectin-Dependent Rolling Step*

The early steps of rolling are mediated by selectins, namely L-selectin expressed in most leukocytes and P-selectin and E-selectin expressed by inflammatory endothelial cells, which interact with P-selectin glycoprotein ligand 1 (PSGL1). L-selectin shedding by proteases during rolling interactions may be physiologically important for limiting leukocyte aggregation and accumulation at sites of inflammation [14]. Following leukocytes activation, L-selectin can be shed by proteolytic cleavage near the cell surface by ADAM-17 [15]. ADAMs (proteins containing A Disintegrin and A Metalloprotease domain) are ubiquitous multidomain metalloproteases that are emerging as key regulators of critical events that occur at the cell surface. The best-characterized in vivo activity of ADAM proteases is as ectodomain sheddase [16]. The biological significance of L-selectin shedding is not well understood, but previous studies have revealed a role in regulating the rolling velocity of PMN. There may be other proteases such as chymotrypsin, stromelysin, and collagenase [17], but not serine proteases [18], involved in L-selectin shedding but they have yet to be fully examined. Selectin ligands like CD43, which has been described as a regulator of leukocyte motility and trafficking [19], can also be cleaved by PMN-derived elastase [20].

2.2 *The Integrin-Dependent Adhesion Step*

The next step will be the integrin-mediated rolling followed by the firm adhesion to endothelium. The $\alpha4\beta7$-integrin [21] and very late antigen 4 (VLA4; also called $\alpha4\beta1$-integrin) [22] on surface of leukocytes interact, respectively, with mucosal vascular addressin cell-adhesion molecule 1 (MADCAM1) and vascular cell-adhesion molecule 1 (VCAM1) on endothelial cells. PMN express four $\beta2$ integrins: macrophage receptor 1 (MAC1 also called CD11b/CD18 and $\alpha M\beta2$-integrin) and lymphocyte function-associated antigen 1 (LFA-1 also known as $\alpha L\beta2$-integrin) are most relevant for slow rolling and PMN arrest in the systemic circulation [23]. Firm adhesion of PMN is largely CD18-dependent since a monoclonal antibody against CD18 inhibits PMN accumulation [24]. MAC-l and LFA-1 bind to endothelial ICAM-1 [25] and LFA-1 also binds to ICAM-2 [26]. VLA4 interacts with VCAM-1 [22] and promote leukocytes arrest [27]. Since truncated $\beta2$-integrin cytoplasmic tail interferes with LFA-1 binding to ICAM-1 [28], it is possible that

LFA-1 cleavage could have significant effect on PMN recruitment. Indeed, the cleavage of LFA-1 by cathepsin X (also known as cathepsin Z or CATZ) a lysosomal cysteine proteases, induces a conformation change, α-actinin-1 binding to LFA-1, thus leading to enhanced T cell migration [29]. Cathepsin X also cleaves αVβ3-integrin thereby inhibiting the attachment of migrating cells to extracellular matrix (ECM) components [30].

Other proteases can regulate leukocyte transmigration through ICAM-1 shedding such as membrane type-1 matrix metalloproteinase (MT1-MMP) [31], PMN-derived cathepsin G [32] or -elastase [33], and calpain [34, 35]. In fact, calpains, which are cytosolic cysteine proteinases activated in response to the accumulation of cytosolic calcium ions [36], have been implicated in leukocytes adhesion through cleavage of numerous integrins. Indeed, integrin-β subunits are susceptible to calpain cleavage [37]; for example, calpain cleaves α2bβ3-integrin thus inhibiting clot retraction [38]. Calpain has been also implicated in leukocyte membrane "expansion" through cleavage of cytoskeletal proteins involved in membrane linkage [36] such as talin and ezrin. In addition, flattening of lymphocytes during adhesion via β2-integrin is dependent on the activity of calpain [39].

2.3 *Migration and Proteolytic Regulation of the Proteins from the Extracellular Matrix*

Leukocytes can traverse the endothelium by the paracellular or the transcellular routes [40]. PMN elastase seems to be important for disrupting cell–cell interactions and thus for transmigration since zymosan-induced leukocyte firm adhesion and transmigration was suppressed in elastase knockout mice [41]. Indeed, α6β1-integrin and elastase appear to facilitate PMN migration through the perivascular basement membrane component, apparently acting in a synergistic way [42]. In support of a role for chymase in inflammatory cell recruitment, it has been reported that human chymase is chemotactic in vitro for neutrophils and monocytes [43]. It has also been demonstrated that intradermal injection of human chymase in mice caused inflammatory cell influx, including eosinophils [44]. One major function of chymase, in an inflammatory context, may be to promote microvascular permeability by degrading components involved in epithelial cell–cell contacts, for instance the tight junction proteins occludin and ZO-1 [45]. Similarly to chymase, a number of studies have reported proinflammatory activities of tryptase and injection of human tryptase into guinea pig skin induced influx of PMN and eosinophils [46]. Using the same model, the authors also demonstrated that the proinflammatory action of tryptase was associated with increased microvascular leakage [47]. Subsequently, two independent studies showed that recombinant mast cell chymase, when injected intraperitoneally to mice, induced an influx of neutrophils [48]. Further, it has been reported that instillation of human recombinant β-tryptase into trachea of mice provokes a neutrophilic inflammation. The mechanisms behind these proinflammatory activities are not known, but it has been suggested that tryptase can stimulate IL-8 and ICAM expression on endothelial cells [48, 49]. After transmigration, leukocytes move through the ECM to reach bacteria, parasites, infected or dead cells, or cell debris. This movement requiring interactions with ECM can be facilitated by α6β1-integrin and possibly proteases, such as MMPs, PMN-derived serine proteases (cathepsin G, elastase, and proteinase 3), or metalloproteinases (gelatinase, collagenase), which cleave ECM components. The location of mast cells within connective tissues and, hence, the release of chymase and tryptase into connective tissues in association with mast cells degranulation is well in line with a role for these proteases in inflammation regulation. Type IV collagen can be cleaved by tryptase from mast cells [50]. Indeed, mast cell proteases affect the inflammatory process by modifying the ECM, either directly by acting on ECM components or indirectly by regulating the activities of ECM-processing enzymes. Chymase and tryptase can degrade fibronectin [51], type IV collagen [50], and gelatin [52]. As a result, ECM protein degradation enhances influx of inflammatory cells into the tissues and potentiates inflammation.

3 Modulation of Effector Mechanisms of Inflammatory Cells by Proteases

3.1 Surface Receptors and Phagocytosis Modulation by Proteases

The surface receptors of PMN and monocytes regulate a range of functions, including cell differentiation, growth and survival, adhesion, migration, phagocytosis, activation, and cytotoxicity. Their ability to recognize a wide range of endogenous and exogenous ligands and to trigger the antimicrobial effect of both PMN and monocytes is a crucial step in innate immunity [2]. Many different receptors recognize microbes and phagocytosis is usually mediated simultaneously by multiple receptors. Some of these receptors are able to transmit intracellular signals that trigger phagocytosis, while others appear primarily to participate in binding or to increase the efficiency of internalization or activation. Several receptors lead to pathogen recognition. Fcγ-receptors are cell surface receptors that bind to the Fc region of IgG thus recognizing IgG-opsonized particles. Phagocytes also express complement receptors (CR): CR1 (also known as CD35), CR3 (αMβ2-integrin or CD11b/CD18 or MAC1), and CR4 (αXβ2-integrin, CD11bc/CD18), which recognize complement-opsonized particles. Proteases can interfere with pathogen recognition by cleavage of these receptors. For example, PMN elastase [53] and MMPs [54] shed FcγR (FcγR), releasing a functional soluble Fcγ receptor, which plays a regulatory role on immune responses with cell binding and antiproliferative capacities [55]. Likewise, PMN-derived elastase and MMP-8 cleave CR1 and release a soluble form that inhibits complement activation [56]. CD14 can bind lipopolysaccharide (LPS) and is expressed predominantly at the surface of monocytes and macrophages [57]. CD14 signaling leads to production of numerous inflammatory mediators. Notably, PMN-derived elastase and cathepsin G can cleave CD14 thus releasing of multiple CD14 fragments and inhibiting LPS-mediated monocyte activation [58]. Moreover, purified PMN-derived elastase and cathepsin G can cleave CD2, CD4, and CD8 on peripheral blood T lymphocytes, thus leading to a clear reduction of cytotoxicity toward target cells and significantly reduced IL-2 and IL-4 production [59]. Likewise, urokinase-type plasminogen activator receptor (uPAR/CD87), which is a multiligand receptor that operates as a key element in physiological processes such as cell migration during inflammation can be cleaved by elastase and cathepsin G, as well. Remarkably, CD87-deficient mice showed impaired leukocyte recruitment in response to local infections [60]. CD87 cleavage decreased urokinase effect and cellular migration during tissue remodeling [61]. Protease-activated receptors (PARs) are seven-transmembrane receptors, which are activated directly by proteolytical cleavage. They are expressed by a variety of immune cells [62]. Several lines of evidence have shown that proteases from inflammatory cells can regulate PAR activities (see the dedicated chapter on Terminating Protease Receptor Signaling).

3.2 Synergy Between Proteases and Other Mediators to Optimize Killing Mechanisms

Degranulation is a crucial mechanism involved in pathogen destruction. Degranulation of vesicles into the phagolysosome or in the extracellular space is a key event for microbicidal activity. This antimicrobial effect depends on two concurrent events occurring in the nascent phagolysosome: the generation of reactive oxygen species (ROS) by assembly and activation of the NADPH-dependent oxidase and the release of antibiotic proteins contained in the granules. The NADPH oxidase is an electron transport chain in "professional" phagocytic cells that transfers electrons from NADPH in the cytoplasm, across the wall of the phagocytic vacuole, to form superoxide. The electron transporting flavocytochrome *b* is activated by the integrated function of four cytoplasmic proteins. It has been proposed that the antimicrobial function of this system involves pumping K+ into the vacuole through potassium channels, the effect of which is to elevate the vacuolar pH and activate neutral serine proteinases [63]. Notably, elastase-deficient mice failed to defend themselves against infection [64], thus underlining the importance of PMN elastase and cathepsin G in microbicidal killing. PMN-derived elastase cleaves the virulence factors and outer membrane proteins of Gram-negative bacteria [65]. Granules contain numerous proteases such as MMP, serine proteases, thiol-proteases, and aspartate proteases, which have direct microbicidal functions [7]. In addition, these proteases can cooperate with antibiotic proteins to enhance the antibactericidal potential of PMN. As an example, proteinase 3 released into the extracellular space can process pro-cathelicidin to generate an active antibiotic peptide [66]. Of note, cathepsin G [67], proteinase 3, and azurocidin [68] have also antimicrobial potential independent of their enzymatic action.

Another means by which PMN can kill microbes are through the generation of neutrophil extracellular traps (NETs) that are structures composed by DNA and proteases such as elastase and cathepsin G [69]. Recently, it has been shown that mast cells may also form structures similar to NETs called mast cell extracellular traps (MCETs) [70]. These extracellular traps are responsible for bacteria killing and lead to cell death [71]. Pathogens are trapped by NETs and are killed by the high local concentration of enzymes. Proteases that are present in these structures play an active role in the elimination of pathogens, thus reinforcing the importance of PMN serine proteases in antimicrobial defense.

4 Modulation of the Immune Response via the Cleavage of Cytokines, Chemokines, and Their Cognate Receptors

Cytokines and chemokines are key players of the inflammatory reaction and proteases from inflammatory cells modulate the bioactivity of chemokine and cytokine networks through proteolytic cleavage. This cleavage might increase their biodisponibility by promoting the processing of an inactive precursor. In contrast, proteolytic processing might also result in cytokine degradation and inactivation. Controlling chemokine activity at sites of infection is important, since excess accumulation of leukocytes may contribute to localized tissue damage. In addition, proteolytic regulation of receptors is one potent way to orientate the inflammatory or the immune response.

4.1 Cleavage of Cytokines and Their Receptors: Potential Modulation of the Immune Response

The inflammatory response is greatly enhanced by macrophage-derived proinflammatory cytokines such as interleukin-1-β (IL-1β) that activates the release of other proinflammatory cytokines such as Tumor Necrosis Factor-α (TNF-α) and interleukin-6 (IL-6) (reviewed in [72]). Modulation of macrophage-derived cytokines by PMN proteases constitutes an important link in the cooperation between these two cell types [73]. The processing of the inactive precursor into the bioactive IL-1β depends on activation of caspase-1 by protein complexes called the inflammasome [74]. Proteinase 3 increases active IL-1β release as well thus suggesting that there might be alternative pathways for the production of this proinflammatory cytokine, particularly in the context of local inflammatory processes, characterized by an overwhelming PMN infiltrate where a minimal role of caspase-1 and more important role of proteinase 3 were observed [75]. In addition, IL-1β can be converted rapidly by human mast cell chymase, a process that might be expected to have a critical role in the initiation of the inflammatory response [76]. IL-6 is one of the most important mediators of the acute phase response. The bioactivity of IL-6 was reduced in inflammatory exudates as evidence by its degradation and subsequent inactivation by the PMN-derived elastase, proteinase 3, and cathepsin G. This inactivation might act as a feedback mechanism to terminate the IL-6-induced PMN activation [77]. Moreover, mast cell chymase can degrade and inactivate IL-6 [78]. As a regulatory cytokine, TNF-α orchestrates communication between immune cells and controls many of their functions [79]. TNF-α is best known for its role in leading immune defenses to protect a localized area from invasion or injury but it is also involved in controlling whether target cells live or die. The transmembrane TNF-α is proteolytically cleaved to yield a soluble biologically active form. Indeed, proteinase 3 has the ability to augment the release of TNF-α from LPS-activated THP-1, a human monocytic cell line [75]. In contrast, proteolytic cleavage of TNF-α, resulting in the loss of its cytotoxic activities has been reported for PMN elastase and cathepsin G, suggesting a negative feedback loop limiting PMN activation [80]. TNF-α can bind to TNF-R1 or TNF-R2, two structurally distinct TNF-α receptors on target cells, also known as p55 or p75, respectively, to activate two separate intracellular signaling pathways to gene transcription. In vitro data [81] provided direct evidence for the involvement of PMN elastase in the TNF-R2 receptor shedding, affecting TNF-mediated cellular activation as well as its cytotoxic effects. Therefore, PMN elastase-catalyzed TNF-R2-shedding is believed to represent an accessory mechanism for controlling the cellular responses to TNF-α at sites of inflammation.

T-cell related cytokines are also the targets of PMN-derived proteases. These latter might thus participate in the regulation of the immune response. Indeed, PMN elastase degrades interleukine-2 (IL-2), a T-cell growth factor that promotes T-cell-dependent immune responses including T cell adhesion to fibronectin [82]. IL-2 receptor is a heterotrimeric protein expressed on the surface of certain immune cells, such as lymphocytes, which binds and responds to IL-2. IL-6 receptor, also known as CD126, is a type I cytokine receptor, which has been shown to interact with IL-6. The potent but differential effects of the three PMN-serine proteases on IL-2 and IL-6 receptors were described [83]. Under in vitro conditions, PMN elastase and, to a lesser extent, proteinase 3 could cleave membrane-bound IL-2 receptor. In contrast, the cleavage of the IL-6 receptor was only due to cathepsin G [83]. The receptor fragments released by the action of these enzymes were found to retain their ligand-binding capacity. These results strongly suggest a biologic role of PMN-derived serine proteases, particularly in the regulation of the cytokine receptor shedding of functional IL-2 and IL-6 receptors at foci of inflammation. Interleukin-32 (IL-32) is a proinflammatory cytokine selectively expressed by T-cells, which has been shown to stimulate TNF-α production by macrophages. Notably, proteinase 3 has been described as a specific IL-32-binding protein, independent of its enzymatic activity. Moreover, cleavage of IL-32 by proteinase 3 enhanced its IL-8 induction activity in monocytes. Therefore, targeting proteinase 3 might by a new mean to modulate IL-32 activity [84].

As it will be discussed at the end of this chapter, resolution of inflammation is dependent on the secretion of anti-inflammatory cytokines such as interleukin-10 or Transforming-Growth factor-β (TGF-β). This latter is a multifunctional cytokine modulating onset and course of autoimmune diseases, which is important in blocking macrophage activation. The most important regulation of TGF-β activity to be reported is based on whether or not TGF-β is biologically active or latent [85]. MMP-9 and MMP-2 play key roles in this activation process [86]. Moreover, proteinase-3 was revealed as a potent activator of latent TGF-β, indicating that TGF-β might serve as a proinflammatory factor in ANCA-associated vasculitis [87]. In addition, PMN elastase releases the active TGF-β and contributes to the tissue remodeling that accompanies inflammation in the lung [88].

4.2 Chemotaxis, Chemokines, and Their Receptors as Targets of Proteases

Various compounds – such as lipid mediators, pathogen-derived products, antimicrobial peptides, and complement products – are chemotactic and regulate leukocyte trafficking. For instance, PMN migration can be induced by a gradient of complement fragment C5a, or platelet-activating factor (PAF), which are released from pathogens, leukocytes, or surrounding cells (reviewed in [1]). Interestingly, chemerin is a chemoattractant for dendritic cells that acts as a ligand for the G-protein-coupled receptor ChemR23. Chemerin is secreted in an inactive form as prochemerin and is activated through cleavage of the C-terminus to stimulate dendritic cells and macrophages to the site of inflammation. Serine proteases factor XIIa and plasmin of the coagulation and fibrinolytic cascades, PMN-derived elastase and cathepsin G, mast cell tryptase are all potent activators of chemerin [89], triggering rapid defenses in the body. However, proteolytic events that negatively control chemerin activity involved proteinase 3 and mast cell chymase. These mechanisms highlight the complex interplay of proteases regulating the bioactivity of this novel mediator during early innate immune responses [90].

Chemokines are key regulators of inflammatory cell migration and activation. They exert their biological effects by interacting with their G protein-coupled transmembrane receptors and to subsequently induce signaling. As already mentioned, chemotactic factors are pivotal to initiate the inflammatory response and CXCL8 (Chemokine (CXC motif) ligand 8 also called interleukine-8, IL-8) produced by a wide range of cells including macrophages and endothelial cells is the most potent chemokine to attract PMN. Significant conversion of IL-8 to more potent, amino-terminally truncated forms was observed upon incubation with PMN granule lysates, indicating that PMN proteases released in inflamed tissues convert IL-8 to enhance its chemotactic activity [91]. CCL3 (Chemokine (C–C motif) ligand 3, also called Macrophage Inflammatory Protein- 1 or MIP-1α) is involved in the recruitment and activation of PMN. MIP-1α proteolysis by cathepsin G, PMN elastase, and proteinase 3 results in loss of chemotactic activity as compared with the parental molecule. Moreover, using PMN lysates from Papillon–Lefevre syndrome patients, containing inactivated serine proteases, it has been demonstrated an absence of degradation of MIP-1α. These findings suggest that severe periodontal tissue destruction may be related to accumulation of intact MIP-1α and dysregulation of the microbial-induced inflammatory response [92]. Likewise, chymase degrades CCL3 activity, dampening the inflammatory reaction [93]. CCL5 (Chemokine (C–C motif) ligand 5, also called RANTES for Regulated upon Activation, Normal T-cell Expressed, and Secreted) plays an active role in recruiting leukocytes into inflammatory sites. N-terminal proteolytic processing modulates the biological activity and receptor specificity of RANTES/CCL5. Cathepsin G was characterized as the enzyme responsible for this processing. This digested variant binds CCR5 but exhibits lower chemotactic and antiviral activities than unprocessed RANTES. These findings suggest that cathepsin G mediates a novel pathway for regulating RANTES activity and may be relevant to the role of RANTES and its analogs in preventing HIV infection [94]. In addition, tryptase degrades RANTES [95] to abrogate its eosinophil chemotactic activity. CCL15 (chemokine (C–C motif) ligand 15 or lungkine) expressed in leukocytes and macrophages of the mouse lung is chemotactic for PMN, monocytes, and lymphocytes and elicits its effects by binding to cell surface chemokine receptors like CCR1 and CCR3. Monocyte infiltration into inflammatory sites is generally preceded by PMN. It has been reported that PMN may support this process by CCL15 activation. Cathepsin G was identified as the principal protease to produce an N-terminal deleted form of CCL15, as well as PMN elastase. Compared with full-length CCL15, truncated CCL15 displayed a significantly increased potency to induce calcium fluxes and chemotactic activity on monocytes and to induce adhesiveness of mononuclear cells to fibronectin [96]. In addition, chymase can also increase the potency of CCL15, suggesting that proteases, released during inflammatory responses in vivo, can convert chemokines into potent chemoattractants [93]. CXCL5 (also called epithelial neutrophil-activating protein 78 or ENA-78) is a potent stimulator of PMN, inducing a variety of biological responses such as chemotaxis, enzyme release, up-regulation of surface receptors, and intracellular calcium mobilization. Proteolysis of ENA-78 by cathepsin G leads to the formation of truncation products with higher potency than native ENA-78 [97]. CXCL7 is a proteolytically processed fragment of platelet basic protein, which binds CXCR2 and chemoattracts and activates PMN. Cathepsin G [98] as well as mast cell chymase [99] have been recognized to cleave the precursor to form the active CXCL7. CXCL12 (also called stromal-derived factor 1 or SDF-1) activates leukocytes and is often induced by proinflammatory stimuli such as LPS, TNF-α, or IL-1β. It is the only known ligand for CXCR4. It has been reported that it generates an SDF-1 fragment, which fails to induce agonistic functions. Furthermore, exposure of CXCR4-expressing cells to PMN elastase resulted in the proteolysis of the extracellular amino-terminal domain of the receptor. Hence, elastase-mediated proteolysis of SDF-1/CXCR4 is part of a mechanism regulating their biological functions in both homeostatic and pathologic processes [100].

5 Role of Proteases in the Resolution of Inflammation

5.1 *The Resolution of Inflammation Is an Active Process*

Over the last few years, many studies have focused their attention on the resolution of inflammation [101]. This is an active and coordinated program involving a switch from pro- to anti-inflammatory mediators, apoptosis of inflammatory cells, and their subsequent engulfment by professional (macrophages and immature dendritic cells) and nonprofessional phagocytes (fibroblasts, endothelial, epithelial and mesenchymal cells). Ultimately, this should result in the release of anti-inflammatory and reparative mediators. Resolution of inflammation is accompanied by an active switch in the mediators that predominate in exudates: from classic prostaglandins and leukotrienes, to genus of specialized proresolving lipid mediators: lipoxins, resolvins, protectins, and maresins [102, 103]. Therefore, it is clear that the resolution of inflammation involves both inhibition of proinflammatory signaling and stimulation of proresolving events and that proteinases from inflammatory cells might influence this process by cleaving intracellular or extracellular signaling molecules. One key event in the resolution of inflammation is the elimination of dead cells and specifically apoptotic PMN. Indeed, PMN apoptosis and their safe clearance provide a mechanism of reducing the number of viable and activated PMN without releasing their potentially harmful enzymes and ROS. However, different type of cell death such as anoikis [104] or autophagy [105] might be involved in inflammation and might be regulated by proteinases.

5.2 *Regulation of PMN Apoptosis by Proteases*

Apoptosis, described as a noninflammatory programmed cell death, is a fundamental physiological process, in which cells die by activating an intrinsic suicide mechanism [105]. Apoptosis is a carefully regulated process with highly controlled proteolytic activation, involving activation of caspases, the traditionally predominant mediators of the death program. Interestingly, roles for noncaspases proteases, such as calpains, cathepsins that are cysteine proteases from the lysosomal compartment, serine proteases, and proteasome in cell death have been reported as well [106]. Serine proteases may function independently of the apoptotic signaling pathway or interact with other mediators, such as caspases. Cleavage of procaspase-7 and -8 by calpain results in their inactivation, whereas cleavage of procaspase-9, -12, and Bcl-xl positively impact upon apoptotic events [107]. Cathepsins are released into the cytosol prior to mitochondrial membrane depolarization and activate caspases [108].

Delayed death in tissues cause unwanted and exaggerated inflammation progressing towards chronic inflammation. PMN apoptosis must be tightly controlled to provide a correct balance between PMN immune functions and their safe removal [109]. If this balance is disturbed, meaning if apoptosis is not tightly regulated, this defect will lead to profound and devastating conditions ranging from cancer to autoimmune diseases [110]. Indeed, if pathogens alter PMN apoptosis by inducing delay in apoptosis or cell lysis, they will be able to survive and disseminate, inducing surrounding tissue destruction and inflammation. Importantly, PMN apoptosis is accompanied by a general decrease in primary cell functions, including chemotaxis, phagocytosis, superoxide production, and degranulation [111]. Indeed, several receptor expressions are decreased, including CD16, CD31, CD50, and CD66, avoiding receptor/ligand interaction and therefore decreasing capacity of PMN to answer to microenvironment signals [112]. At the inflammatory site, PMN are exposed to multiple factors and their fate would ultimately depend on the balance between prosurvival and proapoptotic signals from other blood cells, including endothelial cells [113], red blood cells [114] and platelets [115], and the inflammatory microenvironment [116]. Proinflammatory cytokines, bacterial constituents including IL-1β, IFN-γ [117], LPS [118], G-CSF, GM-CSF [119], CRP, and SAA can rescue PMN from apoptosis, whereas TNF-α, C5a, and Fas-Ligand shorten their lifespan. Mature PMN contain a low number of mitochondria that may have a role restricted to apoptosis [120] and a low content of cytochrome c compensated by the elevated expression of Apaf-1 and caspase-9 [121]. PMN do not express the antiapoptotic factor Bcl-2, but express A1 and Mcl-1, which is degraded by the proteasome in cells undergoing apoptosis [122]. Mcl-1 and A1 are rapidly expressed due to the presence of proinflammatory cytokines in the microenvironment and they have a short half-life indicating that they constitute key regulator factors of the survival/apoptosis balance in PMN. PMN express also the proapoptotic molecules Bax, Bak, Bid, and Bad, which show stable expression overtime and are essential components of the apoptotic machinery. Notably, the calpain system appears to be a key component in the control of PMN survival. During apoptosis, level of calpastatin, a calpain inhibitor, is decreased thus favoring calpain activity [123]. Indeed, pharmacological calpain inhibitors prolong PMN survival; Accordingly, calpastatin belongs to the genes that are strongly induced upon G-CSF-induced PMN survival [124]. Calpains might also contribute to programmed cell death by generating an active form of the proapoptotic factor Bax [125] and by inhibiting the prosurvival molecule XIAP [126]. Notably, calpain mediated the cleavage of Atg5, a protein involved in autophagic death, and provokes apoptotic cell death, therefore, representing a molecular link between autophagy and apoptosis [127].

Cathepsins are other cytosolic proteases that have been involved in apoptosis. The best characterized cytosolic substrates for cathepsins is the proapoptotic factor Bid, but recently, Bak and the antiapoptotic Bcl-xl and Mcl-1 have been identified as additional substrates [128]. Another new issue is the involvement of cathepsin D in the activation of procaspase-8 in PMN [129]. Remarkably, this work suggests a new concept in which the granular protein cathepsin D, stored into the azurophilic granules, can "leak" from this permeabilized granules to activate and cleave procaspase-8. This direct activation of procaspase-8 by cathepsin D seems to be specific to PMN and constitutes a new proapoptotic pathway [130]. Moreover, it has been identified a new function of azurophilic granules that is, in addition to their role in bacterial defense mechanisms, to regulate the life span of PMN and, therefore, the duration of innate immune responses through the release of cathepsin D.

PMN-derived serine proteases can regulate the balance survival/apoptosis in PMN through different mechanisms (Fig. 2). For instance, cathepsin G can cleave procaspase-3 into its active fragment and thus potentiate PMN apoptosis. In contrast, procaspase-3 cleavage by membrane-associated proteinase 3 generates an active 22-kDa fragment restricted to the PMN membrane compartment. Interestingly, this fragment was present only in resting PMN but was absent after apoptosis, strongly suggesting that compartmentalized PR3-induced caspase-3 activation might play specific functions in PMN survival [131]. Finally, PMN elastase is

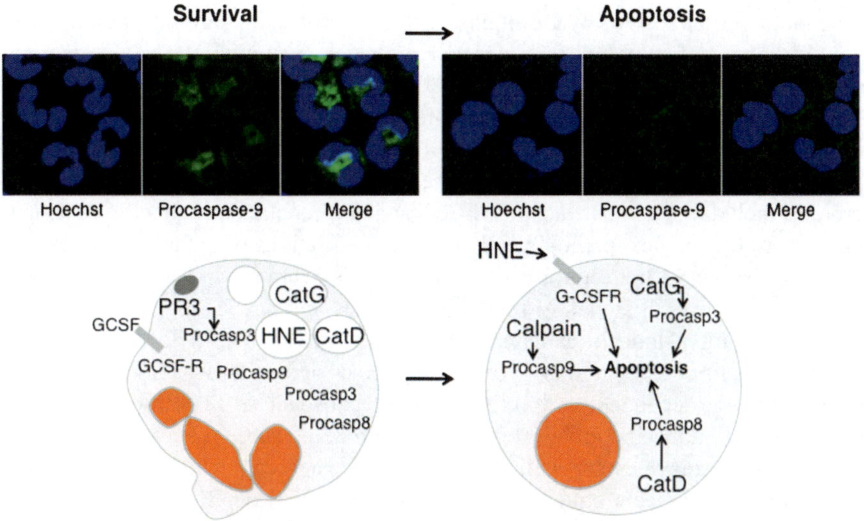

Fig. 2 Proteases modulate the balance survival/apoptosis of neutrophils. *Upper panel*: Visualization of PMN apoptosis. PMN apoptosis induces a morphological change from the characteristic multinucleated cell to a chromatin-condensed morphology typical of apoptosis as evidenced by Hoechst staining. Apoptosis is accompanied by a loss in cytosolic procaspase-9 as shown by indirect immunofluorescence of procaspase-9 in PMN using a rabbit polyclonal antiprocaspase-9 antibody (Witko-Sarsat, unpublished data). *Lower panel*: Cartoon depicting the effects of PMN serine proteases on PMN apoptosis as described in the text. Cleavage of procaspase-3 by proteinase 3 is associated with PMN survival. Proapoptotic effects of proteases include activation of procaspase-3, procaspase-8, and procaspase-9 by cathepsin G, cathepsin D, and calpain, respectively. Moreover, PMN elastase can cleave G-CSF and its receptor thus interfering with prosurvival mechanisms. *catD* Cathepsin D, *catG* cathepsin G, *G-CSFR* G-CSF receptor, *HNE* Human neutrophil elastase, *PR3* proteinase 3, *Procasp*: procaspase

responsible for proteolytic degradation of G-CSF, the most potent cytokine to induce PMN survival and its receptor G-CSFR is also a substrate for this protease. Notably, the decreased G-CSFR surface expression was associated with a reduction in cell viability and proliferation in response to G-CSF [132]. Serine proteases have thus diverse and opposing effects in the regulation of PMN survival and their localization and/or biological activities might also vary during apoptosis, thus adding another level of complexity. Membrane metalloproteinase activity can also be modified by apoptosis and can have access to new substrates. This is illustrated by the proteolysis of the IL-6R by the metalloprotease ADAM17 during apoptosis, thus inducing recruitment of mononuclear cells to the site of infection [133].

5.3 Modulation of Anoikis by PMN-Derived Proteases and Granzyme B

Another type of cell death, called anoikis, relates to the loss of contact of adherent cells, which triggers cell death, having all the feature of apoptosis. Proteolysis of subendothelial adhesive glycoproteins such as fibronectin, thrombospondin, and von Willebrand factor by PMN-derived proteases including PMN elastase and cathepsin G greatly affect the endothelium and promote inflammation [134]. Indeed, it has been reported that the N-terminal domain of thrombospondin can induce anoikis by disrupting cell-to-cell contact [135]. Cathepsin G has been proposed to play an important role in tissue remodeling at sites of injury by cleaving proteins including chemoattractants, ECM, and hormonal factors. In culture neonatal rat cardiomyocytes, pathophysiological concentrations of cathepsin G activate signaling pathways that culminate in myocyte detachment and apoptosis. Some facets of cathepsin G signaling in cardiomyocytes are independent of PAR-1 and PAR-4 activation. Indeed, cathepsin G can transactivate EGFR to induce downstream signaling and anoikis. This paradoxical proapoptotic effect of EGFR appeared to be dependent on protein tyrosine phosphatase SHP2 activation that promotes focal adhesion kinase dephosphorylation and subsequent anoikis [136].

Although granzymes are well known for their capacity to induce apoptosis of their target cells, recent studies have shown that granzyme B possesses a potent ECM remodeling activity during inflammation. Classically, granzymes are described as serine proteases found exclusively in the granules of cytotoxic T lymphocytes and natural killer cells, playing a critical role in eliminating virally infected cells by inducing apoptosis [137]. Cytotoxic granules also contain a pore-forming protein perforin, which mediates the delivery of granzymes into the intermembrane space of the target cells [138]. Granzyme B can directly process and activate procaspase-3 and -7 [139] and cleave Bid, leading to cytochrome c release [140]. Mcl-1 has been identified as a granzyme B substrate, which induces mitochondria membrane depolarization [141]. The role of other granzymes in apoptosis is more controversial and mainly based on in vitro findings. The role of granzymes in the modulation of cell death by anoikis has been more recently uncovered. Cell detachment by granzyme B induced endothelial cell anoikis [142] by cleaving vitronectin, fibronectin, and laminin, three proteins involved in ECM structure and functions. Hence, granzyme B could promote inflammation and tissue destruction, independently of perforin. Recent work has identified the presence of granzyme B in skin but not in lung mast cells in the absence of perforin or granzyme A. Using granzyme B-deficient mice as compared with wild type mice, it has been shown that mast cells contributed to cell death, increased vascular permeability, leukocyte extravasation, and subsequent inflammatory process in affected tissues [143].

5.4 Apoptotic Cells Phagocytosis by Macrophages: A Balance Between "Eat-Me" and "Don't Eat Me"

Both nonprofessional and professional phagocytes, such as macrophages, rapidly eliminate apoptotic PMN. Not surprisingly, impairment of apoptotic cells elimination results in the development of autoimmune diseases [144]. Apoptotic cells expressed "eat-me" signals at their surface to allow their recognition by specific receptors and subsequent phagocytosis by macrophages [145]. Moreover, several soluble molecules called "bridging molecules" have been identified to play a key role in this process, favoring apoptotic cell engulfment [146]. Apoptotic PMN clearance by macrophages leads to the secretion of anti-inflammatory cytokines, such as TGF-β, IL-10, and PGE2 by macrophages and diminution of the release of proinflammatory cytokines including TNF-α and IL-8. After apoptotic cell ingestion, macrophage phenotype switches to an noninflammatory phenotype to induce tissue cicatrization suggesting that phagocytosis of apoptotic cells is involved in the negative regulation of macrophage activation and could be considered as endogenous active anti-inflammatory mechanisms [101]. Phagocytes recognize "eat-me" signals, exposed at the plasma membrane of apoptotic cells [146]. These signals are either modified membrane proteins or molecules newly mobilized at the cell surface. Indeed, many modifications of the apoptotic cell plasma membrane occur, including rearrangement of phospholipid monolayer and relocation of PS from the inner to the outer face of plasma membrane [147], nuclear material exposure [148], modification of the glycosylation profile [149], change of the oxidation status of macromolecules [150], and redistribution of calreticulin at the plasma membrane [151]. Receptors for apoptotic cells include scavenger receptors, integrin, lectin, receptor for the complement and calreticulin. Bridging molecules such as C1q, Gas-6 [152], S protein, and MFG-E8 [153] bind to "eat-me" signals on apoptotic cells and their receptor on macrophages, to induce phagocytosis. Indeed, mice genetically deficient in C1q develop a lupus-like syndrome characterized by systemic inflammation [154].

PS exposure at the plasma membrane is the best-characterized change occurring at the cell surface during early apoptosis and the most characterized "eat-me" signal described. PS has been identified as the ligand for the PS receptor (PSR) on the phagocyte surface [155]. Interestingly, PMN elastase has been shown to cleave PSR on macrophages and therefore, disrupt apoptotic cells phagocytosis [156], contributing to ongoing airway inflammation in cystic fibrosis. However, PSR identification was based on the use of a monoclonal antibody Mab217 that blocked PS-dependent phagocytosis of apoptotic cells. However, the identity of this receptor has been questioned and is the subject of controversy. Indeed, it turned out that this PSR did not mediate apoptotic cell removal and other receptors, which seem to fulfill this function, have been described BAI1 [157], TIM-4 [158], and stabilin 2 [159]. Indeed, it has been demonstrated that PSR is not the protein recognized by Mab217 [160] and that phagocytosis of apoptotic cells is not impaired in this PSR deficient mice. Therefore, the negative effect of PMN elastase on apoptotic cells recognition need to be addressed [156]. Moreover, externalized PS on apoptotic cell surface can be recognized by other receptors including CD36, CD14, and CD68, involved in the uptake of apoptotic cells [161]. In contrast, "don't eat-me" signals namely CD31 (PECAM-1) or CD47 [162] have been described on viable cells; they induce dissociation between the viable cell and macrophage to prevent phagocytosis [163]. A recent study has shown that MMP-2 is a specific protease that cleaves CD47 in vascular smooth muscle cells in presence of glucose [164]. Whether cleavage of CD47 occurs on myeloid cells is an open question and the effect of CD47 cleavage on apoptotic cell recognition is still unknown.

5.5 PMN Serine Proteinases Interfere with the Phagocytosis of Apoptotic PMN by Macrophages

Predigestion of apoptotic PMN with cathepsin G, thrombin, or trypsin dramatically reduced their uptake by macrophages [165] thus suggesting that these enzymes proteolytically inactivate (yet uncharacterized) "eat-me" signals on the PMN cell surface. PMN elastase has been shown to degrade CD14 (also implicated in apoptotic cells recognition [166]) on human monocytes [167] and human fibroblasts [168], impairing apoptotic cells uptake. Indeed, PMN elastase-mediated cleavage of CD14 and impairment of apoptotic cells recognition are reversed by adenovirus-mediated overexpression of elafin, a potent elastase inhibitor [169]. Proteinase 3 membrane expression on PMN has been shown to constitute a proinflammatory factor especially in vasculitis, since proteinase 3 is the target of anti-PMN cytoplasmic antibodies (ANCA) [10]. Remarkably, proteinase 3 expressed at the plasma membrane during apoptosis can hamper the phagocytosis of apoptotic cells by macrophages, thus favoring the persistence of PMN at the site of inflammation [170]. Moreover, it seems that proteinase 3 can act as a "don't eat-me signal" independently of its enzymatic activity thus suggesting that PR3 membrane expression represent a key element in the elimination of apoptotic PMN even in the presence of antiproteinase activity. Elucidation of the molecular basis of proteinase 3 interaction with the plasma membrane or with receptor proteins led to the possibility of targeted therapy [10].

Phagocytosis of apoptotic cells can be modulated by anti-inflammatory and "proresolving" proteins and annexin-A1 is one of them. Annexin-A1 belongs to the annexin family of Ca2+-dependent phospholipid-binding proteins. Since phospholipase A2 is required for the biosynthesis of the potent mediators of inflammation, prostaglandins, and leukotrienes, annexin-A1, which inhibits phospholipase A2, is considered as an anti-inflammatory mediator. Moreover several studies using experimental model of inflammation have shown that administration of annexin-A1 results in potent inhibition of PMN activation and trafficking, such as inhibition of PMN adhesion to the vascular bed and detachment of adherent PMN resulting in a reduction in the number of cells migrating into the subendothelial matrix [171]. In addition, annexin-A1 can stimulate phagocytosis of apoptotic PMN by macrophages [172]. Annexin-A1 appears to be pivotal for apoptotic envelope formation on macrophages infected with *Mycobacterium tuberculosis* [173]. PMN-derived proteinase 3 has been identified as the main enzyme responsible for annexin-A1 cleavage in the amino-terminus bioactive region of the protein [174]. Therefore, because annexin-A1 is an important endogenous anti-inflammatory mediator, blocking this cleavage might augment its homeostatic proresolving actions and could represent an opportunity for innovative anti-inflammatory drug discovery. Indeed, a mutated proteinase 3-resistant form of annexin-A1, so-called super-annexin displayed stronger anti-inflammatory effect over time when compared to the wild type protein, using mouse models of acute inflammation [175]. Thus proteinase 3-mediated cleavage of annexin-A1 might constitutes a proteolytic process impairing apoptotic PMN uptake by macrophages, at the inflammatory site. In the same line of thinking, it has also been shown that other PMN proteases can confuse the recognition of apoptotic cells by macrophages [165]. Therefore, inhibition of proteases involved in inhibition of apoptotic cells clearance might be determinant for the resolution of inflammation and for the design of new therapeutic drugs.

6 Conclusion

A large number of studies have provided evidence that proteases from inflammatory cells are pivotal regulator of all steps of the inflammatory response thus uncovering new potential for therapeutic opportunities. However, further investigations are required to decipher molecular mechanisms involved in the regulation of novel pathways such as those regulating cell death and apoptotic cell clearance.

References

1. Witko-Sarsat V, Rieu P, Descamps-Latscha B, Lesavre P, Halbwachs-Mecarelli L (2000) Neutrophils: molecules, functions and pathophysiological aspects. Lab Invest 80:617–653
2. Kantari C, Pederzoli-Ribeil M, Witko-Sarsat V (2008) The role of neutrophils and monocytes in innate immunity. Contrib Microbiol 15:118–146

3. Nathan C (2006) Neutrophils and immunity: challenges and opportunities. Nat Rev Immunol 6:173–182
4. Owen CA, Campbell EJ (1999) The cell biology of leukocyte-mediated proteolysis. J Leukoc Biol 65:137–150
5. Pham CT (2008) Neutrophil serine proteases fine-tune the inflammatory response. Int J Biochem Cell Biol 40:1317–1333
6. Galli SJ, Grimbaldeston M, Tsai M (2008) Immunomodulatory mast cells: negative, as well as positive, regulators of immunity. Nat Rev Immunol 8:478–486
7. Owen CA (2008) Leukocyte cell surface proteinases: regulation of expression, functions, and mechanisms of surface localization. Int J Biochem Cell Biol 40:1246–1272
8. Hajjar E, Broemstrup T, Kantari C, Witko-Sarsat V, Reuter N (2010) Structures of human proteinase 3 and neutrophil elastase – so similar yet so different. FEBS J 277:2238–2254
9. Klein T, Bischoff R (2010) Physiology and pathophysiology of matrix metalloproteases. Amino Acids
10. Witko-Sarsat V, Reuter N, Mouthon L (2010) Interaction of proteinase 3 with its associated partners: implications in the pathogenesis of Wegener's granulomatosis. Curr Opin Rheumatol 22:1–7
11. Korkmaz B, Moreau T, Gauthier F (2008) Neutrophil elastase, proteinase 3 and cathepsin G: physicochemical properties, activity and physiopathological functions. Biochimie 90:227–242
12. Kahn R, Hellmark T, Leeb-Lundberg LM, Akbari N, Todiras M, Olofsson T, Wieslander J, Christensson A, Westman K, Bader M et al (2009) Neutrophil-derived proteinase 3 induces kallikrein-independent release of a novel vasoactive kinin. J Immunol 182:7906–7915
13. Ley K, Laudanna C, Cybulsky MI, Nourshargh S (2007) Getting to the site of inflammation: the leukocyte adhesion cascade updated. Nat Rev Immunol 7:678–689
14. Rosen SD (2004) Ligands for L-selectin: homing, inflammation, and beyond. Annu Rev Immunol 22:129–156
15. Smalley DM, Ley K (2005) L-selectin: mechanisms and physiological significance of ectodomain cleavage. J Cell Mol Med 9:255–266
16. Murphy G (2008) The ADAMs: signalling scissors in the tumour microenvironment. Nat Rev Cancer 8:929–941
17. Preece G, Murphy G, Ager A (1996) Metalloproteinase-mediated regulation of L-selectin levels on leucocytes. J Biol Chem 271:11634–11640
18. Bazil V, Strominger JL (1994) Metalloprotease and serine protease are involved in cleavage of CD43, CD44, and CD16 from stimulated human granulocytes. Induction of cleavage of L-selectin via CD16. J Immunol 152:1314–1322
19. Sanchez-Mateos P, Campanero MR, del Pozo MA, Sanchez-Madrid F (1995) Regulatory role of CD43 leukosialin on integrin-mediated T-cell adhesion to endothelial and extracellular matrix ligands and its polar redistribution to a cellular uropod. Blood 86:2228–2239
20. Remold-O'Donnell E, Parent D (1995) Specific sensitivity of CD43 to neutrophil elastase. Blood 86:2395–2402
21. Berlin C, Bargatze RF, Campbell JJ, von Andrian UH, Szabo MC, Hasslen SR, Nelson RD, Berg EL, Erlandsen SL, Butcher EC (1995) Alpha 4 integrins mediate lymphocyte attachment and rolling under physiologic flow. Cell 80:413–422
22. Chan JR, Hyduk SJ, Cybulsky MI (2001) Chemoattractants induce a rapid and transient upregulation of monocyte alpha4 integrin affinity for vascular cell adhesion molecule 1 which mediates arrest: an early step in the process of emigration. J Exp Med 193:1149–1158
23. Dunne JL, Ballantyne CM, Beaudet AL, Ley K (2002) Control of leukocyte rolling velocity in TNF-alpha-induced inflammation by LFA-1 and Mac-1. Blood 99:336–341
24. Arfors KE, Lundberg C, Lindbom L, Lundberg K, Beatty PG, Harlan JM (1987) A monoclonal antibody to the membrane glycoprotein complex CD18 inhibits polymorphonuclear leukocyte accumulation and plasma leakage in vivo. Blood 69:338–340
25. Diamond MS, Staunton DE, Marlin SD, Springer TA (1991) Binding of the integrin Mac-1 (CD11b/CD18) to the third immunoglobulin-like domain of ICAM-1 (CD54) and its regulation by glycosylation. Cell 65:961–971

26. Staunton DE, Dustin ML, Springer TA (1989) Functional cloning of ICAM-2, a cell adhesion ligand for LFA-1 homologous to ICAM-1. Nature 339:61–64
27. Hyduk SJ, Chan JR, Duffy ST, Chen M, Peterson MD, Waddell TK, Digby GC, Szaszi K, Kapus A, Cybulsky MI (2007) Phospholipase C, calcium, and calmodulin are critical for alpha4beta1 integrin affinity up-regulation and monocyte arrest triggered by chemoattractants. Blood 109:176–184
28. Hibbs ML, Xu H, Stacker SA, Springer TA (1991) Regulation of adhesion of ICAM-1 by the cytoplasmic domain of LFA-1 integrin beta subunit. Science 251:1611–1613
29. Jevnikar Z, Obermajer N, Pecar-Fonovic U, Karaoglanovic-Carmona A, Kos J (2009) Cathepsin X cleaves the beta2 cytoplasmic tail of LFA-1 inducing the intermediate affinity form of LFA-1 and alpha-actinin-1 binding. Eur J Immunol 39:3217–3227
30. Lechner AM, Assfalg-Machleidt I, Zahler S, Stoeckelhuber M, Machleidt W, Jochum M, Nagler DK (2006) RGD-dependent binding of procathepsin X to integrin alphavbeta3 mediates cell-adhesive properties. J Biol Chem 281:39588–39597
31. Essick E, Sithu S, Dean W, D'Souza S (2008) Pervanadate-induced shedding of the intercellular adhesion molecule (ICAM)-1 ectodomain is mediated by membrane type-1 matrix metalloproteinase (MT1-MMP). Mol Cell Biochem 314:151–159
32. Robledo O, Papaioannou A, Ochietti B, Beauchemin C, Legault D, Cantin A, King PD, Daniel C, Alakhov VY, Potworowski EF et al (2003) ICAM-1 isoforms: specific activity and sensitivity to cleavage by leukocyte elastase and cathepsin G. Eur J Immunol 33:1351–1360
33. Champagne B, Tremblay P, Cantin A, St Pierre Y (1998) Proteolytic cleavage of ICAM-1 by human neutrophil elastase. J Immunol 161:6398–6405
34. Cuzzocrea S, McDonald MC, Mazzon E, Mota-Filipe H, Centorrino T, Terranova ML, Ciccolo A, Britti D, Caputi AP, Thiemermann C (2001) Calpain inhibitor I reduces colon injury caused by dinitrobenzene sulphonic acid in the rat. Gut 48:478–488
35. Sithu SD, English WR, Olson P, Krubasik D, Baker AH, Murphy G, D'Souza SE (2007) Membrane-type 1-matrix metalloproteinase regulates intracellular adhesion molecule-1 (ICAM-1)-mediated monocyte transmigration. J Biol Chem 282:25010–25019
36. Franco SJ, Huttenlocher A (2005) Regulating cell migration: calpains make the cut. J Cell Sci 118:3829–3838
37. Pfaff M, Du X, Ginsberg MH (1999) Calpain cleavage of integrin beta cytoplasmic domains. FEBS Lett 460:17–22
38. Schoenwaelder SM, Yuan Y, Cooray P, Salem HH, Jackson SP (1997) Calpain cleavage of focal adhesion proteins regulates the cytoskeletal attachment of integrin alphaIIbbeta3 (platelet glycoprotein IIb/IIIa) and the cellular retraction of fibrin clots. J Biol Chem 272:1694–1702
39. Dewitt S, Hallett M (2007) Leukocyte membrane "expansion": a central mechanism for leukocyte extravasation. J Leukoc Biol 81:1160–1164
40. Feng D, Nagy JA, Pyne K, Dvorak HF, Dvorak AM (1998) Neutrophils emigrate from venules by a transendothelial cell pathway in response to FMLP. J Exp Med 187:903–915
41. Young RE, Thompson RD, Larbi KY, La M, Roberts CE, Shapiro SD, Perretti M, Nourshargh S (2004) Neutrophil elastase (NE)-deficient mice demonstrate a nonredundant role for NE in neutrophil migration, generation of proinflammatory mediators, and phagocytosis in response to zymosan particles in vivo. J Immunol 172:4493–4502
42. Wang S, Dangerfield JP, Young RE, Nourshargh S (2005) PECAM-1, alpha6 integrins and neutrophil elastase cooperate in mediating neutrophil transmigration. J Cell Sci 118: 2067–2076
43. Tani K, Ogushi F, Kido H, Kawano T, Kunori Y, Kamimura T, Cui P, Sone S (2000) Chymase is a potent chemoattractant for human monocytes and neutrophils. J Leukoc Biol 67:585–589
44. Tomimori Y, Muto T, Fukami H, Saito K, Horikawa C, Tsuruoka N, Saito M, Sugiura N, Yamashiro K, Sumida M et al (2002) Chymase participates in chronic dermatitis by inducing eosinophil infiltration. Lab Invest 82:789–794

45. Scudamore CL, Jepson MA, Hirst BH, Miller HR (1998) The rat mucosal mast cell chymase, RMCP-II, alters epithelial cell monolayer permeability in association with altered distribution of the tight junction proteins ZO-1 and occludin. Eur J Cell Biol 75:321–330
46. He S, Walls AF (1998) Human mast cell chymase induces the accumulation of neutrophils, eosinophils and other inflammatory cells in vivo. Br J Pharmacol 125:1491–1500
47. He S, Peng Q, Walls AF (1997) Potent induction of a neutrophil and eosinophil-rich infiltrate in vivo by human mast cell tryptase: selective enhancement of eosinophil recruitment by histamine. J Immunol 159:6216–6225
48. Huang C, Friend DS, Qiu WT, Wong GW, Morales G, Hunt J, Stevens RL (1998) Induction of a selective and persistent extravasation of neutrophils into the peritoneal cavity by tryptase mouse mast cell protease 6. J Immunol 160:1910–1919
49. Compton SJ, Cairns JA, Holgate ST, Walls AF (1998) The role of mast cell tryptase in regulating endothelial cell proliferation, cytokine release, and adhesion molecule expression: tryptase induces expression of mRNA for IL-1 beta and IL-8 and stimulates the selective release of IL-8 from human umbilical vein endothelial cells. J Immunol 161:1939–1946
50. Kielty CM, Lees M, Shuttleworth CA, Woolley D (1993) Catabolism of intact type VI collagen microfibrils: susceptibility to degradation by serine proteinases. Biochem Biophys Res Commun 191:1230–1236
51. Lazaar AL, Plotnick MI, Kucich U, Crichton I, Lotfi S, Das SK, Kane S, Rosenbloom J, Panettieri RA Jr, Schechter NM et al (2002) Mast cell chymase modifies cell-matrix interactions and inhibits mitogen-induced proliferation of human airway smooth muscle cells. J Immunol 169:1014–1020
52. Fajardo I, Pejler G (2003) Human mast cell beta-tryptase is a gelatinase. J Immunol 171: 1493–1499
53. Tosi MF, Berger M (1988) Functional differences between the 40 kDa and 50 to 70 kDa IgG Fc receptors on human neutrophils revealed by elastase treatment and antireceptor antibodies. J Immunol 141:2097–2103
54. Middelhoven PJ, Van Buul JD, Hordijk PL, Roos D (2001) Different proteolytic mechanisms involved in Fc gamma RIIIb shedding from human neutrophils. Clin Exp Immunol 125:169–175
55. Teillaud C, Galon J, Zilber MT, Mazieres N, Spagnoli R, Kurrle R, Fridman WH, Sautes C (1993) Soluble CD16 binds peripheral blood mononuclear cells and inhibits pokeweed-mitogen-induced responses. Blood 82:3081–3090
56. Sadallah S, Hess C, Miot S, Spertini O, Lutz H, Schifferli JA (1999) Elastase and metalloproteinase activities regulate soluble complement receptor 1 release. Eur J Immunol 29:3754–3761
57. Wright SD, Ramos RA, Tobias PS, Ulevitch RJ, Mathison JC (1990) CD14, a receptor for complexes of lipopolysaccharide (LPS) and LPS binding protein. Science 249:1431–1433
58. Le-Barillec K, Si-Tahar M, Balloy V, Chignard M (1999) Proteolysis of monocyte CD14 by human leukocyte elastase inhibits lipopolysaccharide-mediated cell activation. J Clin Invest 103:1039–1046
59. Doring G, Frank F, Boudier C, Herbert S, Fleischer B, Bellon G (1995) Cleavage of lymphocyte surface antigens CD2, CD4, and CD8 by polymorphonuclear leukocyte elastase and cathepsin G in patients with cystic fibrosis. J Immunol 154:4842–4850
60. Gyetko MR, Sud S, Kendall T, Fuller JA, Newstead MW, Standiford TJ (2000) Urokinase receptor-deficient mice have impaired neutrophil recruitment in response to pulmonary Pseudomonas aeruginosa infection. J Immunol 165:1513–1519
61. Beaufort N, Leduc D, Rousselle JC, Magdolen V, Luther T, Namane A, Chignard M, Pidard D (2004) Proteolytic regulation of the urokinase receptor/CD87 on monocytic cells by neutrophil elastase and cathepsin G. J Immunol 172:540–549
62. Ossovskaya VS, Bunnett NW (2004) Protease-activated receptors: contribution to physiology and disease. Physiol Rev 84:579–621
63. Segal AW (2005) How neutrophils kill microbes. Annu Rev Immunol 23:197–223

64. Belaaouaj A, McCarthy R, Baumann M, Gao Z, Ley TJ, Abraham SN, Shapiro SD (1998) Mice lacking neutrophil elastase reveal impaired host defense against gram negative bacterial sepsis. Nat Med 4:615–618
65. Belaaouaj A (2002) Neutrophil elastase-mediated killing of bacteria: lessons from targeted mutagenesis. Microbes Infect 4:1259–1264
66. Sorensen OE, Follin P, Johnsen AH, Calafat J, Tjabringa GS, Hiemstra PS, Borregaard N (2001) Human cathelicidin, hCAP-18, is processed to the antimicrobial peptide LL-37 by extracellular cleavage with proteinase 3. Blood 97:3951–3959
67. Bangalore N, Travis J, Onunka VC, Pohl J, Shafer WM (1990) Identification of the primary antimicrobial domains in human neutrophil cathepsin G. J Biol Chem 265:13584–13588
68. Gabay JE, Almeida RP (1993) Antibiotic peptides and serine protease homologs in human polymorphonuclear leukocytes: defensins and azurocidin. Curr Opin Immunol 5:97–102
69. Brinkmann V, Reichard U, Goosmann C, Fauler B, Uhlemann Y, Weiss DS, Weinrauch Y, Zychlinsky A (2004) Neutrophil extracellular traps kill bacteria. Science 303:1532–1535
70. von Kockritz-Blickwede M, Goldmann O, Thulin P, Heinemann K, Norrby-Teglund A, Rohde M, Medina E (2008) Phagocytosis-independent antimicrobial activity of mast cells by means of extracellular trap formation. Blood 111:3070–3080
71. Brinkmann V, Zychlinsky A (2007) Beneficial suicide: why neutrophils die to make NETs. Nat Rev Microbiol 5:577–582
72. Philippart F, Cavaillon JM (2007) Sepsis mediators. Curr Infect Dis Rep 9:358–365
73. Soehnlein O, Lindbom L (2010) Phagocyte partnership during the onset and resolution of inflammation. Nat Rev Immunol 10:427–439
74. Martinon F, Burns K, Tschopp J (2002) The inflammasome: a molecular platform triggering activation of inflammatory caspases and processing of proIL-beta. Mol Cell 10:417–426
75. Coeshott C, Ohnemus C, Pilyavskaya A, Ross S, Wieczorek M, Kroona H, Leimer AH, Cheronis J (1999) Converting enzyme-independent release of tumor necrosis factor alpha and IL-1beta from a stimulated human monocytic cell line in the presence of activated neutrophils or purified proteinase 3. Proc Natl Acad Sci USA 96:6261–6266
76. Mizutani H, Schechter N, Lazarus G, Black RA, Kupper TS (1991) Rapid and specific conversion of precursor interleukin 1 beta (IL-1 beta) to an active IL-1 species by human mast cell chymase. J Exp Med 174:821–825
77. Bank U, Kupper B, Reinhold D, Hoffmann T, Ansorge S (1999) Evidence for a crucial role of neutrophil-derived serine proteases in the inactivation of interleukin-6 at sites of inflammation. FEBS Lett 461:235–240
78. Zhao W, Oskeritzian CA, Pozez AL, Schwartz LB (2005) Cytokine production by skin-derived mast cells: endogenous proteases are responsible for degradation of cytokines. J Immunol 175:2635–2642
79. Croft M (2009) The role of TNF superfamily members in T-cell function and diseases. Nat Rev Immunol 9:271–285
80. Scuderi P, Nez PA, Duerr ML, Wong BJ, Valdez CM (1991) Cathepsin-G and leukocyte elastase inactivate human tumor necrosis factor and lymphotoxin. Cell Immunol 135: 299–313
81. Porteu F, Brockhaus M, Wallach D, Engelmann H, Nathan CF (1991) Human neutrophil elastase releases a ligand-binding fragment from the 75-kDa tumor necrosis factor (TNF) receptor. Comparison with the proteolytic activity responsible for shedding of TNF receptors from stimulated neutrophils. J Biol Chem 266:18846–18853
82. Ariel A, Yavin EJ, Hershkoviz R, Avron A, Franitza S, Hardan I, Cahalon L, Fridkin M, Lider O (1998) IL-2 induces T cell adherence to extracellular matrix: inhibition of adherence and migration by IL-2 peptides generated by leukocyte elastase. J Immunol 161:2465–2472
83. Bank U, Reinhold D, Schneemilch C, Kunz D, Synowitz HJ, Ansorge S (1999) Selective proteolytic cleavage of IL-2 receptor and IL-6 receptor ligand binding chains by neutrophil-derived serine proteases at foci of inflammation. J Interferon Cytokine Res 19:1277–1287
84. Novick D, Rubinstein M, Azam T, Rabinkov A, Dinarello CA, Kim SH (2006) Proteinase 3 is an IL-32 binding protein. Proc Natl Acad Sci USA 103:3316–3321

85. Munger JS, Harpel JG, Gleizes PE, Mazzieri R, Nunes I, Rifkin DB (1997) Latent transforming growth factor-beta: structural features and mechanisms of activation. Kidney Int 51:1376–1382
86. Yu Q, Stamenkovic I (2000) Cell surface-localized matrix metalloproteinase-9 proteolytically activates TGF-beta and promotes tumor invasion and angiogenesis. Genes Dev 14:163–176
87. Kekow J, Csernok E, Szymkowiak C, Gross WL (1997) Interaction of transforming growth factor beta (TGF beta) with proteinase 3. Adv Exp Med Biol 421:307–313
88. Taipale J, Lohi J, Saarinen J, Kovanen PT, Keski-Oja J (1995) Human mast cell chymase and leukocyte elastase release latent transforming growth factor-beta 1 from the extracellular matrix of cultured human epithelial and endothelial cells. J Biol Chem 270:4689–4696
89. Zabel BA, Allen SJ, Kulig P, Allen JA, Cichy J, Handel TM, Butcher EC (2005) Chemerin activation by serine proteases of the coagulation, fibrinolytic, and inflammatory cascades. J Biol Chem 280:34661–34666
90. Guillabert A, Wittamer V, Bondue B, Godot V, Imbault V, Parmentier M, Communi D (2008) Role of neutrophil proteinase 3 and mast cell chymase in chemerin proteolytic regulation. J Leukoc Biol 84:1530–1538
91. Padrines M, Wolf M, Walz A, Baggiolini M (1994) Interleukin-8 processing by neutrophil elastase, cathepsin G and proteinase-3. FEBS Lett 352:231–235
92. Ryu OH, Choi SJ, Firatli E, Choi SW, Hart PS, Shen RF, Wang G, Wu WW, Hart TC (2005) Proteolysis of macrophage inflammatory protein-1alpha isoforms LD78beta and LD78alpha by neutrophil-derived serine proteases. J Biol Chem 280:17415–17421
93. Berahovich RD, Miao Z, Wang Y, Premack B, Howard MC, Schall TJ (2005) Proteolytic activation of alternative CCR1 ligands in inflammation. J Immunol 174:7341–7351
94. Lim JK, Lu W, Hartley O, DeVico AL (2006) N-terminal proteolytic processing by cathepsin G converts RANTES/CCL5 and related analogs into a truncated 4-68 variant. J Leukoc Biol 80:1395–1404
95. Pang L, Nie M, Corbett L, Sutcliffe A, Knox AJ (2006) Mast cell beta-tryptase selectively cleaves eotaxin and RANTES and abrogates their eosinophil chemotactic activities. J Immunol 176:3788–3795
96. Richter R, Bistrian R, Escher S, Forssmann WG, Vakili J, Henschler R, Spodsberg N, Frimpong-Boateng A, Forssmann U (2005) Quantum proteolytic activation of chemokine CCL15 by neutrophil granulocytes modulates mononuclear cell adhesiveness. J Immunol 175:1599–1608
97. Nufer O, Corbett M, Walz A (1999) Amino-terminal processing of chemokine ENA-78 regulates biological activity. Biochemistry 38:636–642
98. Cohen AB, Stevens MD, Miller EJ, Atkinson MA, Mullenbach G (1992) Generation of the neutrophil-activating peptide-2 by cathepsin G and cathepsin G-treated human platelets. Am J Physiol 263:L249–L256
99. Schiemann F, Grimm TA, Hoch J, Gross R, Lindner B, Petersen F, Bulfone-Paus S, Brandt E (2006) Mast cells and neutrophils proteolytically activate chemokine precursor CTAP-III and are subject to counterregulation by PF-4 through inhibition of chymase and cathepsin G. Blood 107:2234–2242
100. Valenzuela-Fernandez A, Planchenault T, Baleux F, Staropoli I, Le-Barillec K, Leduc D, Delaunay T, Lazarini F, Virelizier JL, Chignard M et al (2002) Leukocyte elastase negatively regulates Stromal cell-derived factor-1 (SDF-1)/CXCR4 binding and functions by amino-terminal processing of SDF-1 and CXCR4. J Biol Chem 277:15677–15689
101. Serhan CN, Savill J (2005) Resolution of inflammation: the beginning programs the end. Nat Immunol 6:1191–1197
102. Serhan CN, Chiang N, Van Dyke TE (2008) Resolving inflammation: dual anti-inflammatory and pro-resolution lipid mediators. Nat Rev Immunol 8:349–361
103. Serhan CN, Yang R, Martinod K, Kasuga K, Pillai PS, Porter TF, Oh SF, Spite M (2009) Maresins: novel macrophage mediators with potent antiinflammatory and proresolving actions. J Exp Med 206:15–23

104. Chiarugi P, Giannoni E (2008) Anoikis: a necessary death program for anchorage-dependent cells. Biochem Pharmacol 76:1352–1364
105. Maiuri MC, Zalckvar E, Kimchi A, Kroemer G (2007) Self-eating and self-killing: crosstalk between autophagy and apoptosis. Nat Rev Mol Cell Biol 8:741–752
106. Leist M, Jaattela M (2001) Four deaths and a funeral: from caspases to alternative mechanisms. Nat Rev Mol Cell Biol 2:589–598
107. Chua BT, Guo K, Li P (2000) Direct cleavage by the calcium-activated protease calpain can lead to inactivation of caspases. J Biol Chem 275:5131–5135
108. Stoka V, Turk V, Turk B (2007) Lysosomal cysteine cathepsins: signaling pathways in apoptosis. Biol Chem 388:555–560
109. Fox S, Leitch AE, Duffin R, Haslett C, Rossi AG (2010) Neutrophil apoptosis: relevance to the innate immune response and inflammatory disease. J Innate Immun 2:216–227
110. Kennedy AD, DeLeo FR (2009) Neutrophil apoptosis and the resolution of infection. Immunol Res 43:25–61
111. Whyte MK, Meagher LC, MacDermot J, Haslett C (1993) Impairment of function in aging neutrophils is associated with apoptosis. J Immunol 150:5124–5134
112. Dransfield I, Buckle AM, Savill JS, McDowall A, Haslett C, Hogg N (1994) Neutrophil apoptosis is associated with a reduction in CD16 (Fc gamma RIII) expression. J Immunol 153:1254–1263
113. Coxon A, Tang T, Mayadas TN (1999) Cytokine-activated endothelial cells delay neutrophil apoptosis in vitro and in vivo. A role for granulocyte/macrophage colony-stimulating factor. J Exp Med 190:923–934
114. Aoshiba K, Nakajima Y, Yasui S, Tamaoki J, Nagai A (1999) Red blood cells inhibit apoptosis of human neutrophils. Blood 93:4006–4010
115. Andonegui G, Trevani AS, Lopez DH, Raiden S, Giordano M, Geffner JR (1997) Inhibition of human neutrophil apoptosis by platelets. J Immunol 158:3372–3377
116. Luo HR, Loison F (2008) Constitutive neutrophil apoptosis: mechanisms and regulation. Am J Hematol 83:288–295
117. Sakamoto E, Hato F, Kato T, Sakamoto C, Akahori M, Hino M, Kitagawa S (2005) Type I and type II interferons delay human neutrophil apoptosis via activation of STAT3 and up-regulation of cellular inhibitor of apoptosis 2. J Leukoc Biol 78:301–309
118. Murray DA, Wilton JM (2003) Lipopolysaccharide from the periodontal pathogen Porphyromonas gingivalis prevents apoptosis of HL60-derived neutrophils in vitro. Infect Immun 71:7232–7235
119. Kobayashi SD, Voyich JM, Whitney AR, DeLeo FR (2005) Spontaneous neutrophil apoptosis and regulation of cell survival by granulocyte macrophage-colony stimulating factor. J Leukoc Biol 78:1408–1418
120. Maianski NA, Maianski AN, Kuijpers TW, Roos D (2004) Apoptosis of neutrophils. Acta Haematol 111:56–66
121. Murphy BM, O'Neill AJ, Adrain C, Watson RW, Martin SJ (2003) The apoptosome pathway to caspase activation in primary human neutrophils exhibits dramatically reduced requirements for cytochrome C. J Exp Med 197:625–632
122. Moulding DA, Akgul C, Derouet M, White MR, Edwards SW (2001) BCL-2 family expression in human neutrophils during delayed and accelerated apoptosis. J Leukoc Biol 70:783–792
123. Squier MK, Sehnert AJ, Sellins KS, Malkinson AM, Takano E, Cohen JJ (1999) Calpain and calpastatin regulate neutrophil apoptosis. J Cell Physiol 178:311–319
124. Drewniak A, van Raam BJ, Geissler J, Tool AT, Mook OR, van den Berg TK, Baas F, Kuijpers TW (2009) Changes in gene expression of granulocytes during in vivo granulocyte colony-stimulating factor/dexamethasone mobilization for transfusion purposes. Blood 113:5979–5998
125. Altznauer F, Martinelli S, Yousefi S, Thurig C, Schmid I, Conway EM, Schoni MH, Vogt P, Mueller C, Fey MF et al (2004) Inflammation-associated cell cycle-independent block of apoptosis by survivin in terminally differentiated neutrophils. J Exp Med 199:1343–1354

126. Kobayashi S, Yamashita K, Takeoka T, Ohtsuki T, Suzuki Y, Takahashi R, Yamamoto K, Kaufmann SH, Uchiyama T, Sasada M et al (2002) Calpain-mediated X-linked inhibitor of apoptosis degradation in neutrophil apoptosis and its impairment in chronic neutrophilic leukemia. J Biol Chem 277:33968–33977
127. Yousefi S, Perozzo R, Schmid I, Ziemiecki A, Schaffner T, Scapozza L, Brunner T, Simon HU (2006) Calpain-mediated cleavage of Atg5 switches autophagy to apoptosis. Nat Cell Biol 8:1124–1132
128. Droga-Mazovec G, Bojic L, Petelin A, Ivanova S, Romih R, Repnik U, Salvesen GS, Stoka V, Turk V, Turk B (2008) Cysteine cathepsins trigger caspase-dependent cell death through cleavage of bid and antiapoptotic Bcl-2 homologues. J Biol Chem 283:19140–19150
129. Conus S, Perozzo R, Reinheckel T, Peters C, Scapozza L, Yousefi S, Simon HU (2008) Caspase-8 is activated by cathepsin D initiating neutrophil apoptosis during the resolution of inflammation. J Exp Med 205:685–698
130. Conus S, Simon HU (2008) Cathepsins: key modulators of cell death and inflammatory responses. Biochem Pharmacol 76:1374–1382
131. Pederzoli M, Kantari C, Gausson V, Moriceau S, Witko-Sarsat V (2005) Proteinase-3 induces procaspase-3 activation in the absence of apoptosis: potential role of this compartmentalized activation of membrane-associated procaspase-3 in neutrophils. J Immunol 174:6381–6390
132. Hunter MG, Druhan LJ, Massullo PR, Avalos BR (2003) Proteolytic cleavage of granulocyte colony-stimulating factor and its receptor by neutrophil elastase induces growth inhibition and decreased cell surface expression of the granulocyte colony-stimulating factor receptor. Am J Hematol 74:149–155
133. Walcheck B, Herrera AH, St Hill C, Mattila PE, Whitney AR, Deleo FR (2006) ADAM17 activity during human neutrophil activation and apoptosis. Eur J Immunol 36:968–976
134. Bonnefoy A, Legrand C (2000) Proteolysis of subendothelial adhesive glycoproteins (fibronectin, thrombospondin, and von Willebrand factor) by plasmin, leukocyte cathepsin G, and elastase. Thromb Res 98:323–332
135. Pallero MA, Elzie CA, Chen J, Mosher DF, Murphy-Ullrich JE (2008) Thrombospondin 1 binding to calreticulin-LRP1 signals resistance to anoikis. FASEB J 22:3968–3979
136. Rafiq K, Hanscom M, Valerie K, Steinberg SF, Sabri A (2008) Novel mode for neutrophil protease cathepsin G-mediated signaling: membrane shedding of epidermal growth factor is required for cardiomyocyte anoikis. Circ Res 102:32–41
137. Barry M, Bleackley RC (2002) Cytotoxic T lymphocytes: all roads lead to death. Nat Rev Immunol 2:401–409
138. Bots M, Medema JP (2006) Granzymes at a glance. J Cell Sci 119:5011–5014
139. Adrain C, Murphy BM, Martin SJ (2005) Molecular ordering of the caspase activation cascade initiated by the cytotoxic T lymphocyte/natural killer (CTL/NK) protease granzyme B. J Biol Chem 280:4663–4673
140. Waterhouse NJ, Sedelies KA, Trapani JA (2006) Role of Bid-induced mitochondrial outer membrane permeabilization in granzyme B-induced apoptosis. Immunol Cell Biol 84:72–78
141. Han J, Goldstein LA, Gastman BR, Froelich CJ, Yin XM, Rabinowich H (2004) Degradation of Mcl-1 by granzyme B: implications for Bim-mediated mitochondrial apoptotic events. J Biol Chem 279:22020–22029
142. Buzza MS, Zamurs L, Sun J, Bird CH, Smith AI, Trapani JA, Froelich CJ, Nice EC, Bird PI (2005) Extracellular matrix remodeling by human granzyme B via cleavage of vitronectin, fibronectin, and laminin. J Biol Chem 280:23549–23558
143. Pardo J, Wallich R, Ebnet K, Iden S, Zentgraf H, Martin P, Ekiciler A, Prins A, Mullbacher A, Huber M et al (2007) Granzyme B is expressed in mouse mast cells in vivo and in vitro and causes delayed cell death independent of perforin. Cell Death Differ 14:1768–1779
144. Munoz LE, Lauber K, Schiller M, Manfredi AA, Herrmann M (2010) The role of defective clearance of apoptotic cells in systemic autoimmunity. Nat Rev Rheumatol 6:280–289
145. Ravichandran KS, Lorenz U (2007) Engulfment of apoptotic cells: signals for a good meal. Nat Rev Immunol 7:964–974

146. Paidassi H, Tacnet-Delorme P, Arlaud GJ, Frachet P (2009) How phagocytes track down and respond to apoptotic cells. Crit Rev Immunol 29:111–130
147. Fadok VA, Voelker DR, Campbell PA, Cohen JJ, Bratton DL, Henson PM (1992) Exposure of phosphatidylserine on the surface of apoptotic lymphocytes triggers specific recognition and removal by macrophages. J Immunol 148:2207–2216
148. Radic M, Marion T, Monestier M (2004) Nucleosomes are exposed at the cell surface in apoptosis. J Immunol 172:6692–6700
149. Morris RG, Hargreaves AD, Duvall E, Wyllie AH (1984) Hormone-induced cell death. 2. Surface changes in thymocytes undergoing apoptosis. Am J Pathol 115:426–436
150. Buttke TM, Sandstrom PA (1994) Oxidative stress as a mediator of apoptosis. Immunol Today 15:7–10
151. Gardai SJ, McPhillips KA, Frasch SC, Janssen WJ, Starefeldt A, Murphy-Ullrich JE, Bratton DL, Oldenborg PA, Michalak M, Henson PM (2005) Cell-surface calreticulin initiates clearance of viable or apoptotic cells through trans-activation of LRP on the phagocyte. Cell 123:321–334
152. Scott RS, McMahon EJ, Pop SM, Reap EA, Caricchio R, Cohen PL, Earp HS, Matsushima GK (2001) Phagocytosis and clearance of apoptotic cells is mediated by MER. Nature 411:207–211
153. Hanayama R, Tanaka M, Miyasaka K, Aozasa K, Koike M, Uchiyama Y, Nagata S (2004) Autoimmune disease and impaired uptake of apoptotic cells in MFG-E8-deficient mice. Science 304:1147–1150
154. Botto M, Dell'Agnola C, Bygrave AE, Thompson EM, Cook HT, Petry F, Loos M, Pandolfi PP, Walport MJ (1998) Homozygous C1q deficiency causes glomerulonephritis associated with multiple apoptotic bodies. Nat Genet 19:56–59
155. Fadok VA, Bratton DL, Rose DM, Pearson A, Ezekewitz RA, Henson PM (2000) A receptor for phosphatidylserine-specific clearance of apoptotic cells. Nature 405:85–90
156. Vandivier RW, Fadok VA, Hoffmann PR, Bratton DL, Penvari C, Brown KK, Brain JD, Accurso FJ, Henson PM (2002) Elastase-mediated phosphatidylserine receptor cleavage impairs apoptotic cell clearance in cystic fibrosis and bronchiectasis. J Clin Invest 109:661–670
157. Park D, Tosello-Trampont AC, Elliott MR, Lu M, Haney LB, Ma Z, Klibanov AL, Mandell JW, Ravichandran KS (2007) BAI1 is an engulfment receptor for apoptotic cells upstream of the ELMO/Dock180/Rac module. Nature 450:430–434
158. Miyanishi M, Tada K, Koike M, Uchiyama Y, Kitamura T, Nagata S (2007) Identification of Tim4 as a phosphatidylserine receptor. Nature 450:435–439
159. Park SY, Jung MY, Kim HJ, Lee SJ, Kim SY, Lee BH, Kwon TH, Park RW, Kim IS (2008) Rapid cell corpse clearance by stabilin-2, a membrane phosphatidylserine receptor. Cell Death Differ 15:192–201
160. Bose J, Gruber AD, Helming L, Schiebe S, Wegener I, Hafner M, Beales M, Kontgen F, Lengeling A (2004) The phosphatidylserine receptor has essential functions during embryogenesis but not in apoptotic cell removal. J Biol 3:15
161. Somersan S, Bhardwaj N (2001) Tethering and tickling: a new role for the phosphatidylserine receptor. J Cell Biol 155:501–504
162. Gardai SJ, Bratton DL, Ogden CA, Henson PM (2006) Recognition ligands on apoptotic cells: a perspective. J Leukoc Biol 79:896–903
163. Brown S, Heinisch I, Ross E, Shaw K, Buckley CD, Savill J (2002) Apoptosis disables CD31-mediated cell detachment from phagocytes promoting binding and engulfment. Nature 418:200–203
164. Maile LA, Capps BE, Miller EC, Allen LB, Veluvolu U, Aday AW, Clemmons DR (2008) Glucose regulation of integrin-associated protein cleavage controls the response of vascular smooth muscle cells to insulin-like growth factor-I. Mol Endocrinol 22:1226–1237
165. Guzik K, Bzowska M, Smagur J, Krupa O, Sieprawska M, Travis J, Potempa J (2007) A new insight into phagocytosis of apoptotic cells: proteolytic enzymes divert the recognition and clearance of polymorphonuclear leukocytes by macrophages. Cell Death Differ 14:171–182

166. Devitt A, Moffatt OD, Raykundalia C, Capra JD, Simmons DL, Gregory CD (1998) Human CD14 mediates recognition and phagocytosis of apoptotic cells. Nature 392:505–509
167. Sugawara S, Nemoto E, Tada H, Miyake K, Imamura T, Takada H (2000) Proteolysis of human monocyte CD14 by cysteine proteinases (gingipains) from *Porphyromonas gingivalis* leading to lipopolysaccharide hyporesponsiveness. J Immunol 165:411–418
168. Nemoto E, Sugawara S, Tada H, Takada H, Shimauchi H, Horiuchi H (2000) Cleavage of CD14 on human gingival fibroblasts cocultured with activated neutrophils is mediated by human leukocyte elastase resulting in down-regulation of lipopolysaccharide-induced IL-8 production. J Immunol 165:5807–5813
169. Henriksen PA, Devitt A, Kotelevtsev Y, Sallenave JM (2004) Gene delivery of the elastase inhibitor elafin protects macrophages from neutrophil elastase-mediated impairment of apoptotic cell recognition. FEBS Lett 574:80–84
170. Kantari C, Pederzoli-Ribeil M, Amir-Moazami O, Gausson-Dorey V, Moura IC, Lecomte MC, Benhamou M, Witko-Sarsat V (2007) Proteinase 3, the Wegener autoantigen, is externalized during neutrophil apoptosis: evidence for a functional association with phospholipid scramblase 1 and interference with macrophage phagocytosis. Blood 110:4086–4095
171. Perretti M, D'Acquisto F (2009) Annexin A1 and glucocorticoids as effectors of the resolution of inflammation. Nat Rev Immunol 9:62–70
172. Scannell M, Flanagan MB, deStefani A, Wynne KJ, Cagney G, Godson C, Maderna P (2007) Annexin-1 and peptide derivatives are released by apoptotic cells and stimulate phagocytosis of apoptotic neutrophils by macrophages. J Immunol 178:4595–4605
173. Gan H, Lee J, Ren F, Chen M, Kornfeld H, Remold HG (2008) Mycobacterium tuberculosis blocks crosslinking of annexin-1 and apoptotic envelope formation on infected macrophages to maintain virulence. Nat Immunol 9:1189–1197
174. Vong L, D'Acquisto F, Pederzoli-Ribeil M, Lavagno L, Flower RJ, Witko-Sarsat V, Perretti M (2007) Annexin 1 cleavage in activated neutrophils: a pivotal role for proteinase 3. J Biol Chem 282:29998–30004
175. Pederzoli-Ribeil M, Maione F, Cooper D, Al-Kashi A, Dalli J, Perretti M, D'Acquisto F (2010) Design and characterization of a cleavage-resistant Annexin A1 mutant to control inflammation in the microvasculature. Blood 116:4288–4296

Matrix Metalloproteinase Inhibitors as New Anti-inflammatory Drugs

Vincent Lagente, Tatiana Victoni, and Elisabeth Boichot

Abstract Matrix metalloproteinases (MMPs) are a group of proteases known to regulate the turnover of extracellular matrix (ECM) and thus are suggested to be important in the process of several diseases associated with tissue remodeling and inflammation. Degradation of ECM is currently associated with structural and recruited cell activation and release of inflammatory mediators and MMPs. Indeed, a marked increase in their expression is observed associated with a variety of inflammatory diseases. In these conditions, we have to consider MMPs as therapeutic targets which can be inhibited by nonselective and/or selective inhibitors as anti-inflammatory compounds. The present review aims to discuss the potential interest of selective and nonselective MMP inhibitors in several inflammatory diseases including respiratory and cardiovascular pathologies, liver fibrosis, and arthritis. This chapter also includes a special part on macrophage metalloelastase (MMP-12) as a target for inflammatory respiratory diseases.

Keywords Anti-inflammatory compounds • Matrix metalloproteinase • Metalloelastase • TIMP-1

1 Introduction

The increasing family of matrix metalloproteinases (MMPs) has been subject to sustained research and has been widely demonstrated to be important in various fields of medicine including inflammatory process and pathology. MMPs were primarily described to be involved in homeostasis and the turnover of the extracellular matrix (ECM), but there has been numerous evidence suggesting that MMPs

V. Lagente (✉) • T. Victoni • E. Boichot
INSERM UMR 991, Université de Rennes 1, 2 avenue du professeur Léon Bernard, 35043 Rennes cedex, France
e-mail: vincent.lagente@univ-rennes1.fr

N. Vergnolle and M. Chignard (eds.), *Proteases and Their Receptors in Inflammation*, Progress in Inflammation Research, DOI 10.1007/978-3-0348-0157-7_5,

act on cytokines, chemokines, and protein mediators to regulate various aspects of inflammation and immunity [1, 2]. The MMPs form a group of structurally related extracellular zinc endopeptidases known for their ability to cleave one or several constituents of the ECM [3]. Zymogen forms of the MMPs (pro-MMPs) are secreted into the extracellular space from a large number of cell types, where the activation of the pro-MMPs in the local microenvironment can result in discrete alterations in the tissue architecture. MMP synthesis and functions are regulated by transcriptional activation, posttranscriptional processing (release of pro-domain, cell surface shedding), and the control of activity by a family of endogenous inhibitors collectively known as tissue inhibitors of metalloproteinases (TIMP). Upon stimulation, many cell types have been identified as producers of MMPs and TIMPs in the context of inflammatory process, strongly suggesting the involvement of MMPs in numerous inflammatory diseases. Based on this property, MMPs are not only put forward as physiological mediators of the "turnover" of the ECM but are also considered to be critical factors of the remodeling processes in pathological conditions [4]. Indeed, a marked increase in their expression is observed and associated with a variety of inflammatory diseases.

Moreover, some of the MMPs are able to directly activate inflammatory cells leading to an amplification of the inflammatory process. For instance, in vitro studies have demonstrated that MMPs increase the activity of chemokines such as IL-8 [5], but reduce the activity of others, such as ENA78 [6]. Researchers have also demonstrated that MMPs release immobilized chemokine complexes, such as syndecan-1/IL-8 [7], but can convert others, such as monocyte chemoattractant protein-3 (MCP-3), into chemokine receptor antagonists [8].

Consequently, MMPs have been speculated to play a critical role in various inflammatory diseases, such as airway diseases associated with inflammatory process including acute lung injury (ALI) and acute respiratory distress syndrome (ARDS) [9], chronic obstructive pulmonary disease (COPD) [10], pulmonary fibrosis [11], but also liver diseases, rheumatoid arthritis (RA) [12], and cancer [13]. In these conditions, we have to consider MMPs as therapeutic targets which can be inhibited by nonselective and/or selective inhibitors as possible novel anti-inflammatory compounds.

2 MMPs as Modulators of Inflammation

The inflammatory process is characterized by leukocytes trafficking through the tissue barriers, including basement membranes. This is only feasible if the inflammatory cells are able to produce enzymes than can remodel the ECM [14, 15]. Numerous studies have reported that MMPs can either promote or inhibit inflammatory processes through the direct proteolytic processing of inflammatory mediators including chemokines and cytokines to activate, inactivate, or antagonize their functions [16, 17]. Chemokines and cytokines play a central role in the recruitment of leukocytes to the site of infection or injury, thereby influencing the outcome of an inflammatory response. MMP proteolysis can affect the biological functions of chemokines by different processes. Firstly, the proteolysis might inactivate the chemokine. Secondly, processing might produce antagonistic derivatives, which can still bind to the specific receptor but cannot elicit chemotaxis. Thirdly, the truncated chemokine is a more powerful chemotactic agent. Numerous data have been presented in literature (for review see [16]). The influence of MMPs in the progression of the inflammatory process is not limited to leukocyte trafficking but also involves cytokine activation. Indeed, similarly to chemokines, cytokine proteolysis often leads to an altered bioavailability and activity. For example, TNF-α is expressed on T-cells and macrophages as a 26 kDa membrane-bound protein (pro-TNF-α) that is activated by cleavage to a 17 kDa soluble cytokine by TNF-α converting enzyme (TACE), identical to ADAM 17, a member of the disintegrin family of metalloproteinases [18]. IL-1β is another potent pro-inflammatory cytokine that requires proteolytic processing before activation, mainly by caspase-1 but also several MMPs, including MMP-2, MMP-3, and MMP-9. Interestingly, MMP-3 can degrade the mature IL-1β cytokine, suggesting potential dual roles for MMPs in either stimulating or inhibiting IL-1β effects [19]. Nevertheless, it is also important to note that cleavage does not always alter the activities of cytokines and chemokines.

Another mechanism by which MMPs control inflammation is the regulation of chemokine gradients. Indeed, the function of chemokines is partly regulated by proteolytic processing, but also by compartmentalization. This mechanism includes both the immobilization of chemokines to the components of ECM and the generation of chemotactic concentration gradients which provide indications for leukocyte migration. Thus, MMPs can indirectly control the influx of inflammatory cells by cleaving proteins in the pericellular environment that bind chemokines. The MMP7-dependent shedding of syndecan-1 in ACL [20] is a well-known example of this mechanism. In response to lung injury, both CXCL1 (KC) and MMP7 are induced, and MMP7 sheds syndecan-1, a ubiquitous heparin sulfate proteoglycan, that releases the CXCL1-syndecan-1 complex to generate a chemokine gradient. MMP7-KO mice that lack this shedding are unable to create a CXCL1 gradient, and thus, neutrophils fail to efflux into the alveolar space and remain in the perivascular space instead [7].

3 Role of TIMPs in the Inflammatory Process

Tissue inhibitors of the metalloproteinases (TIMPs) are specific endogenous inhibitors that bind to the active site of MMPs in a stoichiometric 1:1 molar ratio, thereby blocking access to ECM substrates. Four TIMPs (TIMP-1, -2, -3, and -4) have been identified in vertebrates, and their expression is regulated during development, tissue remodeling but also inflammation [21]. The mammalian TIMP family presents substantial sequence homology and structural identity on a protein level. TIMPs have basically two structural domains: an N-terminal domain consisting of six conserved cysteine residues forming three disulfide loops, which possesses MMP-inhibitory activity, and a C-terminal domain that also contains six conserved cysteine residues and forms three disulfide loops [22].

Basically, all members of the TIMP family inhibit MMP activity. This is accomplished through the co-ordination of the Zn^{+2} of the MMP active site by the amino and carbonyl groups of the TIMPs N-terminal cysteine residue. Nevertheless, a selective inhibition of some members of the MMP has been reported. For example, although TIMP-1 is the most potent inhibitor for most MMP family members, it is a poor inhibitor of the membrane-type MMPs (MT-MMPs) and MMP-19 [23, 24]. TIMP-3 inhibits members of the A Disintegrin And Metalloproteinase (ADAM) family of proteases, although the mechanism for this inhibition appears to be different from MMP inhibition [25, 26]. TIMP-2 selectively interacts with MT1-MMP to facilitate the cell-surface activation of pro-MMP-2 [23, 24, 27]. Thus, TIMP-2 can both inhibit MMP activity and promote the cell surface activation of pro-MMP-2 by MT1-MMP. TIMPs can be regulated on a transcriptional level by various cytokines and growth factors, resulting in tissue-specific, constitutive, or inducible expression [28].

Given that MMPs degrade various components of the ECM, a tight regulation of the MMP activity is essential to prevent excessive matrix degradation. The primary action of TIMPs is to inhibit MMPs, but numerous studies have reported cell growth-promoting, antiapoptotic, steroidogenic, and antiangiogenic activities (reviewed in [22, 29]), which are in part independent of MMP inhibition. Since the main cellular sources of TIMP-1 are macrophages and fibroblasts, one can easily suggest that TIMP-1 is involved in tissue remodeling associated with the activation of macrophages in the inflammatory process.

Studies have demonstrated that monocytes secrete large quantities of basal levels of TIMP-1, but are unresponsive to LPS, whereas macrophages secrete lower basal levels of TIMP-1, which were found to be upregulated by LPS [29]. It has also been speculated that TIMP-1 may be involved in the modulation of inflammatory responses and may also function to stabilize matrix components deposited in the injured lung.

Moreover, strong evidences imply that TIMPs/MMPs imbalances are an important element in the fibrogenic process: TIMPs, and especially TIMP-1 are upregulated in cases of human pulmonary fibrosis and in bleomycin-induced pulmonary fibrosis. TIMPs, and particularly TIMP-1 induction could lead to a "noncollagenolytic microenvironment," building adequate conditions for a further ECM deposition to occur. Indeed, we previously reported that TIMP-1 was markedly increased in mice's lungs, 24 h after the administration of bleomycin at day 1 [30]. During this period, we were not able to observe collagen deposition, but bleomycin induced an important inflammatory reaction characterized by an influx of neutrophils and probably an increase in macrophage activity. However, the depletion of mice in neutrophils did not modify the level of the TIMP-1 protein in comparison with control mice [30]. We also reported that the nonselective MMP inhibitor, batimastat, reduced the development of bleomycin-induced fibrosis in mice, associated with a decrease in TIMP-1 levels in BAL fluids [31]. This strongly suggests that TIMP-1 may be considered as an available target for tissue remodeling and fibrosis. However, TIMPs which have affinities with the picomolar range seem ideal inhibitors but they do not present selectivity and possess other biological functions, which could lead to side effects [32].

TIMPs have also additional biological activities that are just starting to be recognized and characterized. Despite several evidences suggesting a direct cell signaling capacity for the TIMPs, the connection with MMP-inhibitory activity remains controversial. Although the TIMP-mediated inhibition of the MMP activity is an important component of cell function, the hypothesis of MMP-independent TIMP regulation of cell function, including the promotion of cell growth, anti-apoptotic activity, and growth-inhibitory activity, is now supported by the characterization of specific cell binding partners – and specific signaling events – for TIMP family members. This is particularly relevant to the cell signaling mechanisms mediated by TIMP-1 and TIMP-2. The recognition of these MMP-independent TIMP activities and the understanding of the mechanisms involved have important implications for the development of new therapies for cancer and other chronic diseases. One hypothesis is the presence of receptors for TIMP. For example, the binding of TIMP-2 to the endothelial cell surface and its ability to inhibit endothelial cell proliferation is independent of MMP inhibition, is saturable and reversible [33]. Competitive binding studies have demonstrated that TIMP-2 binding to the surface of human microvascular endothelial cells can be inhibited by anti-β1 and anti-α3 integrin blocking antibodies. The interaction of TIMP-2 with α3β1 cell surface integrin was confirmed by the co-immunoprecipitation of α3β1 integrin using anti-TIMP-2 antibodies, and the loss of TIMP-2 growth suppressive activity in β1-null fibroblasts. This was the first demonstration of TIMP interaction with a specific cell surface protein, identifying this integrin as a TIMP-2 receptor. It has been suggested that TIMP-2 inhibits angiogenesis by inducing endothelial cell differentiation to a quiescent state with G1 cell cycle arrest, enhanced expression of RECK, a membrane-associated inhibitor of MMPs (MMP-2, MMP-9, MT1-MMP) as well as ADAM-10 [34, 35].

It has also been suggested that the TIMP-1 function is also influenced by the cellular context, specifically in that MMPs, in particular MMP-9, may reduce the effective concentration of TIMP-1 and compete with TIMP-1 for binding to the cell surface receptor CD63 [32]. In contrast to TIMP-2, TIMP-1 blood concentrations are increased in cancer patients, particularly in those with breast or colorectal carcinoma, and this increase is negatively associated with patient outcome [36–38]. These recent studies have demonstrated the clinical utility of TIMP-1 as a biomarker and independent prognostic factor in breast, colorectal, and several hematological cancers.

The characterization of receptors for TIMP family members is a first step to understand the MMP-independent, cytokine-like functions of the TIMPs. Hopefully, this can lead to a starting point for the molecular dissection of signaling events associated with the various activities of these proteins and their function in both normal physiologic and pathologic processes. It is clear that the pleotropic activities of the TIMP family members are complex and depend on interactions with other extracellular components, as well as direct interactions with cell binding partners.

Despite the fact that TIMP-mediated effects in the heart are due to their ability to inhibit MMPs, the role of each TIMP during cardiac remodeling is likely to be far more complex than a simple imbalance of MMP activity in the heart. Through their pleiotropic activities, TIMPs may regulate a wide range of cellular responses critical to cardiac remodeling [39]. Research carried out over the last two decades has revealed that the TIMP family members interact with other ECM components and thereby influence cardiac fibroblast phenotype and fibrosis, endothelial cells and angiogenesis, cardiomyocytes proliferation and hypertrophic remodeling and inflammatory cells [39]. The interaction of TIMP-2 with endothelial cells and the consequence on the angiogenesis process has been mentioned previously. Several genetic studies have reported that the imbalance between MMPs and TIMPs influences the infiltration of inflammatory cells into the injured heart after cardiac injury, stress, or infection. For example, TIMP-3 may inhibit TACE, primarily responsible for the bioactivation of the pro-inflammatory cytokine TNF-α, a key inflammatory mediator involved in cardiac remodeling and heart failure [40, 41]. Challenging TIMP-3-deficient mice with lipopolysaccharide (LPS) resulted in an uncontrolled systemic inflammation leading to animal morbidity due to TNF-α-signaling [42]. TIMP-3, however, is not the only metalloproteinase inhibitor that may function during cardiac inflammation. A recent study has demonstrated that TIMP-1 directly protects B-cells from apoptosis through a non-MMP-inhibitory pathway and suggests that this protein may play a pivotal role in the maintenance of B-cell homeostasis [43]. Conversely, recombinant TIMP-2 increases apoptosis in activated human peripheral blood T-cells, whereas unstimulated T-cells are not susceptible. This effect was specific to TIMP-2 and was not observed for TIMP-1 [44]. Other experiments show that the MMP-inhibitory function of TIMP-2 seems to be important in this process. Indeed, TIMP-2 peptide lacking the N-terminal domain, which is critical for MMP inhibition, did not induce apoptosis. Moreover, future research will be essential in order to predict the physiological relevance of the non-MMP-inhibitory functions of the four TIMP species and their involvement in cardiac inflammation. It is clear that a complete and in-depth understanding of the MMP independent TIMP-mediated processes and their modulation during diseases is mandatory to design innovative TIMP-based therapeutic strategies to prevent cardiac remodeling and the progression of heart failure. Understanding these processes and how they are modulated during disease progression will be helpful in the development of novel therapeutic interventions.

4 Anti-inflammatory Properties of Broad Spectrum MMP Inhibitors

Through the importance of ECM remodeling, there is a significant interest in using MMP inhibition as a therapeutic strategy. However, the TIMPs have not proved to be suitable for pharmacological applications due to their short half-life in vivo [45]. Numerous MMP inhibitors are still under development, in spite of extensive efforts by almost all major pharmaceutical companies, indicating that the development of MMP inhibitors is very challenging [46–48]. Most of them are developed as anticancer drugs, but limited studies have reported an anti-inflammatory activity and the development of them in pathologies associated with inflammation is commented below.

The first synthetic broad-spectrum MMP inhibitor includes hydroxamic acid derived inhibitors such as BB-94 (Batimastat), BB-1101, BB-2293, BB-2516 (marimastat), and CT1746. Batimastat and marimastat are competitive MMP inhibitors and Zn^{2+}-chelating mimickers of collagen. Initial results have been promising in cancer research in blocking the progression of tumor growth [49, 50]. We have previously shown that batimastat significantly limits the development of bleomycin-induced pulmonary fibrosis in mice associated with a reduction of levels of TIMP-1 [31]. In vitro studies have shown that an MMP inhibitor can reduce trans-basement membrane neutrophil migration [51]. However, in another study, we have reported that batimastat did not modify either inflammatory cell recruitment or MMP-9 profile in BAL fluids of mice induced by aerosol with LPS [52]. In LPS-stimulated monocytes and macrophages, the MMP inhibitor Bay 17-4003 has no significant effects on the release of various cytokines including TNF-α, IL-6, or IL-8 [53]. Similar results have been reported in vivo in an LPS-driven rat model of airway inflammation but it significantly reduced elastase-induced experimental emphysema [54]. In contrast, Zhang et al. [55] have shown that a dual TACE (tumor necrosis factor-converting enzyme)/MMP inhibitor reduces LPS-induced TNF-α secretion in human monocytes. It is therefore suggested that the inhibition observed by Zhang et al. [55] was due to an effect on TACE activity. This result is consistent with the results of a recent in vitro study showing that hydroxamic acid-based synthetic MMP inhibitors had no significant effect on fMLP (formyl-MET-LEU-PHE)-stimulated neutrophil migration through the endothelial cells and associated basal lamina [56]. The fact that MMP may not be involved in cell migration was reinforced by results of in vivo studies using MMP-9 knockout mice exposed to LPS [57] or to bleomycin [58]. Other studies have reported a reduction in LPS-induced plasma levels of TNF-α after the administration of other MMP inhibitors [59]. We have also reported that marimastat significantly reduced the influx of neutrophils and macrophages induced by an intratracheal administration of MMP-12 in mice [60]. It was reported that a broad spectrum MMP inhibitor, CP-471,474 significantly reduced the extent and severity of inflammation as well as the destructive lesions of the lung in guinea pigs exposed to cigarette smoke at 2 months [61]. In another study [62], two orally bioavailable synthetic inhibitors (RS-113456 and RS-132908) were shown to markedly inhibit the smoke-induced increase in emphysema at every 6-week observation in mice. A dual TACE and MMP inhibitor and a dual MMP and NE inhibitor [63] have both demonstrated to inhibit cellular inflammation. Nevertheless, with the use of these dual inhibitors, it is difficult to determine the simple role of MMPs in cellular inflammation.

The liver is constantly exposed to various endogenous and exogenous compounds and pathogens, which may induce acute and/or chronic injuries. These injuries lead to a complex process of tissue repair which consists in inflammation,

ECM production and remodeling, and cell regeneration. In most cases, this process results in the restoration of the liver architecture without any loss of hepatic functions and obvious clinical signs. However, when liver injuries are repeated, tissue repair may become unsuccessful, a chronic inflammation process takes place and ECM components accumulate in excess with an increase of tissue remodeling which results in the formation of a fibrotic scar.

It has been previously demonstrated that inhibition with marimastat during chronic CCl4 administration resulted in a significantly attenuated hepatic inflammation and necrosis, coupled with a downregulation of genes related to fibrogenesis, but resulted in an increased liver fibrosis [64]. Indeed, the inhibition of MMPs and collagen degradation by marimastat counterbalanced the beneficial anti-inflammatory effect resulting in a positive balance of collagen deposition. Since an effective inhibition of the fibrolytic activity by MMPs accelerates fibrosis progression, these data suggest a note of caution for the use of broad-spectrum MMP inhibitors in patients with chronic, ongoing liver diseases, or for the treatment of liver fibrosis itself. These data are in line with a previous report describing the use of batimastat in the prevention of acute, fulminant hepatitis induced by TNF-α combined with D-(+)-galactosamine [65]. Another study reported a reduction of liver injury following treatment with the MMP-inhibitor CTS-1027 [66]. Using the bile duct ligation model, a decrease in hepatocyte apoptosis and a reduction in markers for HSC activation and fibrogenesis was demonstrated, which is in line with the results that marimastat attenuates hepatic inflammation and necrosis coupled with the downregulation of genes related to fibrogenesis [64].

In both rheumatoid arthritis (RA) and osteoarthritis (OA), the irreversible destruction of the cartilage, tendon, and bone that comprise synovial joints is a hallmark. While cartilage is made up of proteoglycans and type II collagen, tendons and bones are composed primarily of type I collagen. RA is an autoimmune disease afflicting numerous joints throughout the body. In contrast, OA develops in a small number of joints, usually resulting from chronic overuse or injury. In both diseases, inflammatory cytokines such as interleukin-1 β (IL-1β) and tumor necrosis factor-α (TNF-α) stimulate the production of MMPs, enzymes that can degrade all components of the ECM. The collagenases, MMP-1 and MMP-13, have predominant roles in RA and OA because they are rate limiting in the process of collagen degradation. MMP-1 is produced primarily by the synovial cells that line the joints, and MMP-13 is a product of the chondrocytes that reside in the cartilage. In addition to collagen, MMP-13 also degrades the proteoglycan molecule, aggrecan, giving it a dual role in matrix destruction. The expression of other MMPs, such as MMP-2, MMP-3, and MMP-9, is also elevated in arthritis and these enzymes degrade noncollagen matrix components of the joints. Significant efforts have been expended in attempts to design effective inhibitors of MMP activity and/or synthesis with the aim of curbing connective tissue destruction within the joints. To date, however, no effective clinical inhibitors exist. Increasing our knowledge on the crystal structures of these enzymes and on the signal transduction pathways and molecular mechanisms that control MMP gene expression may provide new opportunities for the development of therapeutics to prevent the joint destruction seen in arthritis.

A potent, orally active stromelysin inhibitor, CGS 27023A, has been reported to block the erosion of cartilage matrix and is proposed for the treatment of chronic joint disease [67]. BAY 12-9566 is a developed compound that selectively inhibits MMP-2, MMP-3, and MMP-9 isozymes which suppressed inflammation and cartilage destruction in adjuvant-induced arthritis (AA) in rats [68]. Indeed, the oral treatment of rats for 22 days with 50 mg/kg BAY 12-9566 showed decreased AA, determined by an improvement in body weight gain, arthritic index, and swelling of paws contralateral to the adjuvant injection site. Neutrophil infiltration and collagen degradation were also significantly lower in this treatment group. Cartilage destruction was successfully suppressed in rats treated with 50 mg/kg BAY 12-9566. In vitro, BAY 12-9566 prevented matrix invasion by endothelial cells in a concentration-dependent manner ($IC_{50} = 8.4 \times 10^{-7}$ M), without affecting cell proliferation. In vivo, a daily oral administration of BAY 12-9566 (50–200 mg/kg) inhibited angiogenesis induced by a basic fibroblast growth factor in the Matrigel plug assay, reducing the hemoglobin content of the pellets [69]. AA in rats was also inhibited by the hydroxamate GI168, which resulted in a reduction of paw swelling and a protection against the degradation of bone and cartilage, pannus formation, and abnormal bone deposition78 [70]. GI168 does not inhibit TACE, although TNF inhibitors are currently being used in clinics against rheumatoid arthritis. GW3333, a dual inhibitor of TACE and MMPs, was compared with an anti-TNF antibody to evaluate the importance of soluble TNF and MMPs in rat models of arthritis [71]. In a 21-day AA model, the anti-TNF antibody did not inhibit the ankle swelling or the joint destruction, as assessed by histology or radiology. GW3333, however, showed an inhibition of both ankle swelling and joint destruction. The preclinical efficacy of GW3333 suggests that dual inhibitors of TACE and MMPs may present therapeutic activity as antiarthritic drugs. Another dual TACE/MMP inhibitor TMI-1 has been described with nanomolar IC(50) values in vitro [55]. In cell-based assays such as monocyte cell lines, human primary monocytes, and human whole blood, LPS-induced TNF-α secretion is inhibited. The inhibition of LPS-induced TNF-α secretion is selective because TMI-1 has no effect on the secretion of other proinflammatory cytokines such as interleukin (IL)-1β, IL-6, and IL-8. TMI-1 also potently inhibits TNF-α secretion by human synovium tissue explants of RA patients. In vivo, TMI-1 is highly effective in reducing clinical severity scores in mouse prophylactic collagen-induced arthritis at 5, 10, and 20 mg/kg p.o. b.i.d. and therapeutic model at 100 mg/kg p.o. b.i.d.

5 Macrophage metalloelastase (MMP-12) as a Target for Inflammatory Respiratory Diseases

MMP-12 is a 54-kDa proenzyme, with a 45-kDa NH_2-terminal active form that is processed into a mature 22 kDa form. The human gene, which is designated as human macrophage metalloelastase, produces a 1.8-kb transcript encoding a

470-amino acid protein which is 64% identical to the mouse protein. MMP-12 mRNA and protein are predominantly detected in alveolar macrophages [72, 73], but recently they have been also described in bronchial epithelial cells [74] and airway smooth muscle cells [75].

MMP-12 expression is upregulated by other matrix components such as hyaluronan fragments, cytokines, and growth factors, for example, by transforming growth factor (TGF)-β, interferon (IFN)-gamma and epidermal growth factor (EGF), and serine proteases such as thrombin and plasmin [76, 77]. MMP-12 was initially described as an elastase, but like other metalloproteinases, it has been found to have, in vitro, a wide variety of potential substrates, including type IV collagen, fibronectin, laminin, and gelatin, as well as nonmatrix proteins such as alpha1-antitrypsin and latent tumor necrosis factor (TNF)-α [78–80].

Few studies have reported a role for MMP-12 in asthma or in allergic airway inflammation. In a study demonstrating that MMP-9 deficiency impairs cellular infiltration and bronchial hyperresponsiveness during allergen-induced airway inflammation in mice, increased mRNA levels of MMP-12 were also reported after allergen exposure in the lung extracts of WT mice but not in MMP-9-deficient mice [81]. After using MMP-12-deficient mice, Warner et al. [82] observed a significant reduction in cockroach antigen (CRA)-induced inflammatory injury, which was highlighted by fewer peribronchial leukocytes, significantly less protein in the bronchoalveolar lavage (BAL) fluid, and a significant reduction in the number of infiltrating neutrophils, eosinophils, and macrophages. In another experimental model of allergic bronchial asthma [83], a significant increase in the expression/activity of MMP-12 was found: the peak was observed 12 h after the last antigen challenge. Furthermore, the mRNA expression of MMP-12 had also increased during the early phase (1–3 h) after the last antigen challenge. Immunohistochemical studies revealed that MMP-12 was mainly expressed in airway epithelia and alveolar macrophages. Another study reported that human airway smooth muscle cells express and secrete MMP-12 that is upregulated by interleukin (IL)-1β and TNF-α [75]. Bronchial smooth muscle cells may be an important source of elastolytic activity, thereby contributing to remodeling in airway diseases such as chronic asthma. In a recent study [84], the airway smooth muscle content of different components of ECM, as well as MMPs and TIMPs was analyzed in lung tissue from patients with or without asthma. As regards fatal asthma, the result showed an increased expression of fibronectin, MMP-9 and MMP-12 in the large airways, compared with nonasthma control patients. As a result, the increased expression of fibronectin, MMP-9 and MMP-12 in asthmatic patients could have important consequences for airway smooth muscle functions and excessive airway narrowing in asthma.

Pulmonary fibrosis is characterized by an excessive deposition of ECM in the interstitium resulting in respiratory failure. Metalloproteinases have been described to be involved in the remodeling process and ECM turnover in pulmonary fibrosis [85]. We previously investigated *MMP-12* mRNA levels in lungs of Balb/c and C57BL/6 mice in the classic model of experimental pulmonary fibrosis induced after the administration of bleomycin [86]. Indeed, C57BL/6 mice are known to be "fibrosis prone" strain, whereas Balb/c mice are described as resistant to the development of pulmonary fibrosis [87]. In keeping with these facts, the preliminary hydroxyproline level measurement revealed that C57BL/6 mice developed an important accumulation of collagen 14 days after the administration of bleomycin, whereas a rise in the level of collagen was not observed in the lungs of Balb/c mice [88]. Our experiment showed that *MMP-12* mRNA levels in lung tissue were increased in C57BL/6 mice at day 1 and day 14, whereas a slight raise of *MMP-12* mRNA was detected only at day 14 in Balb/c mice. Through these results, we confirmed that bleomycin elicits *MMP-12* mRNA induction in lung tissue of WT mice, as previously described by Swiderski et al. [89]. We also noted that MMP-$12^{-/-}$ mice responded to the administration of bleomycin by an increase of collagen content in the lung tissue [86]. This raise did not differ significantly from observations in WT mice, suggesting that MMP-12 is not necessary for the development of bleomycin-induced pulmonary fibrosis. Similarly, Lanone et al. [90] observed in IL-13-induced injury in mice that a MMP-12 deficiency did not alter subepithelial fibrosis following inducible IL-13 transgenic expression. Taken together, these results suggest that bleomycin-induced fibrosis and associated inflammation involve different mechanisms from MMP-12 dependent pathways. In contrast, a targeted deletion of MMP-12 protected mice from Fas-induced pulmonary fibrosis, even though the inflammatory responses in the lungs were similar to those of wild-type mice [91]. Compared with wild-type mice, the MMP-$12^{-/-}$ KO mice showed a decreased expression of the profibrotic genes *egr1* and *cyr61*. Several hypotheses suggest that crucial components of fibrogenic processes are due to remodeling disorders involving growth factors such as TGF-β and a nondegrading microenvironment created by a "shield" of protease inhibitors, including TIMP-1 [92]. TGF-β has shown to be a prominent fibrogenic mediator in many organs, including lungs. Moreover, TGF-β presents a pivotal situation by regulating global lung tissue remodeling. Indeed, mice lacking integrin αvβ6 (integrin αvβ6 null mice) fail to activate TGF-β and develop age-related emphysema, which is MMP-12 dependent [93]. MMP-12 in lung has demonstrated to be downregulated by the TGF-β signaling pathway [93, 94]. Consistent with our results, Lanone et al. [90] did not report an alteration in the total of TGF-β in the BAL fluid of MMP-$12^{-/-}$ mice after IL-13 transgene expression. Interestingly, in WT mice, bleomycin elicits the increase of both MMP-12 and TGF-β. Therefore, further investigations are required in order to explain why MMP-12 expression coexists with high TGF-β levels.

COPD is one of the major causes of mortality and morbidity in the developed countries and its prevalence is still increasing [95]. This pathology is also associated with an airway inflammatory process characterized by an accumulation of inflammatory cells such as macrophages and neutrophils. Indeed, it has been shown that cigarette smoke consistently produces an increase in the number of neutrophils in BAL fluid and in lung tissue [96, 97]. Macrophage numbers are also elevated in the lungs of smokers and of patients with COPD where they accumulate in the alveoli, bronchioli, and small airways. Furthermore, there is a positive correlation between the number of macrophages in the alveolar walls and the mild-to-moderate emphysema status in patients with COPD [98]. It is generally believed that the development of emphysema reflects a relative excess of cell-derived proteases that damage the connective tissue of the lung and a relative lack of antiproteolytic defenses. This theory is often referred to as the "protease–antiprotease imbalance" hypothesis and involves mainly serine proteases like neutrophil elastases and matrix MMPs.

MMP-12 seems to play a predominant role in the pathogenesis of chronic lung injury and particularly in emphysema.

Studies using MMP-12$^{-/-}$ KO mice have demonstrated that inflammatory processes and emphysema induced by long-term exposure to cigarette smoke were linked to MMP-12 [99]. In a more recent study, it was reported that inflammatory lesions contained significantly more MMP-12 in macrophages in the lungs of mice after 10, 20, and 30 days of cigarette smoke exposure than in control mice [100]. We recently demonstrated that MMP-12$^{-/-}$ KO mice have a reduced airway inflammatory reaction following an exposure to two cigarettes twice a day during 3 days [101]. In contrast to the observations following cigarette smoke exposure, MMP-12$^{-/-}$ KO mice developed a similar airway neutrophilia as control mice when exposed to LPS. This suggests clear differences between the two models and that the early inflammatory processes following cigarette smoke or LPS exposure, although similar in profile, have different causal mechanisms.

The direct effect of MMP-12 in the development of inflammatory processes in mouse airways has also been evaluated using a recombinant form of human MMP-12 (rhMMP-12) [51]. A single instillation of rhMMP-12 in mouse airways elicited an intense inflammatory response characterized by the development of two successive phases. Indeed, a marked recruitment of neutrophils was observed following an injection of rh-MMP-12 with a maximum increase at 18 h [102]. This cellular recruitment was associated with a very transient increase in cytokines and chemokines and MMP-9 in BAL fluids and in lung parenchyma. From day 4 to day 15, after carrying out the same experiments, we observed an important and stable recruitment of macrophages in BAL fluids in the absence of the inflammatory markers observed during the early phase of inflammation [102]. As this experimental model of lung inflammation partially mimics some features of COPD, we have investigated the effects of a treatment with anti-inflammatory compounds such as the corticosteroid dexamethasone, the phosphodiesterase inhibitor rolipram and a broad-spectrum MMP inhibitor, marimastat [60]. Marimastat, dexamethasone, and rolipram were able to significantly decrease neutrophil recruitment 4 and 24 h after rhMMP-12 instillation, but only marimastat was effective at decreasing the macrophage recruitment which occurred on day 7. Overall, this suggests that dexamethasone and rolipram were able to inhibit the early inflammatory response but were ineffective to limit the macrophage influx. In contrast, marimastat was able to reduce both early and late responses.

As the mechanism by which MMP-12 triggers cell activation and recruitment associated with inflammatory process is not established, we have also examined the effects of the rhMMP-12 catalytic domain on human alveolar type II like epithelial cells (A549) and on human bronchial epithelial cells (Beas-2b). We have shown that rhMMP-12 enhanced the release of several chemokines by A549 cells, in particular MCP-1/CCL2, Gro-α/CXCL1, and IL-8/CXCL8 [103]. In Beas-2b cells, we have also observed a concentration-dependent increase of IL-8/CXCL8 after incubation with rhMMP-12. By focusing our study on IL-8/CXCL8, we were able to show in A549 cells that MMP-12 induced its gene expression and release via EGFR transactivation and further activation of the MAP-kinase ERK1/2 signal transduction pathway (Fig. 1), involving also the AP-1 transcription factor. It was also recently reported that MMP-12 truncates and inactivates ELR + CXC chemokines which contributes to the regulation of CC chemokine activities [105]. These authors demonstrated an early proinflammatory function of MMP-12 in the recruitment of neutrophils and a late anti-inflammatory property that abrogates both the neutrophil and macrophage influx. Because of its ability to induce an inflammatory response and tissue remodeling, it may be possible to consider MMP-12 as an essential component of the process leading to the development of COPD.

COPD is an unmet medical need and the development of selective MMP inhibitors is expected to offer new therapeutic opportunities. However, until now it was difficult to confirm the role of specific MMPs because of the lack of numerous studies involving selective inhibitors. The limited studies include guinea pigs which were exposed to cigarette smoke over 1 month, 2 months, and 4 months,

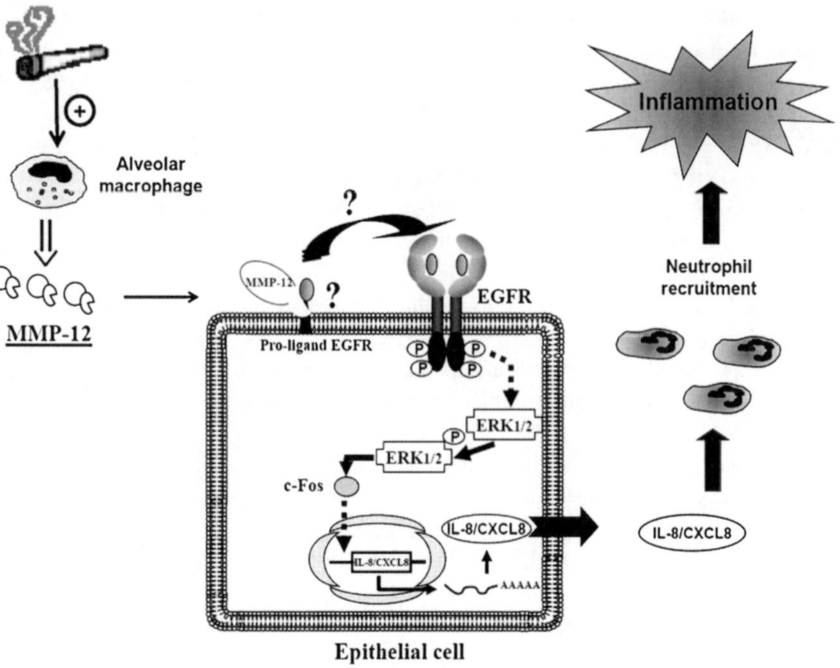

Fig. 1 Proposal mechanism for the effect of MMP-12 on cell activation and recruitment associated with inflammatory process. rhMMP-12-induced IL-8 gene expression and release via EGFR transactivation and further activation of the MAP-kinase ERK1/2 signal transduction pathway, involving also the AP-1 transcription factor (adapted from [104])

and received CP-471,474, a broad spectrum MMP inhibitor [61]. It was reported that CP-471,474 significantly reduced both the extent and severity of inflammation at 2 months. Moreover, the inhibitor significantly decreased the destructive lesions in the lungs, mainly at 2 and 4 months. In another study [62], two orally bioavailable synthetic inhibitors (RS-113456 and RS-132908) were tested in the murine model of smoke-induced emphysema. Both compounds were potent (<1.2 nM) inhibitors of both human and murine MMP-12. RS-132908 markedly inhibited the smoke-induced increase in emphysema at every 6-week observation, and produced 74% inhibition at the end of 6 months of smoke exposure. RS-113456 was also extremely effective since it completely blocked further experimental diseases. Interestingly, both compounds also reduced macrophage accumulation within the lung tissue. In a more recent study, Churg et al. [106] examined the effects of a dual MMP-9/MMP-12 inhibitor, AZ11557272, on the development of anatomic and functional changes associated with experimental COPD in guinea pigs exposed daily to cigarette smoke for up to 6 months. At all times, smoke-induced increases in lavage inflammatory cells and desmosine and in serum TNF-α were completely suppressed by AZ11557272. After 6 months, AZ11557272 reverted a smoke-induced airspace enlargement by about 70%. This study demonstrates that a MMP-9/MMP-12 inhibitor can substantially impede the development of emphysema, small airway remodeling, and improve the functional consequences of these lesions in a nonmurine species. Another relatively selective inhibitor of MMP-12 derived from γ-keto carboxylic acids displayed promising in vivo protection against porcine pancreatic elastase (PPE)-induced emphysema in male golden Syrian hamsters [107]. This result provides additional evidence for the involvement of MMP-12 in the development of emphysema and should stimulate further exploration of MMP-12 as a target for treatment of emphysema.

Two other MMP inhibitors developed by Serono have demonstrated potent anti-inflammatory properties on cell influx induced by cigarette smoke. Indeed, the inhibition of MMP-12 by the selective MMP-12 inhibitor, AS111793, dose-dependently reduced the increase in the number of neutrophils in BAL fluids after 4 days and of macrophages after 11 days [108]. On day 4, AS111793 also significantly reduced all the inflammation markers that had increased after cigarette smoke exposure, including soluble TNF receptors I and II, MIP-1gamma, IL-6 and proMMP-9 activity in BAL fluids, and KC/CXCL1, Fractalkine/CX3CL1, TIMP-1 and I-TAC/CXCL11 in lung parenchyma. In contrast, the inhibition of MMP-12 was ineffective in reducing neutrophil influx, proMMP-9 activity, and KC/CXCL1 release in BAL fluids of mice exposed to LPS [108]. We finally investigated the activity of the MMP-12 inhibitor, AS111793 on the elastolytic and gelatinolytic activities in BAL fluids of patients with COPD [109]. Elastolytic activity was dose-dependently inhibited by AS111793 with an IC_{50} value of 0.37 μM, but AS111793 did not reduce the gelatinolytic activity (mainly MMP-9) as measured by zymography [109].

Another dual inhibitor for MMP-9 and MMP-12, AS112108, has demonstrated to reduce early inflammatory processes in the same experimental model of COPD [110]. Finally, the broad spectrum MMP inhibitor PKF242-484 also reduced the neutrophil influx after a cigarette smoke exposure but did not reduce the number of macrophages when administered systemically and did when dosed topically [111]. Regarding the role of macrophages and their possible production of MMP-12 before the degradation of elastin to produce chemotactic peptides, the next challenge would be the inhibition and/or recruitment of monocytic cells.

Although not yet published, it is possible that several side effects may appear following MMP-inhibitor treatments. Indeed, it was reported that MMP-12 KO mice develop significantly more gross Lewis lung carcinoma pulmonary metastases than their wild-type counterparts, both in spontaneous and experimental metastasis models [112]. The number of micrometastases between the two groups is equivalent; thus, it seems that MMP-12 affects lung tumor growth, and not metastasis formation, per se. Even though no reliable clinical data are available, the results from trials employing broad-spectrum MMP inhibitors were not ineffective as widely believed, but rather beneficial in certain types of tumors and detrimental in others [113]. Experiments still need to be carried out to establish whether the results are the same for selective MMP-12 inhibitors or not.

6 Conclusions and future directions

It has not yet been clearly established which MMP activity needs to be inhibited in order to have an impact on inflammatory diseases. Since numerous reported evidences suggest that different MMPs play an important role in the pathogenesis of tissue remodeling associated with inflammatory processes in several diseases, broad spectrum MMP inhibitors may have therapeutic potential, nevertheless, associated with adverse events [47, 114]. It is therefore possible that a selective MMP inhibitor may have reduced side effects.

One alternative is a gene transfer to overexpress TIMPs which can reduce MMP activity and modulate tissue remodeling. Several preclinical studies of various diseases have reported encouraging data. For example, cartilage degradation and invasion by rheumatoid synovial fibroblasts is inhibited by the gene transfer of TIMP-1 and TIMP-3 [115]. However, expressing wild-type TIMPs could have drawbacks because multiple MMPs may be inhibited. The best route to success is probably the development of engineered TIMPs with altered specificity, to enable the targeting of specific MMPs.

One alternative of selective MMP inhibitors could be the RNA interference therapy development. Regarding the airways, for example, a local administration of siRNA may be easier to achieve than a systemic administration. As shown using siRNA targeting MMP-12 (siMMP-12) [116], the inhibition or degradation of the corresponding mRNA should be an original solution to reduce the development of associated inflammatory diseases. As regards the importance of MMP in many physiopathological processes, the interest in the development of MMP inhibitors remains high, and the therapeutic potential is expected to increase when positive data will be obtained. However, only clinical data will be able to validate the therapeutic potential of this class of compounds.

Acknowledgements The authors thank Florence Jacquet for the corrections of the manuscript.

References

1. Greenlee KJ, Werb Z, Kheradmand F (2007) Matrix metalloproteinases in lung: multiple, multifarious, and multifaceted. Physiol Rev 87:69–98
2. Parks WC, Wilson CL, Lopez-Boado YS (2004) Matrix metalloproteinases as modulators of inflammation and innate immunity. Nat Rev Immunol 4:617–629
3. Vu T, Werb Z (1998) Gelatinase B: structure, regulation, and function. In: Parks WC, Mecham RP (eds) Matrix metalloproteinases. Academic, San Diego, pp 115–148
4. Mott JD, Werb Z (2004) Regulation of matrix biology by matrix metalloproteinases. Curr Opin Cell Biol 16:558–564
5. Van Den Steen PE, Proost P, Wuyts A, Van Damme J, Opdenakker G (2000) Neutrophil gelatinase B potentiates interleukin-8 tenfold by aminoterminal processing, whereas it degrades CTAP-III, PF-4, and GRO-alpha and leaves RANTES and MCP-2 intact. Blood 96:2673–2681
6. Van Den Steen PE, Wuyts A, Husson SJ, Proost P, Van Damme J, Opdenakker G (2003) Gelatinase B/MMP-9 and neutrophil collagenase/MMP-8 process the chemokines human GCP-2/CXCL6, ENA-78/CXCL5 and mouse GCP-2/LIX and modulate their physiological activities. Eur J Biochem 270:3739–3749
7. Li Q, Park PW, Wilson CL, Parks WC (2002) Matrilysin shedding of syndecan-1 regulates chemokine mobilization and transepithelial efflux of neutrophils in acute lung injury. Cell 111:635–646
8. McQuibban GA, Gong JH, Tam EM, McCulloch CA, Clark-Lewis I, Overall CM (2000) Inflammation dampened by gelatinase A cleavage of monocyte chemoattractant protein-3. Science 289:1202–1206
9. Fligiel SE, Standiford T, Fligiel HM, Tashkin D, Strieter RM, Warner RL, Johnson KJ, Varani J (2006) Matrix metalloproteinases and matrix metalloproteinase inhibitors in acute lung injury. Hum Pathol 37:422–430
10. Ohnishi K, Takagi M, Kurokawa Y, Satomi S, Konttinen YT (1998) Matrix metalloproteinase-mediated extracellular matrix protein degradation in human pulmonary emphysema. Lab Invest 78:1077–1087
11. Maisi P, Prikk K, Sepper R, Pirila E, Salo T, Hietanen J, Sorsa T (2002) Soluble membrane-type 1 matrix metalloproteinase (MT1-MMP) and gelatinase A (MMP-2) in induced sputum and bronchoalveolar lavage fluid of human bronchial asthma and bronchiectasis. APMIS 110:771–782
12. Konttinen YT, Ainola M, Valleala H, Ma J, Ida H, Mandelin J, Kinne RW, Santavirta S, Sorsa T, Lopez-Otin C, Takagi M (1999) Analysis of 16 different matrix metalloproteinases (MMP-1 to MMP-20) in the synovial membrane: different profiles in trauma and rheumatoid arthritis. Ann Rheum Dis 58:691–697
13. Urbanski SJ, Edwards DR, Maitland A, Leco KJ, Watson A, Kossakowska AE (1992) Expression of metalloproteinases and their inhibitors in primary pulmonary carcinomas. Br J Cancer 66:1188–1194
14. Pruijt JF, Fibbe WE, Laterveer L, Pieters RA, Lindley IJ, Paemen L, Masure S, Willemze R, Opdenakker G (1999) Prevention of interleukin-8-induced mobilization of hematopoietic progenitor cells in rhesus monkeys by inhibitory antibodies against the metalloproteinase gelatinase B (MMP-9). Proc Natl Acad Sci USA 96:10863–10868

15. Opdenakker G, Fibbe WE, Van Damme J (1998) The molecular basis of leukocytosis. Immunol Today 19:182–189
16. Van Lint P, Libert C (2007) Chemokine and cytokine processing by matrix metalloproteinases and its effect on leukocyte migration and inflammation. J Leuk Biol 82:1375–1381
17. Manicorne AM, McGuire JK (2008) Matrix metalloproteinases as modulators of inflammation. Semin Cell Dev Biol 19:34–41
18. Black RA, Rauch CT, Kozlosky CJ, Peschon JJ, Stack JL, Wolfson MF, Castner BJ, Stocking KL, Reddy P, Srinivasan S, Nelson N, Boiani N, Schooley KA, Gerhart M, Davis R, Fitzner JN, Johnson RS, Paxton RJ, March CJ, Cerretti DP (1997) A metalloproteinase disintegrin that releases tumour-necrosis factor-alpha from cell. Nature 385:729–733
19. Ito A, Mukaiyama A, Itoh H, Nagase H, Thorgersen IB, Enghild JJ, Sasaguri Y, Mori Y (1996) Degradation of interleukin 1 beta by matrix metalloproteinases. J Biol Chem 271: 14657–14660
20. Bernfield M, Gotte M, Park PW, Reizes O, Fitzgerald ML, Lincecum J, Zako M (1999) Functions of cell surface heparan sulfate proteoglycans. Annu Rev Biochem 68:729–777
21. Brew K (2000) Tissue inhibitors of metalloproteinases: evolution, structure and function. Biochim Biophys Acta 1477:267–283
22. Tuuttila A, Morgunova E, Bergmann U, Lindqvist Y, Maskos K, Fernandez-Catalan C, Bode W, Tryggvason K, Schneider G (1998) Three-dimensional structure of human tissue inhibitor ofmetalloproteinases-2 at 2.1 A resolution. J Mol Biol 284:1133–1140
23. Lambert E, Dasse E, Haye B, Petitfrere E (2004) TIMPs as multifacial proteins. Crit Rev Oncol Hematol 49:187–198
24. Baker AH, Edwards DR, Murphy G (2002) Metalloproteinase inhibitors: biological actions and therapeutic opportunities. J Cell Sci 115:3719–3727
25. Brew K, Dinakarpandian D, Nagase H (2000) Tissue inhibitors of metalloproteinases: evolution, structure and function. Biochim Biophys Acta 1477:267–283
26. Amour A, Slocombe PM, Webster A, Butler M, Knight CG, Smith BJ, Stephens PE, Shelley C, Hutton M, Knauper V, Docherty AJ, Murphy G (1998) TNF-alpha converting enzyme (TACE) is inhibited by TIMP-3. FEBS Lett 435:39–44
27. Stetler-Stevenson WG (1999) Matrix metalloproteinases in angiogenesis: a moving target for therapeutic intervention. J Clin Invest 103:1237–1241
28. Clark IM, Swingler TE, Sampieri CL, Edwards DR (2007) The regulation of matrix metalloproteinases and their inhibitors. Int J Biochem Cell Biol 12:006
29. Wong S, Belvisi MG, Birrell MA (2005) Profiling of MMP/TIMP gene expression in human and rodent lung samples, and in models of airway inflammation. Proc Am Thor Soc 2:A73
30. Manoury B, Nenan S, Guénon I, Lagente V, Boichot E (2007) Influence of early neutrophil depletion on MMPs/TIMP-1 balance in bleomycin-induced lung fibrosis. Int Immunopharmacol 7:900–911
31. Corbel M, Caulet-Maugendre S, Germain N, Lagente V, Boichot E (2001) Inhibition of bleomycin-induced pulmonary fibrosis in mice by the matrix metalloproteinase inhibitor, batimastat. J Pathol 193:538–545
32. Chirco R, Liu XW, Jung KK, Kim HR (2006) Novel functions of TIMPs in cell signaling. Cancer Metastasis Rev 25:99–113
33. Seo DW, Li H, Guedez L, Wingfield PT, Diaz T, Salloum R, Wei BY, Stetler-Stevenson WG (2003) TIMP-2 mediated inhibition of angiogenesis: an MMP-independent mechanism. Cell 114:171–180
34. Oh J, Seo DW, Diaz T, Wei B, Ward Y, Ray JM, Morioka Y, Shi S, Kitayama H, Takahashi C, Noda M, Stetler-Stevenson WG (2004) Tissue inhibitors of metalloproteinase 2 inhibits endothelial cell migration through increased expression of RECK. Cancer Res 64:9062–9069
35. Muraguchi T, Takegami Y, Ohtsuka T, Kitajima S, Chandana EP, Omura A, Miki T, Takahashi R, Matsumoto N, Ludwig A, Noda M, Takahashi C (2007) RECK modulates Notch signaling during cortical neurogenesis by regulating ADAM10 activity. Nature Neurosci 10:838–845

36. Holten-Andersen MN, Fenger C, Nielsen HJ, Rasmussen AS, Christensen IJ, Brunner N, Kronborg O (2004) Plasma TIMP-1 in patients with colorectal adenomas: a prospective study. Eur J Cancer 40:2159–2164
37. McCarthy K, Maguire T, McGreal G, McDermott E, O'Higgins N, Duffy MJ (1999) High levels of tissue inhibitor of metalloproteinase-1 predict poor outcome in patients with breast cancer. Int J Cancer 84:44–48
38. Nakopoulou L, Giannopoulou I, Stefanaki K, Panayotopoulou E, Tsirmpa I, Alexandrou P, Mavrommatis J, Katsarou S, Davaris P (2002) Enhanced mRNA expression of tissue inhibitor of metalloproteinase-1 (TIMP-1) in breast carcinomas is correlated with adverse prognosis. J Pathol 197:307–313
39. Vanhoutte D, Heymans S (2010) TIMPs and cardiac remodeling: 'Embracing the MMP-independent-side of the family'. J Mol Cell Cardiol 48:445–453
40. Kassiri Z, Oudit GY, Sanchez O, Dawood F, Mohammed FF, Nuttall RK et al (2005) Combination of tumor necrosis factor-alpha ablation and matrix metalloproteinase inhibition prevents heart failure after pressure overload in tissue inhibitor of metalloproteinase-3 knock-out mice. Circ Res 97:380–390
41. Wisniewska M, Goettig P, Maskos K, Belouski E, Winters D, Hecht R, et al. Structural determinants of the ADAM inhibition by TIMP-3: crystal structure of the TACE-N–TIMP-3 complex. *J Mol Biol* 381:1307–19.
42. Smookler DS, Mohammed FF, Kassiri Z, Duncan GS, Mak TW, Khokha R (2006) Tissue inhibitor of metalloproteinase 3 regulates TNF-dependent systemic inflammation. J Immunol 176:721–725
43. Guedez L, Stetler-Stevenson WG, Wolff L, Wang J, Fukushima P, Mansoor A et al (2002) In vitro suppression of programmed cell death of B cells by tissue inhibitor of metalloproteinases-1. J Clin Invest 102:2002–2010
44. Lim MS, Guedez L, Stetler-Stevenson WG, Stetler-Stevenson M (1999) Tissue inhibitor of metalloproteinase-2 induces apoptosis in human T lymphocytes. Ann NY Acad Sci 878: 522–523
45. Skiles JW, Gonnella NC, Jeng AY (2001) The design, structure, and therapeutic application of matrix metalloproteinase inhibitors. Curr Med Chem 8:425–474
46. Skotnicki JS, Levin JI, Killar LM (1999) Matrix metalloproteinase inhibitors. In: Bottomley KMK, Bradshaw D, Nixon JS (eds)"Metalloproteinases as targets for anti-inflammatory drugs" Progress in Inflammation Research. pp 17–57
47. Dorman G, Cseh S, Hdju I, Barna L, Konya D, Kupai K, Kovacs L, Ferdinandy P (2010) Matrix metalloproteinase Inhibitors. Drugs 70:949–964
48. Hu J, Van den Steen E, Sang QXA, Opdenaker G (2007) Matrix metalloproteinase inhibitors as therapy for inflammatory and vascular diseases. Nat Rev Drug Discover 6:480–498
49. Wang X, Fu X, Brown PD, Crimmin MJ, Hoffman RM (1994) Matrix metalloproteinase inhibitor BB-94 (batimastat) inhibits human colon tumor growth and spread in a patient-like orthotopic model in nude mice. Cancer Res 54:4726–4728
50. Watson SA, Morris TM, Collins HM, Bawden LJ, Hawkins K, Bone EA (1999) Inhibition of tumour growth by marimastat in a human xenograft model of gastric cancer: relationship with levels of circulating CEA. Br J Cancer 81:19–23
51. Delclaux C, Delacourt C, D'Ortho MP, Boyer V, Lafuma C, Harf A (1997) Role of gelatinase B and elastase in human polymorphonuclear neutrophil migration across basement membrane. Am J Respir Cell Mol Biol 14:288–295
52. Corbel M, Lanchou J, Germain N, Boichot E, Lagente V (2001) Modulation of airway remodeling-associated mediators by the antifibrotic compound pirfenidone and the matrix metalloproteinase inhibitor batimastat during acute lung injury in mice. Eur J Pharmacol 426:113–121
53. Birrell MA, Wong S, Dekkak A, De Alba J, Haj-Yahia S, Belvisi MG (2006) Role of matrix metalloproteinases in the inflammatory response in human airway cell-based assays and in rodent models of airway disease. J Pharmacol Exp Ther 318:741–750

54. Birrell MA, Wong S, Hele DJ, McCluskie K, Hardaker E, Belvisi MG (2005) Steroid-resistant inflammation in a rat model of chronic obstructive pulmonary disease is associated with a lack of nuclear factor-kappaB pathway activation. Am J Respir Crit Care Med 172:74–84
55. Zhang Y, Xu J, Levin J, Hegen M, Li G, Robertshaw H, Brennan F, Cummons T, Clarke D, Vansell N, Nickerson-Nutter C, Barone D, Mohler K, Black R, Skotnicki J, Gibbons J, Feldmann M, Frost P, Larsen G, Lin LL (2004) Identification and characterization of 4-[[4-(2-butynyloxy)phenyl]sulfonyl]-N-hydroxy-2,2-dimethyl-(3S)thiomorpholinecarboxamide (TMI-1), a novel dual tumor necrosis factor-alpha-converting enzyme/matrix metalloprotease inhibitor for the treatment of rheumatoid arthritis. J Pharmacol Exp Ther 309:348–355
56. Mackarel AJ, Cottell DC, Russell KJ, FitzGerald MX, O'Connor CM (1999) Migration of neutrophils across human pulmonary endothelial cells is not blocked by matrix metalloproteinase or serine protease inhibitors. Am J Respir Cell Mol Biol 20:1209–1219
57. Betsuyaku T, Shipley JM, Liu Z, Senior RM (1999) Gelatinase B deficiency does not protect against lipopolysaccharide-induced acute lung injury. Chest 116:17S–18S
58. Betsuyaku T, Fukuda Y, Parks WC, Shipley JM, Senior RM (2000) Gelatinase B is required for alveolar bronchiolization after intratracheal bleomycin. Am J Pathol 157:525–535
59. Solorzano CC, Ksontini R, Pruitt JH, Auffenberg T, Tannahill C, Galardy RE, Schultz GP, MacKay SL, Copeland EM, Moldawer LL (1997) A matrix metalloproteinase inhibitor prevents processing of tumor necrosis factor alpha (TNF alpha) and abrogates endotoxin-induced lethality. Shock 7:427–431
60. Nenan S, Lagente V, Planquois JM, Hitier S, Berna P, Bertrand CP, Boichot E (2007) Metalloelastase (MMP-12) induced inflammatory response in mice airways: effects of dexamethasone, rolipram and marimastat. Eur J Pharmacol 559:75–81
61. Selman M, Cisneros-Lira J, Gaxiola M, Ramirez R, Kudlacz EM, Mitchell PG, Pardo A (2003) Matrix metalloproteinases inhibition attenuates tobacco smoke-induced emphysema in guinea-pigs. Chest 123:1633–1641
62. Martin RL, Shapiro SD, Tong SE, Van Wart HE (2001) Macrophage metalloelastase inhibitors. Prog Respir Res 31:177–180
63. Trifilieff A, Walker C, Keller T, Kottirsch G, Neumann U (2002) Pharmacological profile of PKF242-484 and PKF241-466, novel dual inhibitors of TNF-alpha converting enzyme and matrix metalloproteinases, in models of airway inflammation. Br J Pharmacol 135: 1655–1664
64. de Meijer V, Sverdlov DY, Popov D, Le HD, Meisel JD, Nose V, Schuppan D, Puder M (2010) Broad-spectrum matrix metalloproteinase inhibition curbs inflammation and liver injury but aggravates experimental liver fibrosis in mice. PLoS One 5:e11256
65. Wielockx B, Lannoy K, Shapiro SD, Itoh T, Itohara S et al (2001) Inhibition of matrix metalloproteinases blocks lethal hepatitis and apoptosis induced by tumor necrosis factor and allows safe antitumor therapy. Nat Med 7:1202–1208
66. Kahraman A, Bronk SF, Cazanave S, Werneburg NW, Mott JL et al (2009) Matrix metalloproteinase inhibitor, CTS-1027, attenuates liver injury and fibrosis in the bile duct-ligated mouse. Hepatol Res 39:805–813
67. MacPherson LJ, Bayburt EK, Capparelli MP, Carroll BJ, Goldstein R, Justice MR, Zhu L, Hu S, Melton RA, Fryer L, Goldberg RL, Doughty JR, Spirito S, Blancuzzi V, Wilson D, O'Byrne EM, Ganu V, Parker DT (1997) Discovery of CGS 27023A, a non-peptidic, potent, and orally active stromelysin inhibitor that blocks cartilage degradation in rabbits. J Med Chem 40:2525–2532
68. Hamada T, Arima N, Shindo M, Sugama K, Sasaguri Y (2000) Suppression of adjuvant arthritis of rats by a novel matrix metalloproteinase-inhibitor. Br J Pharmacol 131: 1513–1520
69. Gatto C, Rieppi M, Borsotti P, Innocenti S, Ceruti R, Drudis T, Scanziani E, Casazza AM, Taraboletti G, Giavazzi R (1999) BAY 12–9566, a novel inhibitor of matrix metalloproteinases with antiangiogenic activity. Clin Cancer Res 5:3603–3607

70. Conway JG, Wakefield JA, Brown RH, Marron BE, Sekut L, Stimpson SA, McElroy A, Menius JA, Jeffreys JJ, Clark RL (1995) Inhibition of cartilage and bone destruction in adjuvant arthritis in the rat by a matrix metalloproteinase inhibitor. J Exp Med 182:449–457
71. Conway JG, Andrews RC, Beaudet B, Bickett DM, Boncek V, Brodie TA, Clark RL, Crumrine RC, Leenitzer MA, McDougald DL, Han B, Hedeen K, Lin P, Milla M, Moss M, Pink H, Rabinowitz MH, Tippin T, Scates PW, Selph J, Stimpson SA, Warner J, Becherer JD (2001) Inhibition of tumor necrosis factor-α (TNF-α) production and arthritis in the rat by GW3333, a dual inhibitor of TNF-α-converting enzyme and matrix metalloproteinases. J Pharmacol Exp Ther 298:900–908
72. Shapiro SD, Kobayashi DK, Ley TJ (1993) Cloning and characterization of a unique elastolytic metalloproteinase produced by human alveolar macrophages. J Biol Chem 268:23824–23829
73. Shapiro SD, Griffin GL, Gilbert DJ, Jenkins NA, Copeland NG, Welgus HG, Senior RM, Ley TJ (1992) Molecular cloning, chromosomal localization, and bacterial expression of a murine macrophage metalloelastase. J Biol Chem 267:4664–4671
74. Lavigne MC, Thakker P, Gunn J, Wong A, Miyashiro JS, Wasserman AM, Wei SQ, Pelker JW, Kobayashi M, Eppihimer MJ (2004) Human bronchial epithelial cells express and secrete MMP-12. Biochem Biophys Res Commun 324:534–546
75. Xie S, Issa R, Sukkar MB, Oltmanns U, Bhavsar PK, Papi A, Caramori G, Adcock I, Chung KF (2005) Induction and regulation of matrix metalloproteinase-12 in human airway smooth muscle cells. Respir Res 6:148
76. Horton MR, Shapiro S, Bao C, Lowenstein CJ, Noble PW (1999) Induction and regulation of macrophage metalloelastase by hyaluronan fragments in mouse macrophages. J Immunol 162:4171–4176
77. Raza SL, Nehring LC, Shapiro SD, Cornelius LA (2000) Proteinase-activated receptor-1 regulation of macrophage elastase (MMP-12) secretion by serine proteinases. J Biol Chem 275:41243–41250
78. Gronski TJ Jr, Martin RL, Kobayashi DK, Walsh BC, Holman MC, Huber M, Van Wart HE, Shapiro SD (1997) Hydrolysis of a broad spectrum of extracellular matrix proteins by human macrophage elastase. J Biol Chem 272:12189–12194
79. Gearing AJH, Beckett P, Christodoulou M, Churchill M, Clements J, Davidson H, Drummond H, Galloway WA, Gilbert R, Gordon JL, Leber TM, Mangan M, Miller K, Nayee P, Owen K, Patel S, Thomas W, Wells GL, Wood M, Woolley K (1994) Processing of tumour necrosis factor- precursor by metalloproteinases. Nature 370:555–557
80. Banda MJ, Rice AG, Griffin GL, Senior RM (1988) Alpha 1-proteinase inhibitor is a neutrophil chemoattractant after proteolytic inactivation by macrophage elastase. J Biol Chem 263:4481–4484
81. Cataldo DD, Tournoy KG, Vermaelen K, Munaut C, Foidart JM, Louis R, Noël A, Pauwels RA (2002) Matrix metalloproteinase-9 deficiency impairs cellular infiltration and bronchial hyperresponsiveness during allergen-induced airway inflammation. Am J Pathol 161:491–498
82. Warner RL, Lukacs NW, Shapiro SD, Bhagarvathula N, Nerusu KC, Varani J, Johnson KJ (2004) Role of metalloelastase in a model of allergic lung responses induced by cockroach allergen. Am J Pathol 165:1921–1930
83. Chiba Y, Yu Y, Sakai H, Misawa M (2007) Increase in the expression of matrix metalloproteinase-12 in the airways of rats with allergic bronchial asthma. Biol Pharm Bull 30:318–323
84. Araujo BB, Dolhnikoff M, Silva LFF, Elliot J, Lindeman JNN, Ferreira DS, Mulder A, Gomes HAP, Fernezlian SM, James A, Mauad T (2008) Extracellular matrix components and regulators in the airway smooth muscle in asthma. Eur Respir J 32:61–69
85. Lagente V, Manoury B, Nénan S, Le Quément C, Boichot E, Martin-Chouly C (2005) Role of matrix metalloproteinases (MMPs) in the development of airway inflammation and remodeling. Brazil J Biol Med Res 38:1521–1530

86. Manoury B, Nenan S, Guenon I, Boichot E, Planquois JM, Bertrand CP, Lagente V (2006) Macrophage metalloelastase (MMP-12) deficiency does not alter bleomycin-induced pulmonary fibrosis in mice. J Inflamm 22:2
87. Kolb M, Bonniaud P, Galt T, Sime PJ, Kelly MM, Margetts PJ et al (2002) Differences in the fibrogenic response after transfer of active transforming growth factor-beta1 gene to lungs of "fibrosis-prone" and "fibrosis-resistant" mouse strains. Am J Respir Cell Mol Biol 27: 141–150
88. Manoury B, Caulet-Maugendre S, Guénon I, Lagente V, Boichot E (2006) TIMP-1 is a key factor of fibrogenic response to bleomycin on lung of mice. Int J Immunopathol Pharmacol 19:471–487
89. Swiderski RE, Dencoff JE, Floerchinger CS, Shapiro SD, Hunninghake GW (1998) Differential expression of extracellular matrix remodeling genes in a murine model of bleomycin-induced pulmonary fibrosis. Am J Pathol 152:821–828
90. Lanone S, Zheng T, Zhu Z, Liu W, Lee CG, Ma B et al (2002) Overlapping and enzyme-specific contributions of matrix metalloproteinases-9 and -12 in IL-13-induced inflammation and remodeling. J Clin Invest 110:463–474
91. Matute-Bello G, Wurfel MM, Lee JS, Park DR, Frevert CW, Madtes DK, Shapiro SD, Martin TR (2007) Essential role of MMP-12 in Fas-induced lung fibrosis. Am J Respir Cell Mol Biol 37:210–221
92. Selman M, Ruiz V, Cabrera S, Segura L, Ramirez R, Barrios R et al (2000) TIMP-1, -2, -3, and -4 in idiopathic pulmonary fibrosis. A prevailing nondegradative lung microenvironment? Am J Physiol Lung Cell Mol Physiol 279:L562–L574
93. Morris DG, Huang X, Kaminski N, Wang Y, Shapiro SD, Dolganov G et al (2003) Loss of integrin alpha(v)beta6-mediated TGF-beta activation causes Mmp12-dependent emphysema. Nature 422:169–173
94. Bonniaud P, Kolb M, Galt T, Robertson J, Robbins C, Stampfli M et al (2004) Smad3 null mice develop airspace enlargement and are resistant to TGF-beta-mediated pulmonary fibrosis. J Immunol 173:2099–2108
95. Pauwels R (2000) COPD: the scope of the problem in Europe. Chest 117:332S–335S
96. Saetta M (1999) Airway inflammation in chronic obstructive pulmonary disease. Am J Respir Crit Care Med 160:S17–S20
97. Finkelstein R, Fraser RS, Ghezzo H, Cosio MG (1995) Alveolar inflammation and its relation to emphysema in smokers. Am J Respir Crit Care Med 152:1666–1672
98. Tetley TD (2002) Macrophages and the pathogenesis of COPD. Chest 121:156S–159S
99. Hautamaki RD, Kobayashi DK, Senior RM, Shapiro SD (1997) Requirement for macrophage elastase for cigarette smoke-induced emphysema in mice. Science 277:2002–2004
100. Valença SS, Da Hora K, Castro P, Gonçalves de Moraes V, Carvalho L, Moraes Sobrino Porto LC (2004) Emphysema and metalloelastase expression in mouse lung induced by cigarette smoke. Tox Pathol 32:351–356
101. Leclerc O, Lagente V, Planquois JM, Berthelier C, Artola M, Eichholtz T, Bertrand CP, Schmidlin F (2006) Involvement of MMP-12 and phosphodiesterase type 4 in cigarette smoke-induced inflammation in mice. Eur Respir J 27:1102–1109
102. Nenan S, Planquois JM, Berna P, De Mendez I, Hitier S, Shapiro SD, Boichot E, Lagente V, Bertrand CP (2005) Analysis of the inflammatory response induced by rhMMP-12 catalytic domain instilled in mouse airways. Int Immunopharmacol 5:511–524
103. Le Quément C, Guénon I, Gillon JY, Lagente V, Boichot E (2008) MMP-12 induces IL-8/CXCL8 secretion through EGFR and ERK1/2 activation in epithelial cells. Am J Physiol 294:L1076–L1084
104. Le Quement C (2008) PhD Thesis, University of Rennes 1
105. Dean RA, Cox JH, Bellac CL, Doucet A, Starr AE, Overall CM (2008) Macrophage-specific metalloelastase (MMP-12) truncates and inactivates ELR + CXC chemokines and generates CCL2, -7, -8, and -13 antagonists: potential role of the macrophage in terminating polymorphonuclear leukocyte influx. Blood 112:3455–3464

106. Churg A, Wang R, Wang X, Onnervik PO, Thim K, Wright JL (2007) An MMP-9/-12 inhibitor prevents smoke-induced emphysema and small airway remodeling in guinea pigs. Thorax 62:706–713
107. Ma D, Jiang Y, Chen F, Gong LK, Ding K, Xu Y, Wang R, Ge A, Ren J, Li J, Li J, Ye Q (2006) Selective inhibition of matrix metalloproteinase isozymes and in vivo protection against emphysema by substituted gamma-keto carboxylic acids. J Med Chem 49:456–458
108. Le Quément C, Guénon I, Gillon JY, Valença S, Cayron-Elizondo V, Lagente V, Boichot E (2008) Involvement of MMP-12 in airway inflammation of mice exposed to cigarette smoke. Br J Pharmacol 154:1206–1215
109. Lisbonne M, Belleguic C, Le Quement C, Brinhault G, Cayron-Elizondo V, Papasioulotis O, Delaval P, Braunstein G, Gillon JY, Lagente V (2007) Inhibition of elastolytic activity by the MMP inhibitors, AS111793 and marimastat in bronchoalveolar lavage fluids of patients with chronic obstructive pulmonary disease. Rev Mal Respir 24:1198
110. Le Quément C, Lagente V, Guenon I, Muzio V, Gillon JY, Boichot E (2008) Anti-inflammatory properties of MMP inhibitors in experimental models of chronic obstructive pulmonary disease and lung inflammation In: "Matrix metalloproteinases in tissue remodelling and inflammation", Progress in Inflammation Research, Birkauser, pp 57–69
111. Morris A, Kinnear G, Wan WY, Wyss D, Bahra P, Stevenson CS (2008) Comparison of cigarette smoke-induced acute inflammation in multiple strains of mice and the effect of an MMP inhibitor on these responses. J Pharmacol Exp Ther 327:851–862
112. Houghton AM, Grisolano JL, Baumann ML, Kobayashi DK, Hautamaki RD, Nehring LC, Cornelius LA, Shapiro SD (2006) Macrophage elastase (matrix metalloproteinase-12) suppresses growth of lung metastases. Cancer Res 66:6149–6155
113. Coussens LM, Fingleton B, Matrisian L (2002) Matrix metalloproteinase inhibitors and cancer: Trials and tribulations. Science 295:2387–2392
114. Pavlaki M, Zucker S (2003) Matrix metalloproteinase inhibitors (MMPIs): The beginning of phase I or the termination of phase III clinical trials. Cancer Metastasis Rev 22:177–203
115. Van der Laan WH, Quax PH, Seemayer CA, Huisman LG, Pieterman EJ, Grimbergen JM, Verheijen JH, Breedveld FC, Gay RE, Gay S, Huizinga TW, Pap T (2003) Cartilage degradation and invasion by rheumatoid synovial fibroblasts is inhibited by gene transfer of TIMP-1 and TIMP-3. Gene Ther 10:234–242
116. Garbacki N, Di Valentin E, Piette J, Cataldo D, Crahay C, Colige A (2009) Matrix metalloproteinase 12 silencing: a therapeutic approach to treat pathological lung tissue remodeling? Pulm Pharmacol Ther 22:267–278

Dual Role for Proteases in Lung Inflammation

Giuseppe Lungarella, Eleonora Cavarra, Silvia Fineschi, and Monica Lucattelli

Abstract Proteases identified in the lung were initially thought to be involved in extracellular matrix destruction. Today, several lines of evidence suggest that proteases have a multitude of regulatory functions in local inflammation processes, innate immunity, and infections. In particular, enzymes belonging to serine and metalloproteases can modulate many biological functions by promoting chemokine and cytokine activation and degradation, cytokine receptor shedding, proteolysis of cytokine binding proteins, and the activation of different specific cell surface receptors.

Inflammatory processes are essential in host defense, but when excessive they can contribute to tissue injury, organ dysfunction, and lung diseases. Individual protease can be both beneficial and/or detrimental in inflammatory reactions. However, the positive or negative contribution of a certain enzymes may depend on biological context such as location, substrate availability, inhibitors, cell type, and disease state.

This review summarizes the current knowledge on the implication of proteases in inflammation and focuses on the dual role of certain proteases in both pro- and anti-inflammatory pathways.

Further exploration and understanding of protease functions in lung inflammation will have important implication in health and disease.

Keywords Cell surface receptors • Chemokine activation • Chemokine degradation • Cytokine activation • Cytokine binding proteins • Cytokine degradation • Cytokine receptors shedding • Extracellular matrix proteolysis • Lung inflammation • Metalloproteases • Proteinase-mediated signaling • Serine proteases

---

G. Lungarella (✉) • E. Cavarra • S. Fineschi • M. Lucattelli
Department of Physiopathology and Experimental Medicine, University of Siena, via Aldo Moro n.6, 53100 Siena, Italy
e-mail: lungarella@unisi.it; cavarra@unisi.it; fineschi2@unisi.it; lucattelli@unisi.it

N. Vergnolle and M. Chignard (eds.), *Proteases and Their Receptors in Inflammation*, Progress in Inflammation Research, DOI 10.1007/978-3-0348-0157-7_6,

1 Introduction

The lung, with its enormous surface area, is continuously exposed to a variety of potential toxic agents, which include microbial pathogens, particulate pollutants, and allergens. The respiratory system has evolved several important mechanisms that minimize the impact of inhaled toxicants, protect from injury, and preserve its homeostasis. The major pulmonary defense mechanisms include cough, mucus production, mucociliary clearance, bronchoconstriction, and cellular events such as phagocyte and immunocompetent cell responses [1]. The cells involved in the defense mechanisms include both structural cells (such as fibroblasts, epithelial, and endothelial cells) as well as a wide spectrum of inflammatory/immune effector cells of hematopoietic origin [such as neutrophils (PMNs), monocyte/macrophages (AMs), various lymphocytes subsets, dendritic cells, mast cells] that sometimes take up residence in the lung. Other cells involved in inflammatory processes are eosinophils, basophils, and platelets [1].

Two different components of the defense system are recognized: the innate immune system (evolutionary conserved and "nonspecific") and the acquired (adaptive) immune system that is able to mount a highly specific response against precise components of molecules or pathogenic microorganisms. Together these lung systems are highly effective in protecting the host against microbiological infections. Although the two components of the defense system work in concert, they have several distinct features and involve different cell populations [2].

The innate immune response (inflammation) represents the first line of defense against bacteria and pollutants, and provides immediate host defense in a nonspecific manner by means of a variety of structural and inflammatory cells. These cells are able to phagocytosize bacteria, to secrete various soluble factors (that are directly or indirectly microbicidal), and to modulate the inflammatory response by influencing the function of diverse cell types involved in the lung inflammatory processes [2].

Cells of the inflammatory system, particularly neutrophils and mononuclear phagocytes, are able of elaborating and secreting into the extracellular milieu, proteases, oxidants, eicosanoids, chemokines, cytokines and neuropeptides, which are capable of interacting synergistically, or sometimes antagonistically, to modulate the intensity of inflammatory reactions on a local and systemic level [2]. Inflammation is essential for host defense and tissue repair processes, but when dysregulated or excessive, it can contribute to ongoing tissue injury, organ dysfunction, and chronic disease [3]. Unfortunately, our understanding of the interactive effects of the above-mentioned inflammatory mediators is still rudimentary.

Among these factors proteinases are prominent, and exert both physiological and pathological effects in the respiratory system. The activity of proteases is counterbalanced by endogenous inhibitors. Altered expression of protease/antiprotease family members has been associated with lung diseases characterized by a loss or an abnormal accumulation of extracellular matrix (ECM) (such as emphysema, asthma and pulmonary fibrosis) or lung tissue damage after cardiopulmonary bypass, reperfusion injury after unilateral ischemia, smoke inhalation, and adult respiratory distress syndrome [4].

This chapter deals with the role of proteases in inflammatory responses and summarizes the present knowledge regarding the dual role (pro-/anti-inflammatory, protective/destructive) of these enzymes in pulmonary inflammation. The emphasis will be on proteases originating in neutrophils, macrophages and some lymphocyte subsets, although mast cells, platelets, eosinophils, fibroblasts, and bacteria are also important potential sources of proteases in lung chronic inflammation.

2 Serine Proteinases

2.1 Neutrophil Serine Proteinases

Neutrophils represent one of the first lines of host defense against infection achieving their antimicrobial function through a series of coordinate responses (intracellular sequestration of microorganisms into phagosome, fusion of neutrophil granules with the phagolysosome, release of proteases and antimicrobial peptides into the phagolysosomes, release of large quantities of reactive oxygen species generated by the activation of the membrane-bound NADPH oxidase system) that culminate in destruction of pathogens [5]. Neutrophils are vital components of the immune defense. The importance of these cells in the inflammatory response is demonstrated by studies carried out on several experimental animal models and in humans. Depletion of neutrophils abrogates inflammation in animal models [6], and disorders of neutrophil number and function lead to increased risk of infection in patients with severe chronic neutropenia [7].

Neutrophils express a series of neutral serine proteases including neutrophil elastase (NE), cathepsin G (CG), and proteinase 3 (PR3) in their azurophil granules. These cells achieve their antimicrobial activity through a series of coordinated responses that culminate in phagocytosis and destruction of pathogens. The bactericidal activity of neutrophils depends partly on the presence of these granule-associated proteases whose relative importance in the killing of gram-positive and gram-negative bacteria was confirmed in genetically manipulated mice deficient of CG and NE [8–10]. Additionally, an increased susceptibility to recurrent bacterial infections (with bleeding defects and partial albinism) characterizes the patients with Chediak–Higashi syndrome [11], or *beige* mice [12], whose neutrophils show a severe defect in exocytosis and a significantly reduced serine protease activity. Similarly, neutrophils of patients with Papillon–Lefevre syndrome show minimal residual activity of all three serine proteinases [6] that may explain the susceptibility to chronic infection (such as periodontal diseases) and defective antimicrobial activity against common microorganisms [13]. On the contrary, a series of studies suggest that neutrophil serine proteases may also interfere with the host normal defense and promote viral infection [6].

Although the role in host defense has been acknowledged, neutrophil serine proteases have also been implicated in various noninfectious inflammatory lung processes. Following limited granule exocytosis, these enzymes may be released extracellularly playing crucial roles in extracellular proteolytic processes at sites of inflammation [14]. This event is generally considered as being potentially injurious to airway and alveolar function, but the presence of a sophisticated network of extracellular antiproteases – such as α2-macroglobulin (α2M), α1-antitrypsin (α1AT), serine leukocyte protease inhibitor (SLPI), elafin (also known as skin-derived antileukoprotease), and α1-antichymotrypsin (α1ACT) – provides protection from both endogenous and exogenous serine proteases. Under normal circumstances, elafin antagonizes NE activity, whereas α1AT, SLPI, and α2M all inhibit CG and NE. With regard to PR3, α1AT, α2M, α1ACT but not SLPI inhibit its enzyme activity. However, an excess of proteases may overwhelm the lung's antiprotease defense and cause lung damage [15].

The combined action of all three neutrophil serine proteinases can degrade every extracellular matrix (ECM) component of the lung that leads to compromise the alveolar structures. In particular, NE is believed to play a particularly important role in ECM proteolysis. It is able to degrade most of the components of the pulmonary ECM, including elastin, type I–IV collagens, proteoglycans, fibronectin, and laminin [16]. In addition to the proteolytic effects on the ECM, NE has other direct destructive effects on other pulmonary components inducing apoptosis of epithelial cells via activation of surface receptors such as proteinase-activated receptor 1 (PAR-1) [17], or causing detachment and death (anoikis) of endothelial [18] and airway epithelial cells [19]. Due to these capacities, NE has been involved in the pathogenesis of several lung conditions including pulmonary emphysema [20], acute lung injury [21], acute respiratory distress syndrome [21], and chronic inflammation airway diseases [22].

However, over the past few years it has become evident that NE [23] and the other neutrophil serine proteases [6] have active regulatory functions in tissue destruction and in repair during inflammation. It has become clear that various active molecules (i.e., cytokines and chemokines), specific cell receptors, and cytokine binding proteins are also natural substrates of NE, CG, and PR3 [6, 23].

Recent findings indicate that NE, CG, and PR3 can modulate the activity of important mediators involved in (a) inflammation (such as TNF-α, IL-6, IL-8, SDF-1α, MIP-1α, MIP-2, RANTES, etc.), (b) adaptive immune responses (such as IL-2, IL-18, IL-32, chemerin, etc.), or (c) repair processes (such as TGF-β, TGF-α, EGF, IGF, TNF-α, etc.).

In some cases a limited proteolysis of chemokines or cytokines (Table 1) can enhance their biological activity, or stability; in others, proteolysis causes inactivation providing a negative feedback mechanism to terminate immunostimulating, or to block immunoinhibiting signals. Thus, neutrophil serine proteases may participate in the regulation of various biological pathways that fine-tune the local inflammatory response. Neutrophil serine proteases may enhance the inflammatory response by modifying the activity of several chemokines through a limited truncation obtained by enzymatic proteolysis. Chemokines are a family of chemoattractant proteins that play an important role in recruiting leukocytes at the

Table 1 Overview of serine proteases and their products involved in modulation of inflammation

Enzymes	Targets	Activity	Biological effects
Neutrophil elastase (NE)	CXCL8 (IL-8)	Proteolysis	Abrogation of PMN chemotaxis
	CXCR1 (IL-8R)	Proteolysis	Abrogation of PMN bacterial killing
	TNF receptor II (TNFRII)	Proteolysis	Abrogation of PMN adhesion
	CD43	Proteolysis	Abrogation of PMN spreading/aggregation
	CD4 and CD8	Proteolysis	Reduction of cytotoxicity and cytokine production by T cells
	Chemerin	Limited proteolysis	Pro-chemerin activation: activation of antigen-presenting cells, chemoattraction
	CCL3 (MIP-1α)	Proteolysis	Abrogation of macrophage chemotaxis and activation
	CXCL12 (SDF-1)	Proteolysis	Inhibition of T-lymphocyte migration
	CXCR4 (SDF-1R)	Proteolysis	Inhibition of T-lymphocyte migration
	IL-18 (IFN-γ)	Proteolysis	Decreased biological activity
	TNF-α	Proteolysis	Loss of activity
	TNF-α	Limited proteolysis	Activation
	CD43 (Sialophorin)	Proteolytic shedding	Neutrophil spreading
	CD11b/CD18 (CR3, MAC-1)	Reversible binding	Cellular attachment/detachment
	TLR4	unknown	Receptor activation: modulation of inflammatory responses
	PAR-1	Proteolysis	Inactivation
	PAR-2	Proteolysis of distinct sites	Activation/inactivation
	PAR-3	Proteolysis	Inactivation
	CD87 (Urokinase-type plasminogen activator receptor)	Proteolytic cleavage	Cell migration, chemotaxis
	CD14	Proteolysis	Inhibition of LPS-mediated cell activation
	CR1 (complement receptor 1)	Proteolysis	Inhibition of complement activation and phagocytosis
	C5	Limited proteolysis	C5b generation: Complement cascade activation
	C5a	Proteolysis	Inhibition of chemotaxis
	CD23	Limited proteolysis	Monocytes stimulation

(continued)

Table 1 (continued)

Enzymes	Targets	Activity	Biological effects
	Perivascular basement membrane	Proteolysis	Enhancement of neutrophil migration
	E-cadherins	Limited proteolysis	Enhancement of neutrophil migration
	ICAM-1 (intercellular adhesion molecule-1)	Limited proteolysis	Enhancement of neutrophil migration
	VCAM-1 (vascular cell adhesion molecule-1)	Limited proteolysis	Enhancement of neutrophil migration
	Immunoglobulins	Proteolysis	Decreased phagocytosis and killing of bacteria
	Annexin I	Proteolytic degradation	Inactivation: loss of anti-inflammatory properties
	MMP-2 and cathepsins	Limited proteolysis	Pro-enzymes activation: cleavage of defensins and lactoferrin
	MMP-9	Limited proteolysis	Pro-enzyme activation
	TIMP-1	Proteolysis	Inactivation: imbalance MMP/TIMP
	Surfactant protein (SP) – A and D	Proteolytic degradation	Impairment of lung innate host defense
Cathepsin G (CG)	CXCL5 (ENA-78)	Limited proteolysis	PMN chemotaxis: enhanced efficiency
	CCL15 (MIP-1δ)	Limited proteolysis	Monocytes chemotaxis: enhanced efficiency
	CXCL7 (NAP-2)	Limited proteolysis of connective tissue activating peptide (CTAP-III)	PMN chemotaxis and activation
	Chemerin	Limited proteolysis of pro-chemerin	Antigen-presenting cells and chemoattraction
	CCL5 (RANTES)	Limited proteolysis	Monocytes chemotaxis: low efficiency
	CCL3 (MIP-1α)	Proteolysis	Abrogation of macrophage chemotaxis and activation
	CXCL2 (MIP-2)	Modulation of integrin clustering	PMN migration and chemotaxis
	CXCL12 (SDF-1)	Proteolysis	Inhibition of T-lymphocyte migration
	CXCR4 (SDF-1R)	Proteolysis	Inhibition of T-lymphocyte migration
	TNF-α	Proteolysis	Loss of activity
	TNF-α	Limited proteolysis	Activation
	Formyl peptide receptor (FPR)	Receptor binding	PMN chemotaxis and activation
	PAR-1	Proteolysis	Inactivation
	PAR-2	Proteolysis of distinct sites	Activation/inactivation

(continued)

Table 1 (continued)

Enzymes	Targets	Activity	Biological effects
	PAR-3	Proteolysis	Inactivation
	PAR-4	Limited proteolysis	Activation. Neutrophil/ platelet interaction
	CD87 (urokinase-type plasminogen activator receptor)	Proteolytic cleavage	Cell migration, chemotaxis
	CD14	Proteolysis	Inhibition of LPS-mediated cell activation
	C5a	Proteolysis	Inhibition of chemotaxis
	C3	Limited proteolysis	Generation of active fragment
	CD23	Limited proteolysis	Monocytes stimulation
	E-cadherins	Limited proteolysis	Enhancement of neutrophil migration
	ICAM-1 (intercellular adhesion molecule-1)	Limited proteolysis	Enhancement of neutrophil migration
	VCAM-1 (vascular cell adhesion molecule-1)	Limited proteolysis	Enhancement of neutrophil migration
Proteinase 3 (PR3)	CXCL8 (IL8)	Limited proteolysis	PMN chemotaxis: enhanced efficiency
	CCL3 (MIP-1α)	Proteolysis	Abrogation of macrophage chemotaxis and activation
	IL-18 (IFN-γ)	Limited proteolysis	Pro-IL-18 activation: activation of T-cells functions
	TNF-α	Limited proteolysis	Pro-TNF-α activation
	IL-1β	Limited proteolysis	Pro-IL-1β activation
	IL-32	Limited proteolysis	Enhanced activity
	IL-6	Proteolysis	Inactivation: limiting or pro-inflammatory consequence
	PAR-1	Proteolysis	Inactivation
	PAR-2	Proteolysis of distinct sites	Activation/inactivation
	PAR-3	Proteolysis	Inactivation
Mast cell β-tryptase	Endothelial cells	unknown	Secretion of neutrophil chemokines
Mast cell chymase	IL-6	Proteolysis	Inactivation: limiting or pro-inflammatory consequence
	IL-13	Proteolysis	Inactivation
	MMP-9	Limited proteolysis	Pro-MMP-9 activation
	PAR-2	Limited proteolysis	Activation
	Angiotensin-1	Limited proteolysis	Angiotensin II generation
	MMP-1	Limited proteolysis	Pro-MMP-1 activation
	CXCL7 (NAP-2)	Limited proteolysis of CTAP-III	PMN chemotaxis and activation

(continued)

Table 1 (continued)

Enzymes	Targets	Activity	Biological effects
Granzyme A	Immune cells	unknown	Secretion of pro-inflammatory cytokines: IL-1β, TNF-α, IL-6
Granzyme A,B,H	Immune cells	unknown	Cytokine secretion, autoantigen production and autoimmunity

sites of inflammation. The proteolytic modification of these molecules enhances their biological activity as the truncated variants show higher potency and efficiency for inflammatory cell chemotaxis than the full-length forms.

The N-terminal processing of CXCL8 (also known as IL-8) and CXCL-5 (also known as ENA-78) by PR3 [24] or CG [25], respectively, increase neutrophil chemotaxis. Similarly, the N-terminal modification of CCL15 (also known as MIP-1δ) by CG increases the chemotactic activity of this molecule for monocytes [26]. CG also converts CTAP-III (connective tissue-activating peptide III) into CXCL7 (also known as NAP-2) by limited proteolysis [27]. NE and CG are also able to initiate a more specific defensive mechanism by converting pro-chemerin [28], through the removal of a C-terminal peptide, to chemerin an important chemoattractant factor for antigen-presenting cells.

In other cases, the proteolytic processing of chemokines by neutrophil serine proteases results in inactivation or decreased activity of these molecules. For example, the removal of N-terminal residues of CCL5 (also known as RANTES) by CG [29] or proteolysis of CCL3 (also known as MIP-1) by all three neutrophil serine proteases [30] results in a form exhibiting a lower chemotactic activity. Likewise, the amino-terminal processing of CXCL12 (also known SDF-1) and its cognate receptor by NE and CG [31–33] abrogates its T lymphocyte chemotactic activity.

These data all together suggest that serine proteases released from neutrophils at the site of inflammation may play an important role in orchestrating the subsequent recruitment of different subsets of immune cells. Unfortunately, most of the current knowledge derives from in vitro studies and the in vivo relevance of these protease-dependent feedback mechanisms still remains poorly understood and a matter for further investigations.

During the inflammatory process, several important cytokines require proteolytic processing to be released from an inactive precursor form. The pro-form of TNF-α is bound to the cell surface membrane and is proteolytically activated by the membrane-bound metalloproteinase TACE (TNF-α-converting enzyme) [34]. However studies indicate that this pro-cytokine may be alternatively processed by PR3 [35], whereas CG and NE can degrade TNF-α resulting in a loss of activity [36]. Similarly, IL-1β (another important pro-inflammatory cytokine) requires proteolytic activation from its inactive precursor form (pro-IL-1β) by caspase 1 (also known as interleukin-1-converting enzyme or ICE) [37]. However, it can be also processed into its biologically active form by PR3 [35]. This serine protease can also activate interleukin-18 (IL-18), also known as interferon-gamma (IFN-γ)-inducing factor, into its active mature form [38]. IL-18 is constitutively secreted by a variety of cells and activated by caspase-1 [39].

Although the enzyme activity of PR-3 closely resembles that of NE, the latter enzyme can proteolyse mature IL-18 leading to diminished biological activity [40]. In contrast, all three neutrophil serine proteases have been shown to degrade and inactivate IL-6 [41], an important cytokine recently implicated in the pathogenesis of chronic obstructive pulmonary disease (COPD) [42]. Unfortunately, it is unclear whether inactivation of IL-6 has pro-inflammatory or limiting consequences since its role in inflammation is still controversial [6].

In addition to the proteolytic modulation and release of chemokines and cytokines, neutrophil serine proteases can also contribute to the modulation of the lung inflammatory reaction through the specific activation of cell surface receptors or their proteolytic shedding [6].

Neutrophil proteases can activate or disarm cell surface receptors implicated in inflammatory cell adhesion, spreading, or migration.

Studies indicate that integrins (a large family of cell surface receptors that mediate the interaction between cells and the environment), proteoglycans and other surface molecules expressed in leukocytes and in other inflammatory or structural cells may provide binding sites for neutrophil serine proteinases. There is strong evidence that membrane-associated serine proteinases on leukocytes play a critical role in inflammation, fibrinolysis, coagulation, wound healing, and ECM remodeling [43].

Only to mention few examples, CG is able to regulate adhesion-dependent neutrophil effector function by modulating integrin clustering [44] and this action does not require its proteolytic activity. The modulation of neutrophil spreading requires the shedding of CD43 (also known as leukosialin) that can be done by NE [45]. NE also binds directly CD11b/CD18 regulating integrin-mediated cellular attachment and detachment [46].

Leukocyte chemotaxis can be modulated also by CG through the binding of the G-protein-coupled formyl peptide receptor (FPR) [47]. FPR is present on the surface of neutrophils and macrophages and has been involved in the pathogenesis of COPD [48]. Subcutaneous injection of CG induces the recruitment of monocytes and neutrophils that can be abolished by neutrophil serine protease inhibitors [49]. This suggests that CG is a potent chemoattractant that can modulate in vivo the chemotaxis of different leukocyte populations.

Another family of surface receptor that can be activated, or inactivated, by neutrophil serine proteases are protease-activated receptors (PARs) [50]. These receptors may play important role in lung inflammation mediating a series of pro- or anti-inflammatory cellular responses. NE, CG, and PR3 can display opposite effects on PARs, activating or disarming them through different cleavage sites (this topic is covered in Chapter "Proteolytic Enzymes and Cell Signaling: Pharmacological Lessons" by Hollenberg et al.).

Shedding of cell surface receptors by proteases may represent an important mechanism for controlling cytokine bioactivity at the sites of inflammation [43].

As reported in Table 1, NE and CG may cleave urokinase-type plasminogen activator receptor (CD87), involved in cell migration and tissue remodeling, generating soluble chemotactic fragments that can amplify the inflammatory response [51]. NE and CG can also cleave CD14 the main surface receptor for LPS. Shedding of this receptor inhibits LPS-induced cell activation and CXCL8 production [52, 53]. This mechanism can down-modulate the inflammatory reaction through a negative feedback. NE can also influence the recognition and/or removal of apoptotic cells by cleaving CD14 [54] and/or the phosphatidylserine receptor [55], respectively. The impairment of apoptotic cell removal may contribute to the chronicity of the inflammatory reaction.

The interaction between complement and neutrophils is a crucial event in acute inflammatory reaction. The generation of biologically active fragments from complement cascade can induce a series of important phenomena (vasodilation, inflammatory cell migration, activation of coagulation cascade, plasmin activation, etc.) that may amplify the inflammatory reaction and may culminate in the destruction of pathogens. Several studies suggest that serine proteases from neutrophils have the potential to control the complement cascade by inhibiting complement activation (and/or the responsiveness of neutrophils) or by generating active fragments after the cleavage of single components of the cascade. In particular, NE is able to cleave the transmembrane complement receptor 1 (CR1, also known as CD35 or C3b/C4b receptor), present on many hematopoietic cells, releasing a functional soluble fragment (sCR1) that inhibits complement activation, thus preventing further amplification of the inflammatory response [56, 57]. NE and CG have also been shown to down-regulate the acute inflammatory reaction by inhibiting the responsiveness of neutrophils to C5a in terms of enzyme release and chemotaxis [58]. In contrast, important pro-inflammatory active fragments can be generated from the proteolysis of C3 and C5 by membrane bound CG and NE, respectively [59, 60]. Further evidence for a role of neutrophil serine proteases in the regulation of acute inflammation comes from studies demonstrating that CG and NE can release CD23 soluble fragments stimulating monocytes to produce oxidative burst and pro-inflammatory cytokines [61].

Studies suggest that neutrophil serine proteases may play specific functions in leukocyte adhesion and transmigration during inflammation, however how these proteases participate remains controversial. Recently, it has been demonstrated that (1) NE plays an important role in neutrophil migration in response to zymosan particles in vivo [62], (2) NE cooperates with platelet/endothelial-cell adhesion molecule 1 (PECAM-1) and alpha6 integrins in mediating neutrophil migration through the peri-vascular environment [63], and (3) CG and NE are able to cleave vascular endothelial cadherins (E-cadherins) [64, 65], vascular cell adhesion molecule 1 (VCAM-1) [66], and intercellular adhesion molecule 1 (ICAM-1) [67, 68]. The extracellular cleavage of these molecules may induce formation of gaps through which neutrophils transmigrate.

These studies altogether suggest that, rather than degradative enzymes, neutrophil serine proteases may represent important regulators of the local immune response and potential targets for therapeutic interventions.

2.2 Serine Proteinases from Other Immune Cells

Mast cells (MCs) produce and secrete large amounts of peptidases, in particular tryptases and chymases, which are stored in dense granules. Recent works suggest that these peptidases contribute to the innate as well as adaptive arms of host defenses, protecting lung and airways from certain infections, while in some cases augment the deleterious pathological responses. MC peptidases can be pro- or anti-inflammatory depending on timing and context [69].

MC contributes to recruit neutrophils to sites of inflammation by releasing TNF-α and other factors [69]. In particular β tryptase (but not α tryptase) produces neutrophilic inflammation (neutrophilic pneumonitis) when injected intratracheally in experimental animals [70, 71]. The mechanism of neutrophil recruitment has been related to the stimulation of endothelial cells to secrete neutrophil chemokines [69]. Like CG, the chymotryptic chymases belong to the same family of immune serine peptidases whose human members include granzymes A, B, H, and M of cytolytic lymphocytes and natural killer cells, elastase, proteinase 3 of neutrophils and monocytes, and factor D of the complement system. The role of chymases in lung inflammation is controversial. These enzymes differ very much in substrate preferences. It has been reported that they are capable to cleave endogenous immune proteins such as IL-6 and IL-13 [72].

A more balanced picture of their role in lung physiologic and pathological processes now replaces the old notion that MC peptidases are deleterious, pro-inflammatory enzymes. Their roles in innate immunity and host defense have been carefully reviewed by Trivedi and Caughey [69].

Granzymes (GRs) are granule-associated enzymes that are predominantly expressed by *CD8+ T lymphocytes* and are stored in their lytic granules. These enzymes markedly differ in their substrate specificity and have been involved in processes ranging from cell death to ECM cleavage to viral inactivation [73]. It has been known for years that granzymes accumulate in the extracellular milieu in many chronic inflammatory settings [74]. The accumulation of active granzyme has been widely considered to be a consequence of inflammation rather than a cause of disease [75]. The interest for granzymes in lung inflammation derives from studies carried out in COPD demonstrating the shift to a predominant type I immune response that induces chronic airway inflammation with an increase in the number of CD8+ cells and release of perforin and granzymes [76, 77]. A similar picture was observed in smoking mice that develop pulmonary emphysema [76]. Recent studies carried out in COPD patients demonstrated that CD8+ cells produce more perforin [78] and that CD8+ cells, CD57+ NK cells, type II pneumocytes, and alveolar macrophages express granzyme A and B (GRA and GRB) [74, 79]. Additionally, preliminary studies suggest that GRB expression is associated with increased COPD severity and GRA may promote inflammatory cytokine production [80]. Unfortunately, little is known about the contribution of other members of the same family of granzymes to lung immune reactions. Research in this area has only just begun, and additional basic and clinical studies are necessary to increase our current understanding of granzyme biology.

3 Lung Metalloproteinases

The matrix metalloproteinases (MMPs) belong to a family of secreted and cell surface-bound zinc-dependent endopeptidases, which process a large array of extracellular and cell surface proteins under normal and pathological conditions [81]. These enzymes can be subdivided into several groups according to substrate specificity [43]. These include: (a) collagenases (MMP-1, MMP-8, and MMP-13); (b) stromelysins (MMP-3, MMP-10, and MMP-11) which have a broad spectrum of activity against ECM proteins; (c) gelatinases (MMP-2 and MMP-9) which degrade gelatins (denatured collagens), elastin, and basement membrane proteins; (d) metalloelastases (MMP-7 and MMP-12) which also have a broad spectrum of susceptible ECM substrates; and finally (e) "membrane-type" MMPs, also known "integral membrane" MMPs (MT-MMPs) which include a series of MT-MMPs (MT1-MMP, MT2-MMP, MT3-MMP, MT5-MMP, MT6-MMP) and ADAMs, a family of at least 35 members which have a disintegrin as well as a metalloproteinase domain [43]. MMPs are synthesized from cells as latent pro-enzymes, whose activation in vitro can be achieved by various proteinases and reactive oxygen species [81]. Unfortunately, the mechanism of extracellular activation of pro-MMPs in vivo is not clear. In the lung, under in vivo conditions, some reactive oxygen species may inactivate rather than activate MMPs [82].

The main function of MMPs in vivo was initially thought to be in ECM remodeling. Today, several lines of evidence suggest that MMPs have evolved to serve broad functions in defense, injury, inflammation, or repair [83]. In particular, MMPs are now being recognized as one of the critical mediators for host defense [84], either by promoting or repressing inflammation, through the direct proteolytic processing of inflammatory cytokines, chemokines, clotting factors, regulators of angiogenesis and defensins, to either increase or decrease their biological functions (Table 2) [85–87].

Although not all members of the MMP family are found within the pulmonary tissue, an increase in the expression of MMPs has been observed in most inflammatory lung diseases and has been involved in the development and progression of a number of pulmonary diseases such as COPD, emphysema, and asthma [81]. It should be emphasized that, under various conditions, different types of inflammatory and resident lung cells express MMPs, which regulate cell function via the alteration of ECM composition and the direct modulation of the bioavailability of mediators of inflammation, or indirectly by changing the biological properties of proteins after a limited proteolysis [43]. For example, an increase of pro-inflammatory mediators is obtained by the proteolytic shedding of CXCL1 from syndecans on the surface of lung epithelial cells by MMP-7 [88], by the enzymatic activation of CXCL-5 by MMP-8 [89] and MMP-9 [84], or by shedding and activation of pro-

Table 2 Overview of metalloproteases and their products involved in modulation of inflammation.

Enzymes	Targets	Activity	Biological effects
MMP-12	Elastin	Proteolysis	Chemotactic fragments for blood monocytes
	TNF-α	Limited proteolysis	Pro-TNF-α activation
	Plasminogen	Limited proteolysis	Angiostatin generation
MMP-7	CXCL1(IL-8, KC)	Limited proteolysis	CXCL1 shedding from syndecans, PMN migration
	Defensins	Limited proteolysis	Activation. Promotion of host defense
	E-Cadherin	Limited proteolysis	Epithelial repair
MMP-8	CCL3 (MIP-1α)	Proteolysis	Abrogation of macrophage chemotaxis and activation
	CXCL5 (ENA-78)	Limited proteolysis	PMN chemotaxis: enhanced efficiency
	CXCL5-6 (LIX)	Limited proteolysis	PMN chemotaxis: enhanced efficiency
MMP-1	TIMP-2	Proteolysis	Inactivation
MMP-9	CXCL8	Limited proteolysis	Fragments with more potent chemoattractant activity
	CXCL5 (ENA-78)	Limited proteolysis	PMN chemotaxis: enhanced efficiency
MMP-2	CCL7 (MCP3)	Limited proteolysis	Truncated form functioning as CC chemokine receptor antagonist
MMP-1, MMP-9	ECM	Proteolysis	Release of *N*-acetyl Pro-Gly-Pro (PGP peptide) promoting PMN influx
MMP-2, MMP-3, MMP-9	IL-1β	Limited proteolysis	Pro-IL-1β activation
MMP-1, MMP-2, MMP-3, MMP-9	IL-1β	Proteolysis	Degradation
MMP-8, MMP-9	Necrotic tissue	Proteolysis	Removal of debris during wound healing
MMP-1, MMP-2, MMP-3, MMP-9, MMP-13, MMP-14	CXCL12 (SDF-1)	Proteolysis	Inhibition of T-lymphocyte migration, loss of chemokine ability to bind CXCR4
MMP-1, MMP-3, MMP-13, MMP-14	CCL2 (MCP1)	Limited proteolysis	Truncated form functioning as CC chemokine receptor antagonist
	CCL8 (MCP2)	Limited proteolysis	Truncated form functioning as CC chemokine receptor antagonist
	CCL13 (MCP4)	Limited proteolysis	Truncated form functioning as CC chemokine receptor antagonist

(continued)

Table 2 (continued)

Enzymes	Targets	Activity	Biological effects
MMP-2, MMP-7, MMP9, MMP-12	α-1 proteinase inhibitor	Proteolysis	Inactivation. Generation of fragment with PMN chemoattractant activity
ADAM 17	TNF-α	Limited proteolysis	Activation
ADAM 10, ADAM 17	Fractalkine (CX_3CL1)	Limited proteolysis	Activation. Monocytes chemotaxis
	L-selectin	Limited proteolysis	Neutrophil migration

TNF-α by MMP-12 from macrophage surfaces of mice exposed to cigarette smoke [90]. These mediators involved in chemotaxis and neutrophil activation can amplify lung neutrophilic inflammation. In contrast, the same or other MMPs can dampen lung inflammation by cleaving and activating defensins (MMP-7) [91, 92] or by cleaving plasminogen to generate angiostatin (MMP-12) [89, 93].

The potential dual role played by MMPs in either promoting or repressing inflammation derives from investigations into the MMP regulation of interleukin-1β (IL-1β) activity. Similar to TNF-α, IL-1β is a potent pro-inflammatory cytokine that requires proteolytic processing for activation. MMP2, MMP3, and MMP9 can generate the active 17-kDa pro-inflammatory form from IL-1β precursor [94], whereas MMP3, MMP1, and MMP9 can proteolytically inactivate the mature form of this cytokine [95].

Surface-bound MMPs have been thought to degrade ECM proteins, as efficiently as the soluble form of proteinases, and to participate to the remodeling and destruction of tissue in several lung diseases. However, surface-bound MMP-8 and MMP9 on activated neutrophils have been reported to contribute to resolution of lung inflammatory responses by removing debris during wound healing and/or by inactivating pro-inflammatory mediators [96, 97]. In particular, MMP-8 may limit lung inflammation and alveolar capillary barrier injury during acute lung injury by inactivating MIP-1α [98].

Different cell types including epithelial and endothelial cells, fibroblasts, smooth muscle cells, and leukocytes express the enzymes belonging to MT-MMP family. Among inflammatory cells, MT1-MMP is expressed by monocytes, macrophages, and dendritic cells, MT4-MMP by eosinophils, whereas MT6-MMP is expressed only in neutrophils [43]. Like other MMPs, MT-MMPs contribute to ECM remodeling, angiogenesis, cell migration, and tumor invasion and metastasis. Additionally, MT-MMPs may also modulate inflammatory responses by degrading proteinase inhibitors [99] or by cleaving and activating TNF-α [100].

Enzymes belonging to the ADAM (a disintegrin and a metalloproteinase) family are expressed in a variety of mammalian tissues. ADAM-8, -10, -15, -17, and -28 are expressed in leukocytes and their level of expression is regulated by various agonists including cytokines, chemokines, phorbol esters, and growth factors [43]. In particular, ADAM-10 and -17 are up-regulated by IL-1α and TNF-α [101–103], whereas ADAM-8 is stored as a preformed proteinase in neutrophil-specific and gelatinase granules. This enzyme translocates to the cell surface under stimulation of neutrophil with phorbol esters [104].

The major biological activities of ADAMs are linked to their multi-domain structure, and include roles for shedding cell surface protein (metalloproteinase domain), regulating cell adhesion and migration (disintegrin domain), promoting cell fusion (cysteine-rich and disintegrin domains), and finally for signaling intracellular events (cytoplasmic tail) [43].

The role of ADAMs in regulating leukocyte adhesion and migration, and in transduction of intracellular signaling events in leukocyte or other immune cells is not clear. It has become evident that the sheddase activities of ADAMs play a role in regulating lung inflammation by the direct proteolytic processing of inflammatory cytokines and chemokines, or by cleaving cytokine and chemokine receptors [43] or by inducing mucin production [105, 106].

The best-known example of sheddase activity relevant for inflammation is that of ADAM-17 (also known as TNF-α convertase or TACE). Pro-TNF-α (26 kDa protein), expressed on the surface of macrophages, PMNs, and other cells, is activated by cleavage to a soluble 17 kDa cytokine by ADAM-17 on the cell surfaces [34, 107]. Other membrane-associated proteins cleaved by ADAMs include a series of other cytokines and their receptors, adhesion molecules, growth factors and their receptors (see Table 2), which have been involved in inflammation and in repair processes.

Of note, ADAM-10 and ADAM-17 cleave the membrane-anchored adhesion molecule fractalkine (CX_3CL1) [108, 109], that functions as chemoattractant for monocytes [110], and other adhesion molecules including L-selectin, constitutively expressed by neutrophil and implicated in directing these cells to sites of inflammation [111, 112]. In lung diseases, ADAM-8 and ADAM-33 have been implicated in asthma [113–115], however, the mechanisms by which these enzymes contribute to the pathogenesis have not been elucidated.

Like serine proteases, the activity of metalloproteases can be regulated and balanced by α2M and the four members of tissue inhibitors of metalloproteases family (TIMPs), which are synthesized by leukocytes and connective tissue cells (this topic is covered by other contributors in this book).

4 Concluding Remarks

An increased expression of members of different classes of proteases is seen in almost every human tissue in which inflammation is present, including the lung. Proteases have evolved as important regulatory enzymes in both pro- and anti-inflammatory pathways. There is growing evidence that these enzymes function in lung inflammatory processes primarily modulating neutrophil influx either through regulation of barrier function, cytokine/chemokine activation/inactivation, or regulating leukocyte function through cell surface binding sites.

Although inflammatory processes are essential for host defense, when excessive they can contribute to tissue injury, organ dysfunction and lung diseases such as COPD, emphysema and asthma only to mention a few. Because experimental evidence suggests that proteases can be both beneficial in regulating host defense and detrimental in inflammatory disease, it is necessary to define the specific molecular mechanisms by which individual proteases function in normal and unregulated inflammatory processes.

In this short review, we outlined the biology of the proteases potentially important in lung inflammatory processes and their roles (in most cases dual) in regulating lung inflammation. Despite the fact that new important insights have been done into the mechanisms of action of several proteases in inflammatory conditions of human lung, it is not still clear whether protease inhibition is beneficial or harmful, and in which situation it works, or may be of clinical help.

Broadly speaking, proteases contribute to inflammatory processes; however, the positive or negative contribution of a certain enzymes may depend on biological context such as location, substrate availability, cell type, and disease state.

References

1. Cross CE, Eiserich JP, Halliwell B (1997) General biological consequences of inhaled environmental toxicants. In: Crystal RG, West JB, Weibel ER, Barnes PJ (eds) The lung, 2nd edn. Lippincott-Raven, Philadelphia, pp 2421–2437
2. Suzuki T, Chow CW, Downey GP (2008) Role of innate immune cells and their products in lung immunopathology. Int J Biochem Cell Biol 40:1348–1361
3. Parsons PE, Wothen GS, Henson PM (1997) Injury from inflammatory cells. In: Crystal RG, West JB, Weibel ER, Barnes PJ (eds) The lung, 2nd edn. Lippincott-Raven, Philadelphia, pp 2439–2449
4. Greene CM, McElvaney NG (2009) Protease and antiproteases in chronic neutrophilic lung disease-relevance to drug discovery. Br J Pharmacol 158:1048–1058
5. Nathan C (2006) Neutrophils and immunity: challenges and opportunities. Nat Rev Immunol 6:173–182
6. Pham CT (2008) Neutrophil serine proteases fine-tune the inflammatory response. Int J Biochem Cell Biol 40:1317–1333
7. Newburger PE (2006) Disorders of neutrophil number and function. Hematology Am Soc Hematol Educ Program:104–110
8. Belaaouaj A, Kim KS, Shapiro SD (2000) Degradation of outer membrane protein A in *Escherichia coli* killing by neutrophil elastase. Science 289:1185–1188
9. Belaaouaj A, McCarthy R, Baumann M, Gao Z, Ley TJ, Abraham SN et al (1998) Mice lacking neutrophil elastase reveal impaired host defense against Gram negative bacterial sepsis. Nat Med 4:615–618
10. Reeves EP, Lu H, Jacobs HL, Messina CG, Bolsover S, Gabella G et al (2002) Killing activity of neutrophils is mediated through activation of proteases by K + flux. Nature 416:291–297
11. Holt OJ, Gallo F, Griffiths GM (2006) Regulating secretory lysosomes. J Biochem 140:7–12
12. Kaplan J, De Domenico I, Ward DM (2008) Chediak-Higashi syndrome. Curr Opin Hematol 15:22–29
13. de Haar SF, Hiemstra PS, van Steenbergen MT, Everts V, Beertsen W (2006) Role of polymorphonuclear leukocyte-derived serine proteases in defense against *Actinobacillus actinomycetemcomitans*. Infect Immun 74:5284–5291

14. Mc Elvaney NG, Crystal RG (1997) Antiproteases and Lung Defense. In: Crystal RG, West JB, Weibel ER, Barnes PJ (eds) The lung, 2nd edn. Lippincott-Raven, Philadelphia, pp 2219–2235
15. Owen CA, Campbell EJ (1995) Neutrophil proteinases and matrix degradation. The cell biology of pericellular proteolysis. Semin Cell Biol 6:367–376
16. Ginzberg HH, Cherapanov V, Dong Q, Cantin A, McCulloch CA, Shannon PT, Downey GP (2001) Neutrophil-mediated epithelial injury during transmigration: role of elastase. Am J Physiol Gastrointest Liver Physiol 281:G705–G717
17. Suzuki T, Moraes TJ, Vachon E, Ginzberg HH, Huang TT, Matthay MA et al (2005) Proteinase-activated receptor-1 mediates elastase-induced apoptosis of human lung epithelial cells. Am J Respir Cell Mol Biol 33:231–247
18. Pendergraft WF 3rd, Rudolph EH, Falk RJ, Jahn JE, Grimmler M, Hengst L, Jennette JC, Preston GA (2004) Proteinase 3 sidesteps caspases and cleaves p21(Waf1/Cip1/Sdi1) to induce endothelial cell apoptosis. Kidney Int 65:75–84
19. Gilmore AP (2005) Anoikis. Cell Death Differ 12(Suppl 2):1473–1477
20. Snider GL (1989) Chronic obstructive pulmonary disease: risk factors, pathophysiology and pathogenesis. Annu Rev Med 40:411–429
21. Lee WL, Downey GP (2001) Leukocyte elastase: physiological functions and role in acute lung injury. Am J Respir Crit Care Med 164:896–904
22. Fischer BM, Voynow JA (2002) Neutrophil elastase induces MUC5AC gene expression in airway epithelium via a pathway involving reactive oxygen species. Am J Respir Cell Mol Biol 26:447–452
23. Lungarella G, Cavarra E, Lucattelli M, Martorana PA (2008) The dual role of neutrophil elastase in lung destruction and repair. Int J Biochem Cell Biol 40:1287–1296
24. Padrines M, Wolf M, Walz A, Baggiolini M (1994) Interleukin-8 processing by neutrophil elastase, cathepsin G and proteinase-3. FEBS Lett 352:231–235
25. Nufer O, Corbett M, Walz A (1999) Amino-terminal processing of chemokine ENA-78 regulates biological activity. Biochemistry 38:636–642
26. Richter R, Bistrian R, Escher S, Forssmann WG, Vakili J, Henschler R, Spodsberg N, Frimpong-Boateng A, Forssmann U (2005) Quantum proteolytic activation of chemokine CCL15 by neutrophil granulocytes modulates mononuclear cell adhesiveness. J Immunol 175:1599–1608
27. Schiemann F, Grimm TA, Hoch J, Gross R, Lindner B, Petersen F, Bulfone-Paus S, Brandt E (2006) Mast cells and neutrophils proteolytically activate chemokine precursor CTAP-III and are subject to counterregulation by PF-4 through inhibition of chymase and cathepsin G. Blood 107:2234–2242
28. Wittamer V, Bondue B, Guillabert A, Vassart G, Parmentier M, Communi D (2005) Neutrophil-mediated maturation of chemerin: a link between innate and adaptive immunity. J Immunol 175:487–493
29. Lim JK, Lu W, Hartley O, DeVico AL (2006) N-terminal proteolytic processing by cathepsin G converts RANTES/CCL5 and related analogs into a truncated 4–68 variant. J Leukoc Biol 80:1395–1404
30. Ryu OH, Choi SJ, Firatli E, Choi SW, Hart PS, Shen RF, Wang G, Wu WW, Hart TC (2005) Proteolysis of macrophage inflammatory protein-1alpha isoforms LD78beta and LD78alpha by neutrophil-derived serine proteases. J Biol Chem 280:17415–17421
31. Delgado MB, Clark-Lewis I, Loetscher P, Langen H, Thelen M, Baggiolini M, Wolf M (2001) Rapid inactivation of stromal cell-derived factor-1 by cathepsin G associated with lymphocytes. Eur J Immunol 31:699–707
32. Rao RM, Betz TV, Lamont DJ, Kim MB, Shaw SK, Froio RM, Baleux F, Arenzana-Seisdedos F, Alon R, Luscinskas FW (2004) Elastase release by transmigrating neutrophils deactivates endothelial-bound SDF-1alpha and attenuates subsequent T lymphocyte transendothelial migration. J Exp Med 200:713–724

33. Valenzuela-Fernández A, Planchenault T, Baleux F, Staropoli I, Le-Barillec K, Leduc D, Delaunay T, Lazarini F, Virelizier JL, Chignard M, Pidard D, Arenzana-Seisdedos F (2002) Leukocyte elastase negatively regulates stromal cell-derived factor-1 (SDF-1)/CXCR4 binding and functions by amino-terminal processing of SDF-1 and CXCR4. J Biol Chem 277:15677–15689
34. Black RA, Rauch CT, Kozlosky CJ, Peschon JJ, Slack JL, Wolfson MF, Castner BJ, Stocking KL, Reddy P, Srinivasan S, Nelson N, Boiani N, Schooley KA, Gerhart M, Davis R, Fitzner JN, Johnson RS, Paxton RJ, March CJ, Cerretti DP (1997) A metalloproteinase disintegrin that releases tumour-necrosis factor-alpha from cells. Nature 385:729–733
35. Coeshott C, Ohnemus C, Pilyavskaya A, Ross S, Wieczorek M, Kroona H, Leimer AH, Cheronis J (1999) Converting enzyme-independent release of tumor necrosis factor alpha and IL-1beta from a stimulated human monocytic cell line in the presence of activated neutrophils or purified proteinase 3. Proc Natl Acad Sci USA 96:6261–6266
36. Robache-Gallea S, Morand V, Bruneau JM, Schoot B, Tagat E, Réalo E, Chouaib S, Roman-Roman S (1995) In vitro processing of human tumor necrosis factor-alpha. J Biol Chem 270:23688–23692
37. Monack DM, Hersh D, Ghori N, Bouley D, Zychlinsky A, Falkow S (2000) Salmonella exploits caspase-1 to colonize Peyer's patches in a murine typhoid model. J Exp Med 192:249–258
38. Sugawara S, Uehara A, Nochi T, Yamaguchi T, Ueda H, Sugiyama A, Hanzawa K, Kumagai K, Okamura H, Takada H (2001) Neutrophil proteinase 3-mediated induction of bioactive IL-18 secretion by human oral epithelial cells. J Immunol 167:6568–6575
39. Gu Y, Kuida K, Tsutsui H, Ku G, Hsiao K, Fleming MA, Hayashi N, Higashino K, Okamura H, Nakanishi K, Kurimoto M, Tanimoto T, Flavell RA, Sato V, Harding MW, Livingston DJ, Su MS (1997) Activation of interferon-gamma inducing factor mediated by interleukin-1beta converting enzyme. Science 275:206–209
40. Robertson SE, Young JD, Kitson S, Pitt A, Evans J, Roes J, Karaoglu D, Santora L, Ghayur T, Liew FY, Gracie JA, McInnes IB (2006) Expression and alternative processing of IL-18 in human neutrophils. Eur J Immunol 36:722–731
41. Bank U, Küpper B, Ansorge S (2000) Inactivation of interleukin-6 by neutrophil proteases at sites of inflammation. Protective effects of soluble IL-6 receptor chains. Adv Exp Med Biol 477:431–437
42. He JQ, Foreman MG, Shumansky K, Zhang X, Akhabir L, Sin DD, Man SF, DeMeo DL, Litonjua AA, Silverman EK, Connett JE, Anthonisen NR, Wise RA, Paré PD, Sandford AJ (2009) Associations of IL6 polymorphisms with lung function decline and COPD. Thorax 64:698–704
43. Owen CA (2008) Leukocyte cell surface proteinases: regulation of expression, functions, and mechanisms of surface localization. Int J Biochem Cell Biol 40:1246–1272
44. Raptis SZ, Shapiro SD, Simmons PM, Cheng AM, Pham CT (2005) Serine protease cathepsin G regulates adhesion-dependent neutrophil effector functions by modulating integrin clustering. Immunity 22:679–691
45. Nathan C, Xie QW, Halbwachs-Mecarelli L, Jin WW (1993) Albumin inhibits neutrophil spreading and hydrogen peroxide release by blocking the shedding of CD43 (sialophorin, leukosialin). J Cell Biol 122:243–256
46. Cai TQ, Wright SD (1996) Human leukocyte elastase is an endogenous ligand for the integrin CR3 (CD11b/CD18, Mac-1, alpha M beta 2) and modulates polymorphonuclear leukocyte adhesion. J Exp Med 184:1213–1223
47. Sun R, Iribarren P, Zhang N, Zhou Y, Gong W, Cho EH, Lockett S, Chertov O, Bednar F, Rogers TJ, Oppenheim JJ, Wang JM (2004) Identification of neutrophil granule protein cathepsin G as a novel chemotactic agonist for the G protein-coupled formyl peptide receptor. J Immunol 173:428–436
48. Stockley RA, Grant RA, Llewellyn-Jones CG, Hill SL, Burnett D (1994) Neutrophil formyl-peptide receptors. Relationship to peptide-induced responses and emphysema. Am J Respir Crit Care Med 149:464–468

49. Chertov O, Ueda H, Xu LL, Tani K, Murphy WJ, Wang JM, Howard OM, Sayers TJ, Oppenheim JJ (1997) Identification of human neutrophil-derived cathepsin G and azurocidin/CAP37 as chemoattractants for mononuclear cells and neutrophils. J Exp Med 186:739–747
50. Coughlin SR, Camerer E (2003) Participation in inflammation. J Clin Invest 111:25–27
51. Beaufort N, Leduc D, Rousselle JC, Magdolen V, Luther T, Namane A, Chignard M, Pidard D (2004) Proteolytic regulation of the urokinase receptor/CD87 on monocytic cells by neutrophil elastase and cathepsin G. J Immunol 172:540–549
52. Le-Barillec K, Si-Tahar M, Balloy V, Chignard M (1999) Proteolysis of monocyte CD14 by human leukocyte elastase inhibits lipopolysaccharide-mediated cell activation. J Clin Invest 103:1039–1046
53. Le-Barillec K, Pidard D, Balloy V, Chignard M (2000) Human neutrophil cathepsin G down-regulates LPS-mediated monocyte activation through CD14 proteolysis. J Leukoc Biol 68:209–215
54. Henriksen PA, Devitt A, Kotelevtsev Y, Sallenave JM (2004) Gene delivery of the elastase inhibitor elafin protects macrophages from neutrophil elastase-mediated impairment of apoptotic cell recognition. FEBS Lett 574:80–84
55. Vandivier RW, Fadok VA, Hoffmann PR, Bratton DL, Penvari C, Brown KK, Brain JD, Accurso FJ, Henson PM (2002) Elastase-mediated phosphatidylserine receptor cleavage impairs apoptotic cell clearance in cystic fibrosis and bronchiectasis. J Clin Invest 109:661–670
56. Hamacher J, Sadallah S, Schifferli JA, Villard J, Nicod LP (1998) Soluble complement receptor type 1 (CD35) in bronchoalveolar lavage of inflammatory lung diseases. Eur Respir J 11:112–119
57. Sadallah S, Hess C, Miot S, Spertini O, Lutz H, Schifferli JA (1999) Elastase and metalloproteinase activities regulate soluble complement receptor 1 release. Eur J Immunol 29:3754–3761
58. Tralau T, Meyer-Hoffert U, Schröder JM, Wiedow O (2004) Human leukocyte elastase and cathepsin G are specific inhibitors of C5a-dependent neutrophil enzyme release and chemotaxis. Exp Dermatol 13:316–325
59. Maison CM, Villiers CL, Colomb MG (1991) Proteolysis of C3 on U937 cell plasma membranes. Purification of cathepsin G. J Immunol 147:921–926
60. Vogt W (2000) Cleavage of the fifth component of complement and generation of a functionally active C5b6-like complex by human leukocyte elastase. Immunobiology 201:470–477
61. Brignone C, Munoz O, Batoz M, Rouquette-Jazdanian A, Cousin JL (2001) Proteases produced by activated neutrophils are able to release soluble CD23 fragments endowed with proinflammatory effects. FASEB J 15:2027–2029
62. Young RE, Thompson RD, Larbi KY, La M, Roberts CE, Shapiro SD, Perretti M, Nourshargh S (2004) Neutrophil elastase (NE)-deficient mice demonstrate a nonredundant role for NE in neutrophil migration, generation of proinflammatory mediators, and phagocytosis in response to zymosan particles in vivo. J Immunol 172:4493–4502
63. Wang S, Dangerfield JP, Young RE, Nourshargh S, Wang S, Dangerfield JP, Young RE, Nourshargh S (2005) PECAM-1, alpha6 integrins and neutrophil elastase cooperate in mediating neutrophil transmigration. J Cell Sci 118:2067–2076
64. Hermant B, Bibert S, Concord E, Dublet B, Weidenhaupt M, Vernet T, Gulino-Debrac D (2003) Identification of proteases involved in the proteolysis of vascular endothelium cadherin during neutrophil transmigration. J Biol Chem 278:14002–14012
65. Mayerle J, Schnekenburger J, Krüger B, Kellermann J, Ruthenbürger M, Weiss FU, Nalli A, Domschke W, Lerch MM (2005) Extracellular cleavage of E-cadherin by leukocyte elastase during acute experimental pancreatitis in rats. Gastroenterol 129:1251–1267
66. Lévesque JP, Takamatsu Y, Nilsson SK, Haylock DN, Simmons PJ (2001) Vascular cell adhesion molecule-1 (CD106) is cleaved by neutrophil proteases in the bone marrow following hematopoietic progenitor cell mobilization by granulocyte colony-stimulating factor. Blood 98:1289–1297

67. Champagne B, Tremblay P, Cantin A, St PY (1998) Proteolytic cleavage of ICAM-1 by human neutrophil elastase. J Immunol 161:6398–6405
68. Robledo O, Papaioannou A, Ochietti B, Beauchemin C, Legault D, Cantin A, King PD, Daniel C, Alakhov VY, Potworowski EF, St-Pierre Y (2003) ICAM-1 isoforms: specific activity and sensitivity to cleavage by leukocyte elastase and cathepsin G. Eur J Immunol 33:1351–1360
69. Trivedi NN, Caughey GH (2010) Mast cell peptidases: chameleons of innate immunity and host defense. Am J Respir Cell Mol Biol 42:257–267
70. McNeil HP, Adachi R, Stevens RL (2007) Mast cell-restricted tryptases: structure and function in inflammation and pathogen defense. J Biol Chem 282:20785–20789
71. Huang C, De Sanctis GT, O'Brien PJ, Mizgerd JP, Friend DS, Drazen JM, Brass LF, Stevens RL (2001) Evaluation of the substrate specificity of human mast cell tryptase beta I and demonstration of its importance in bacterial infections of the lung. J Biol Chem 276:26276–26284
72. Zhao W, Oskeritzian CA, Pozez AL, Schwartz LB (2005) Cytokine production by skin-derived mast cells: endogenous proteases are responsible for degradation of cytokines. J Immunol 175:2635–2642
73. Chowdhury D, Lieberman J (2008) Death by a thousand cuts: granzyme pathways of programmed cell death. Annu Rev Immunol 26:389–420
74. Hendel A, Hiebert PR, Boivin WA, Williams SJ, Granville DJ (2010) Granzymes in age-related cardiovascular and pulmonary diseases. Cell Death Differ 17:596–606
75. Boivin WA, Cooper DM, Hiebert PR, Granville DJ (2009) Intracellular versus extracellular granzyme B in immunity and disease: challenging the dogma. Lab Invest 89:1195–1220
76. Maeno T, Houghton AM, Quintero PA, Grumelli S, Owen CA, Shapiro SD, Maeno T, Houghton AM, Quintero PA, Grumelli S, Owen CA, Shapiro SD (2007) CD8+ T Cells are required for inflammation and destruction in cigarette smoke-induced emphysema in mice. J Immunol 178:8090–8096
77. Chrysofakis G, Tzanakis N, Kyriakoy D, Tsoumakidou M, Tsiligianni I, Klimathianaki M, Siafakas NM (2004) Perforin expression and cytotoxic activity of sputum CD8+ lymphocytes in patients with COPD. Chest 125:71–76
78. Hodge S, Hodge G, Nairn J, Holmes M, Reynolds PN (2006) Increased airway granzyme b and perforin in current and ex-smoking COPD subjects. COPD 3:179–187
79. Granville DJ (2010) Granzymes in disease: bench to bedside. Cell Death Differ 17:565–566
80. Ngan DA, Vickerman SV, Granville DJ, Man SF, Sin DD (2009) The possible role of granzyme B in the pathogenesis of chronic obstructive pulmonary disease. Ther Adv Respir Dis 3:113–129
81. Greenlee KJ, Werb Z, Kheradmand F (2007) Matrix metalloproteinases in lung: multiple, multifarious, and multifaceted. Physiol Rev 87:69–98
82. Owens MW, Milligan SA, Jourd'heuil D, Grisham MB (1997) Effects of reactive metabolites of oxygen and nitrogen on gelatinase A activity. Am J Physiol 273:L445–L450
83. Parks WC, Wilson CL, López-Boado YS (2004) Matrix metalloproteinases as modulators of inflammation and innate immunity. Nat Rev Immunol 4:617–629
84. Manicone AM, McGuire JK (2008) Matrix metalloproteinases as modulators of inflammation. Semin Cell Dev Biol 19:34–41
85. McQuibban GA, Butler GS, Gong JH, Bendall L, Power C, Clark-Lewis I, Overall CM (2001) Matrix metalloproteinase activity inactivates the CXC chemokine stromal cell-derived factor-1. J Biol Chem 276:43503–43508
86. McQuibban GA, Gong JH, Tam EM, McCulloch CA, Clark-Lewis I, Overall CM (2000) Inflammation dampened by gelatinase A cleavage of monocyte chemoattractant protein-3. Science 289:1202–1206
87. McQuibban GA, Gong JH, Wong JP, Wallace JL, Clark-Lewis I, Overall CM (2002) Matrix metalloproteinase processing of monocyte chemoattractant proteins generates CC chemokine receptor antagonists with anti-inflammatory properties in vivo. Blood 100:1160–1167

88. Li Q, Park PW, Wilson CL, Parks WC (2002) Matrilysin shedding of syndecan-1 regulates chemokine mobilization and transepithelial efflux of neutrophils in acute lung injury. Cell 111:635–646
89. Balbín M, Fueyo A, Tester AM, Pendás AM, Pitiot AS, Astudillo A, Overall CM, Shapiro SD, López-Otín C (2003) Loss of collagenase-2 confers increased skin tumor susceptibility to male mice. Nat Genet 35:252–257
90. Churg A, Wang RD, Tai H, Wang X, Xie C, Dai J, Shapiro SD, Wright JL (2003) Macrophage metalloelastase mediates acute cigarette smoke-induced inflammation via tumor necrosis factor-alpha release. Am J Respir Crit Care Med 167:1083–1089
91. Wilson CL, Heppner KJ, Rudolph LA, Matrisian LM (1995) The metalloproteinase matrilysin is preferentially expressed by epithelial cells in a tissue-restricted pattern in the mouse. Mol Biol Cell 6:851–869
92. Wilson CL, Ouellette AJ, Satchell DP, Ayabe T, López-Boado YS, Stratman JL, Hultgren SJ, Matrisian LM, Parks WC (1999) Regulation of intestinal alpha-defensin activation by the metalloproteinase matrilysin in innate host defense. Science 286:113–117
93. Cornelius LA, Nehring LC, Harding E, Bolanowski M, Welgus HG, Kobayashi DK, Pierce RA, Shapiro SD (1998) Matrix metalloproteinases generate angiostatin: effects on neovascularization. J Immunol 161:6845–6852
94. Schönbeck U, Mach F, Libby P (1998) Generation of biologically active IL-1 beta by matrix metalloproteinases: a novel caspase-1-independent pathway of IL-1 beta processing. J Immunol 161:3340–3346
95. Ito A, Mukaiyama A, Itoh Y, Nagase H, Thogersen IB, Enghild JJ, Sasaguri Y, Mori Y (1996) Degradation of interleukin 1beta by matrix metalloproteinases. J Biol Chem 271:14657–14660
96. Gueders MM, Balbin M, Rocks N, Foidart JM, Gosset P, Louis R, Shapiro S, Lopez-Otin C, Noël A, Cataldo DD (2005) Matrix metalloproteinase-8 deficiency promotes granulocytic allergen-induced airway inflammation. J Immunol 175:2589–2597
97. Owen CA, Hu Z, Lopez-Otin C, Shapiro SD (2004) Membrane-bound matrix metalloproteinase-8 on activated polymorphonuclear cells is a potent, tissue inhibitor of metalloproteinase-resistant collagenase and serpinase. J Immunol 172:7791–7803
98. Quintero PA, Knolle MD, Cala LF, Zhuang Y, Owen CA (2010) Matrix metalloproteinase-8 inactivates macrophage inflammatory protein-1 alpha to reduce acute lung inflammation and injury in mice. J Immunol 184:1575–1588
99. Maquoi E, Frankenne F, Baramova E, Munaut C, Sounni NE, Remacle A, Noël A, Murphy G, Foidart JM (2000) Membrane type 1 matrix metalloproteinase-associated degradation of tissue inhibitor of metalloproteinase 2 in human tumor cell lines. J Biol Chem 275:11368–11378
100. English WR, Puente XS, Freije JM, Knauper V, Amour A, Merryweather A, Lopez-Otin C, Murphy G (2000) Membrane type 4 matrix metalloproteinase (MMP17) has tumor necrosis factor-alpha convertase activity but does not activate pro-MMP2. J Biol Chem 275:14046–14055
101. Bandyopadhyay S, Hartley DM, Cahill CM, Lahiri DK, Chattopadhyay N, Rogers JT (2006) Interleukin-1alpha stimulates non-amyloidogenic pathway by alpha-secretase (ADAM-10 and ADAM-17) cleavage of APP in human astrocytic cells involving p38 MAP kinase. J Neurosci Res 84:106–118
102. Bzowska M, Jura N, Lassak A, Black RA, Bereta J (2004) Tumour necrosis factor-alpha stimulates expression of TNF-alpha converting enzyme in endothelial cells. Eur J Biochem 271:2808–2820
103. Walcheck B, Herrera AH, St Hill C, Mattila PE, Whitney AR, Deleo FR (2006) ADAM17 activity during human neutrophil activation and apoptosis. Eur J Immunol 36:968–976
104. Gómez-Gaviro M, Domínguez-Luis M, Canchado J, Calafat J, Janssen H, Lara-Pezzi E, Fourie A, Tugores A, Valenzuela-Fernández A, Mollinedo F, Sánchez-Madrid F, Díaz-González F (2007) Expression and regulation of the metalloproteinase ADAM-8 during

human neutrophil pathophysiological activation and its catalytic activity on L-selectin shedding. J Immunol 178:8053–8063

105. Basbaum C, Li D, Gensch E, Gallup M, Lemjabbar H (2002) Mechanisms by which gram-positive bacteria and tobacco smoke stimulate mucin induction through the epidermal growth factor receptor (EGFR). Novartis Found Symp 248:171–176
106. Lemjabbar H, Li D, Gallup M, Sidhu S, Drori E, Basbaum C (2003) Tobacco smoke-induced lung cell proliferation mediated by tumor necrosis factor alpha-converting enzyme and amphiregulin. J Biol Chem 278:26202–26207
107. Moss ML, Jin SL, Milla ME, Bickett DM, Burkhart W, Carter HL, Chen WJ, Clay WC, Didsbury JR, Hassler D, Hoffman CR, Kost TA, Lambert MH, Leesnitzer MA, McCauley P, McGeehan G, Mitchell J, Moyer M, Pahel G, Rocque W, Overton LK, Schoenen F, Seaton T, Su JL, Becherer JD et al (1997) Cloning of a disintegrin metalloproteinase that processes precursor tumour-necrosis factor-alpha. Nature 385:733–736
108. Garton KJ, Gough PJ, Blobel CP, Murphy G, Greaves DR, Dempsey PJ, Raines EW (2001) Tumor necrosis factor-alpha-converting enzyme (ADAM17) mediates the cleavage and shedding of fractalkine (CX3CL1). J Biol Chem 276:37993–38001
109. Hundhausen C, Misztela D, Berkhout TA, Broadway N, Saftig P, Reiss K, Hartmann D, Fahrenholz F, Postina R, Matthews V, Kallen KJ, Rose-John S, Ludwig A (2003) The disintegrin-like metalloproteinase ADAM10 is involved in constitutive cleavage of CX3CL1 (fractalkine) and regulates CX3CL1-mediated cell-cell adhesion. Blood 102: 1186–1195
110. Blaschke S, Koziolek M, Schwarz A, Benöhr P, Middel P, Schwarz G, Hummel KM, Müller GA (2003) Proinflammatory role of fractalkine (CX3CL1) in rheumatoid arthritis. J Rheumatol 30:1918–1927
111. Le Gall SM, Bobé P, Reiss K, Horiuchi K, Niu XD, Lundell D, Gibb DR, Conrad D, Saftig P, Blobel CP (2009) ADAMs 10 and 17 represent differentially regulated components of a general shedding machinery for membrane proteins such as transforming growth factor alpha, L-selectin, and tumor necrosis factor alpha. Mol Biol Cell 20:1785–1794
112. Wang Y, Herrera AH, Li Y, Belani KK, Walcheck B (2009) Regulation of mature ADAM17 by redox agents for L-selectin shedding. J Immunol 182:2449–2457
113. Van Eerdewegh P, Little RD, Dupuis J, Del Mastro RG, Falls K, Simon J, Torrey D, Pandit S, McKenny J, Braunschweiger K, Walsh A, Liu Z, Hayward B, Folz C, Manning SP, Bawa A, Saracino L, Thackston M, Benchekroun Y, Capparell N, Wang M, Adair R, Feng Y, Dubois J, FitzGerald MG, Huang H, Gibson R, Allen KM, Pedan A, Danzig MR, Umland SP, Egan RW, Cuss FM, Rorke S, Clough JB, Holloway JW, Holgate ST, Keith TP (2002) Association of the ADAM33 gene with asthma and bronchial hyperresponsiveness. Nature 418:426–430
114. Shapiro SD, Owen CA (2002) ADAM-33 surfaces as an asthma gene. N Engl J Med 347:936–938
115. Knolle MD, Owen CA (2009) ADAM8: a new therapeutic target for asthma. Expert Opin Ther Targets 13:523–540

Proteases and Fibrosis

Melissa Heightman, Tatiana Ort, Lawrence de Garavilla, Ken Kilgore, and Geoffrey J. Laurent

Abstract Protein turnover, orchestrated by a large array of proteases, protein complexes, and organelles consumes one-fifth of the human energy. In many diseases this homeostasis is lost resulting in loss of tissue function. One such group of diseases are the fibrotic disorders in which there is progressive fibroproliferation and extracellular matrix deposition. The causes of these diseases are sometimes identified, but often they are cryptic. In any case it is clear that there are shared pathways and processes leading to fibrosis in different tissues such as liver, lung, and kidney. There are also intense efforts to discover ways to inhibit the pathways that lead to fibrosis. Proteases of all classes are known to play roles in fibrosis. The matrix metalloproteases (MMPs) and their inhibitors (the tissue inhibitors of metalloproteinases, TIMPS) can degrade all extracellular matrix components and are thought to be central to pathogenesis. However, cysteine and serine proteases (including caspases, cathepsins, coagulation cascade proteases, and mast cell proteases) are also thought to regulate key processes and cell functions in fibrosis, including extracellular matrix gene expression, cell proliferation, cell apoptosis, epithelial mesenchymal transduction, and fibrocyte recruitment. The multiple pathways and range of proteases involved clearly provides challenges for therapy but it is not surprising that regulation of protease activity and the cell functions they control are being explored as potential targets to treat fibrotic disorders.

M. Heightman • G.J. Laurent (✉)
Centre for Respiratory Research, University College London, 5 University Street, London WC1E 6JF, UK
e-mail: g.laurent@ucl.ac.uk

T. Ort • K. Kilgore
Immunology Research, Centocor R&D Inc., 145 King of Prussia Road, Radnor, PA 19087, USA

L. de Garavilla
Johnson and Johnson Pharmaceutical Research and Development, L.L.C., Welsh & McKean Roads, Spring House, PA 19477-0776, USA

N. Vergnolle and M. Chignard (eds.), *Proteases and Their Receptors in Inflammation*, Progress in Inflammation Research, DOI 10.1007/978-3-0348-0157-7_7,

Keywords Angiogenesis • Apoptosis • Caspases • Cathepsins • Chymase • Collagen • Cysteine proteases • Cytokines • Fibroblast activation protein • Fibroproliferation • Fibrosis • Growth factors • Inflammation • Matrix metalloproteinases • Proteinase activated receptor • Serine proteases • Tryptase

1 Introduction

Rudolf Shoenheimer [1] was the first person to demonstrate that body proteins were continuously turning over and that a considerable amount of human energy is devoted to this process. Subsequent studies recognized the diversity of the rates of turnover with some proteins having turnover rates (half-lives) measured in seconds whilst others, including highly cross-linked matrix protein, were apparently inert in disease-free adults. Many of the proteases and anti-proteases responsible for this process have now been characterized and grouped according to their structure and function.

Lysosomes and proteosomes – the specialized organelles and protein complexes for degradation – are well characterized but proteases are known to be active in all compartments of the cell and extracellular milieu. There are also a myriad of cell membrane-bound proteases such as the ADAM family of metalloproteases that direct cellular interactions between cells and their environment. Other proteases acting extracellularly have been known for many years to play key roles in regulation turnover of a large and diverse family of extracellular matrix proteins. More recently, other functions have been recognized that have revolutionized our concept of proteases (Fig. 1). Proteases can activate families of cell surface receptors initiating signaling cascade that have dramatic effects on cell function. They can also activate, or inactivate cytokines, chemokines and growth factors and thus modify their functions. The degradation of extracellular molecules often generates fragments that possess biological activity and finally proteases activate or deactivate breakdown of other proteases.

Proteases are also key constituents of the myriad of biochemical cascades that regulated cell and tissue function. This includes the coagulation, complement, and signal transduction cascades. These processes are vital for survival controlling blood clotting, wound healing and playing key roles in regulating vision, smell and pain.

Proteases secreted by parasites and microorganisms have also been recognized as effector molecules in disease and it is thought that they often act by disabling or activating the targets of the host proteases. For example, a neutrophil elastase like protease produced by the lung pathogen pseudomonas has been proposed to play roles in cystic fibrosis by disabling receptors or rendering antiproteases ineffective. Proteases derived from house dust mite are reported to contribute to the pathogenesis of asthma by mimicking serine proteases of the coagulation cascade that activate pro-inflammatory receptors [2].

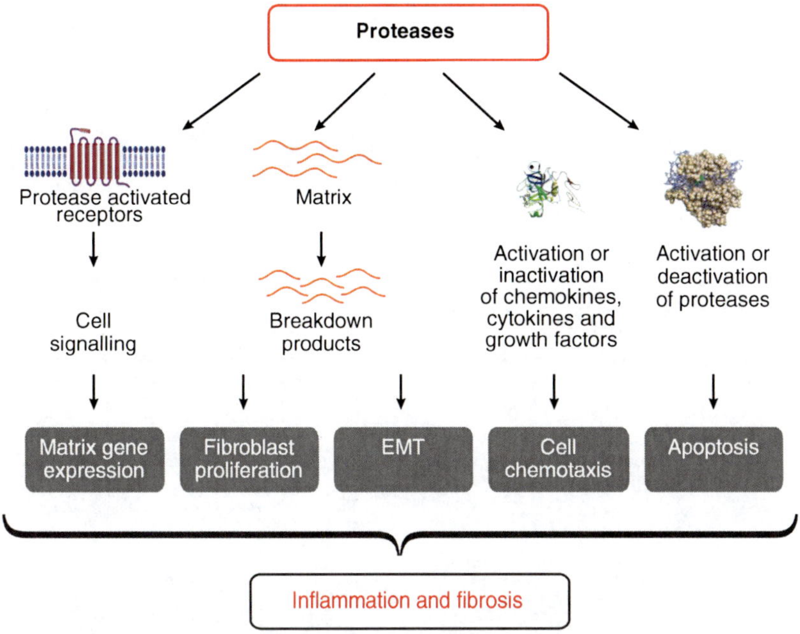

Fig. 1 Proteases influence a range of processes that are thought to be central to the development of tissue fibrosis

The importance of proteases as therapeutic targets is already established. Proteosome inhibitors are used in the treatment of multiple myeloma and are showing promise in other cancers. Protease inhibitors are key to therapy targeting HIV infection and there are other antiproteases being used in antiretroviral therapy. Some of these agents are also being used with success to treat malaria and gastrointestinal protozoal infections. There is also abundant evidence in animal studies that protease inhibitors may provide novel therapies to treat inflammatory and immune disorders. This chapter reviews the roles proteases play in fibrosis. We commence with a short description of organ fibrosis and then review in some detail the current literature on the role of proteases in fibrosis.

1.1 Fibrosis

Fibrosis refers to the process of disordered and progressive scarring that can occur in all tissues. It often follows trauma or infections but in many clinical settings the cause is not known. It is frequently associated with inflammation and can be driven by pathways of innate and acquired immunity. The consequences of fibrosis represent a huge medical problem whether it occurs in livers (e.g., alcohol induced cirrhosis), kidney (e.g., renal sclerosis) or lung [e.g., asbestosis, idiopathic

Fig. 2 Fibroproliferation and extracellular matrix deposition leading to loss of fine structure are the hallmarks of lung, renal, and liver fibrosis

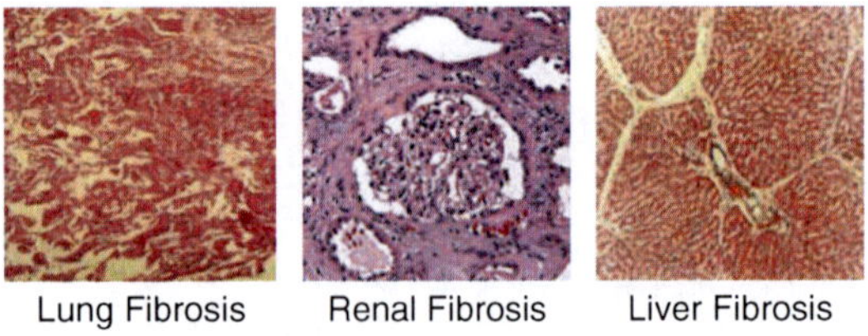

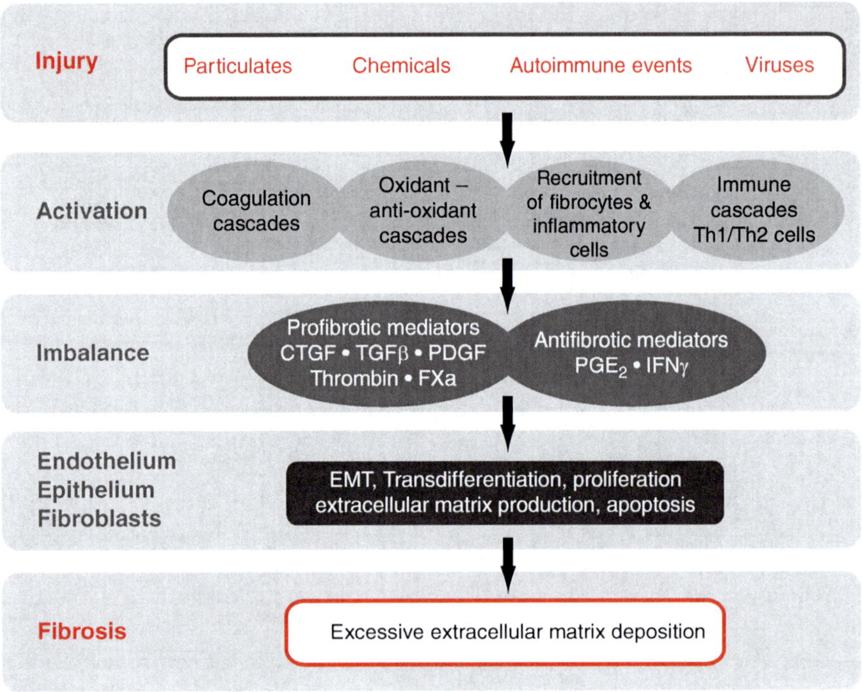

Fig. 3 The pathogenesis of fibrosis. Various forms of injury as well as pathogens can initiate multiple cascades that can drive the development of fibrosis (Figure adapted [3], EMT: epithelial mesenchymal transition)

pulmonary fibrosis (IPF), Fig. 2]. This group of diseases represent a large unmet medical need due to the lack of therapies to either prevent progression or reverse fibrosis. This need is made particularly urgent by the rapid deterioration of the progressive fibrotic disorders which often have survival times of a few years after diagnosis, similar to many cancers.

The key overlapping features of pathogenesis in fibrosis are described in Fig. 3. They include enhanced expression of matrix (particularly collagens), proliferation of fibroblasts and myofibrosis, epithelial mesenchymal transition (EMT), fibrocyte recruitment from blood and altered apoptosis. Matrix production and fibroproliferation are clearly important and probably the key processes. However, the latter three processes are more controversial and require further explanation. EMT is clearly possible in cell culture and it can be driven by the same mediators (e.g., TGFβ) that have been implicated in fibrosis. Fibrocytes are collagen expressing cells in blood that have been proposed to target sites of tissue damage where fibrosis occurs. The importance of apoptosis is also uncertain. Reduced apoptosis rates of fibroblasts have been reported in fibrosis which could explain the increased numbers of those cells but other cells appear to have enhanced rates of apoptosis (epithelial cells, some inflammatory cells) and this may also represent an important aberration in fibrosis.

There are a variety of cascades that can lead to fibrosis (Fig. 3), but in almost all of them proteases play a role and in one example (coagulation cascade), proteases and their receptors are clearly key.

It is also clear that there are downstream cytokines and growth factors that are commonly associated with different proteases and that these same pathways are ubiquitous in fibrotic disorders in different tissues. For example, TGFβ is an archetypical profibrotic molecule that appears to play roles in all settings where fibrosis occurs.

2 Matrix Metalloproteases and Fibrosis

A final common feature of fibrotic disorders is the abnormal accumulation of extracellular matrix (ECM), including fibrillar collagens (types I and II), fibronectin, elastic fibers, and proteoglycans. The consequence is extensive structural disorganization with eventual loss of tissue function. We have known from early tracer studies that the rate of turnover of ECM components is often rapid [4] and is mainly dependent on the activity of the matrix metalloproteinase (MMP) family of 26 zinc-dependent endopeptidases, which are collectively capable of degrading all types of extracellular matrix proteins. They are also involved in the processing of a large number of bioactive molecules sequestered by ECM components or on cell surfaces, such as growth factors, apoptotic ligands, chemokines and cytokines, or cell surface receptors. MMPs therefore have a fundamental involvement in physiological cellular behavior and play a crucial role in morphogenesis, angiogenesis, and tissue repair. They also influence many of the processes thought to be important in fibroproliferative disease, as outlined in Fig. 4, where different members of the family can have both profibrotic and anti-fibrotic actions depending on their precise substrate specificity and the environment in which they are acting.

2.1 MMP Structure and Function

MMPs share a common structure, with a propeptide domain, catalytic domain and haemopexin-like C-terminus. A cleft in the catalytic domain contributes

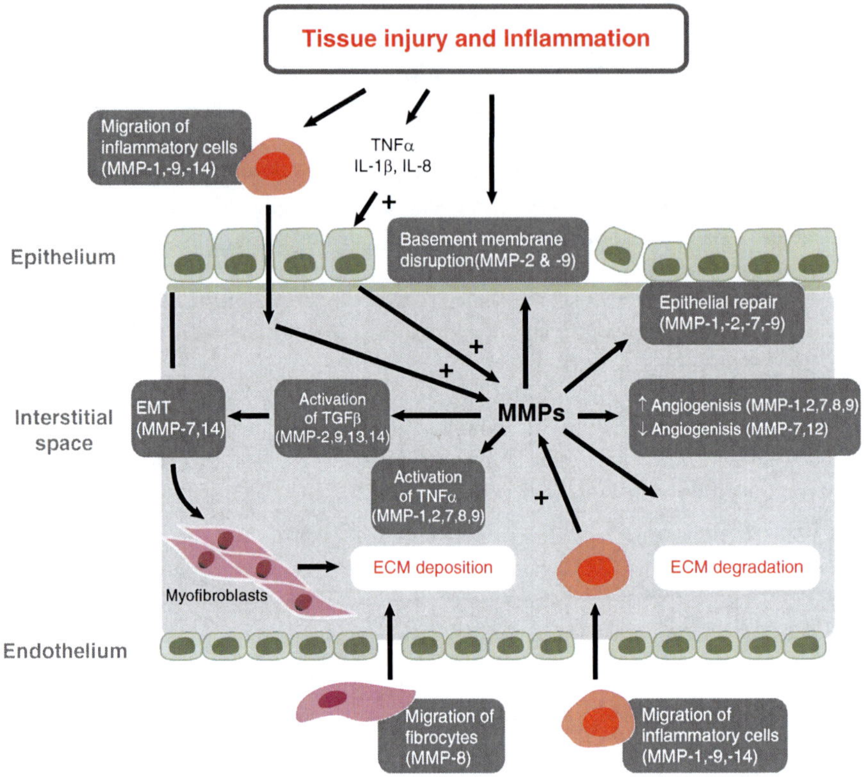

Fig. 4 Matrix metalloproteinases influence rate of turnover of extracellular matrix and can influence many processes that play a part in fibrotic diseases via their nonmatrix substrates. (EMT: epithelial–mesenchymal transition)

to proteases specificity, as does the C-terminal domain, which is thought to be involved in a number of other protein–protein interactions. Membrane-type MMPs also have a transmembrane domain. There is some redundancy in terms of ECM substrate between MMP family members and they are often grouped in this way. However, non-ECM substrates can vary greatly between members of the same family, highlighting their often very different roles in influencing cellular behavior. For example, comparing the substrate degradomes of the gelatinases MMP-2 and MMP-9 among fibroblast-secreted proteins revealed 201 cleavage products for MMP-2 and only 19 for the homologous MMP-9.

The majority of MMPs are secreted, although there is recent evidence that MMPs-1, -2, and -11 may act on intracellular proteins [5]. As with many proteases, MMPs are synthesized as inactive zymogens in which a cysteine residue interacts with the zinc in the active site preventing enzymatic activity. Activation within the extracellular space requires removal of the pro-peptide domain containing this cysteine residue (the "cysteine switch") which can occur through the action of plasmin or other serine proteases or previously activated MMPs. MMPs are tightly regulated following activation by specific endogenous protease inhibitors, the "Tissue Inhibitors of Metalloproteinases" (TIMPs). There are four TIMPs, which differ in spatiotemporal expression. MMP/TIMP balance is critical for determining proteolytic activity.

2.2 *The Role of MMPs in Fibrosis*

There is a large body of literature investigating the roles of the MMPs in fibrosis. In this chapter we describe their role in processes considered to be important in fibrosis pathogenesis: epithelial injury, basement membrane disruption, migration of inflammatory and matrix producing cells, ECM remodeling, dysregulated apoptosis, angiogenesis, and modulation of growth factors. This is also summarized in Fig. 4.

2.3 *Inflammation and Injury*

Tissue injury is a common precursor to fibrosis and can be mediated by a wide range of factors ranging from viral infection to cigarette smoke. MMPs have been shown to play a role in the early immune response to injury. For example, following stimulation of mesenchymal cells with TNFα, IL-1 β, or IFN γ, TLR3-mediated upregulation of TIMP-1 is observed within hours, followed by later upregulation MMP-9 [6]. Inhibition of MMP-9 has been shown to prevent trans-migration of neutrophils in ventilator-induced lung injury, highlighting its role in inflammation [7].

2.4 *Basement Membrane Disruption*

Disruption of the basement membrane is often an early finding in fibrosis and is likely to play an important part in pathogenesis, either by corrupting epithelial repair, or by allowing inappropriate migration of fibroblasts and myofibroblasts. For example, in IPF, both MMP-2 and MMP-9 are up-regulated in bronchoalveolar lavage fluid and both have been shown to be synthesized by sub-epithelial myofibroblasts coinciding in some areas with denuded basement membranes [8]. The substrates of MMPs-2 and -9 include type IV collagen which is an important component of basement membrane.

2.5 Epithelial Repair

This is an essential part of recovery from tissue injury which requires migration of epithelial cells over denuded basement membrane and their subsequent differentiation and reconstitution of the basement membrane. MMPs-1, -2, -7, and -9 have been shown to contribute to this process in vitro [8]. Treatment of alveolar epithelial cells with exogenous MMP-1 increased cell migration on type I collagen, proposed to occur via alteration of cell adhesion sites in degraded collagen [9]. The expression of MMP-7 is known to be upregulated in migrating airway epithelial cells and MMP-7 facilitates shedding of E-cadherin ectodomain and syndecan-1 from injured airway epithelial cells [10].

2.6 ECM Accumulation and Remodeling/Changes in MMP: TIMP Balance

Although it might be predicted that a non-matrix-degrading environment prevails in fibrosis and that the MMP/TIMP balance would be shifted in favor of the TIMPs, studies on the relative levels of expression of specific MMPs and TIMPs both in experimental models and human disease give mixed results. Many studies show that ongoing turnover of ECM persists even in established fibrosis and that there is upregulation of the entire "degradome." For example, in IPF transcriptional profiling identified that there was upregulation of a large number of different proteases (particularly MMPs-1, -2, -7, -9) suggesting that, IPF involves ongoing intensive remodeling even with advanced fibrosis [11].

In hepatic cirrhosis a shift of the MMP/TIMP balance in favor of MMP inhibition is seen, with TIMP1 expression by hepatic stellate cells (HSCs) increasing in parallel with fibrosis as analyzed by hydroxyproline levels, whilst MMP-1 levels remain unchanged [12, 13]. There has also been shown to be reduced MMP activation in cirrhosis due to the production of PAI-1 by HSCs, which inhibits plasmin synthesis [14].

Studies examining specific MMP and TIMP levels in pulmonary fibrosis, however, can be conflicting and challenging to interpret in terms of the overall consequence for ECM turnover. For example, in the bleomycin model, greatest upregulation is seen with MMP-13 and TIMP1 [15], whereas in human interstitial lung diseases an increase in MMP-1 [16] and MMPs-7, -2, and -9 is described, with TIMPs 2, 3, and 4 being more highly expressed than TIMP1 [16]. MMP-1 has typically been associated with inflammatory conditions associated with ECM degradation such as rheumatoid arthritis, rather than fibrosis, but its microlocalization in IPF in epithelial cells rather than in areas of fibrogenesis points to a likely role regulating epithelial/mesenchymal interactions, rather than ECM degradation. The elevation of MMP-7 in a number of types of fibrotic lung disease [17] is of particular interest as it has a broad substrate affinity for ECM molecules (collagen IV, aggrecan, fibronectin, geltin, vitronectin, elastin, osteopontin) and many bioactive substrates. MMP-7 knockout mice are protected against fibrosis, supporting the human data [18]. This raises the question whether MMP-7 is directly profibrotic, or whether its increased expression represents a compensatory mechanism to limit matrix accumulation.

2.7 *Accumulation of ECM Producing Cells*

Matrix secreting cells seen in fibrosis of all organs are phenotypically similar mesenchymal cells, such as interstitial fibroblasts, glomerular mesangial cells, vascular smooth muscle cells, or HSCs. These cell types accumulate in fibrosis via several different mechanisms: by proliferation of a resident cell population, by increasing resistance to apoptosis, by epithelial or endothelial-to-mesenchymal transition or by recruitment of circulating mesenchymal progenitor cells such as fibrocytes. MMPs have been shown to influence some of these mechanisms.

2.8 *Altered Apoptosis*

MMPs can promote or protect against apoptosis via proteolysis of a number of biologically active molecules. The end effect depends on the cellular context and MMP/TIMP balance. For example, MMP-7 has been shown to release soluble FAS ligand, which is pro-apoptotic, yet MMP-7 also releases EGF from ECM binding, which promotes cell survival via erb-B4 receptor tyrosine kinase [19]. Matrix degradation can also result in the loss of critical survival signals, e.g., cadherins and integrins. This has been most clearly shown for epithelial cells with degradation of basement membrane [8], but it has also been shown that HSCs cultured on matrigel (a basement membrane-type matrix) are more resistant to apoptosis than those cultured on collagen I. Apoptosis of HSCs appears to shift the balance of TIMP/MMP expression in favor of matrix degradation in models of liver fibrosis [20].

2.9 *Epithelial to Mesenchymal Transition*

Epithelial to Mesenchymal Transition (EMT) involves a sequence of changes that lead to the expression of migratory and invasive properties by epithelial cells. Several MMPs are regulated by transcription factors (snail, ETS, β-catenin), known to regulate EMT, for example, MMP-7 and MT1-MMP are targets of the β-catenin/TCF pathway and TIMP3 has been shown to have direct effects on β-catenin. Consequently, MMPs are often considered as target genes of EMT pathways, with MMP expression occurring as a late event. However, some MMPs have been shown to initiate EMT changes directly, for example, via cleavage of E-cadherin, thereby inducing E-cadherin complex fragility and EMT changes [21], or via proteolytic cleavage of latent TGF-β_1, a crucial mediator of EMT by MMP-2 [22, 23]. Over-expression of MMP-2 in mice results in changes consistent with chronic kidney disease, perhaps highlighting the importance of EMT in this organ, where it has been estimated to contribute to almost 40% of FSP-1 positive cells (a marker of newly acquired fibroblast like features) [24].

2.10 Cellular Migration

MT-MMPs (membrane-type MMPs), particularly MT1-MMP, have been shown to play a key role in the pericellular proteolysis associated with cell migration and invasion (MT1-MMP can cleave several specific substrates including collagens (I, II, III), laminin and fibronectin). MMPs have also been shown to modulate chemokines, for example, the action of MMP-1 and MMP-9 on collagen exposes normally hidden N-acetyl Pro-Gly-Pro (PGP) containing strands, which can act as a partial agonist at chemokine receptors due to molecular similarity to an important motif on alpha chemokines [25]. Fibrocytes have become increasingly recognized as important cell types in hepatic, pulmonary, and renal fibrosis [26–28]. They have also been shown to produce MMPs-2, -7, -8, and -9 and MMP-8 can promote fibrocyte migration through basement membrane and interstitial matrix [29].

2.11 Angiogenesis

In Hepatic fibrosis aberrant angiogenesis is thought to play a key role in pathogenesis [30]. The role of angiogensis in pulmonary fibrosis was first identified by Turner-Warwick [31], who demonstrated neovascularization leading to anastomoses between the systemic and pulmonary microvasculature in patients with IPF. There has also been shown to be reduced or delayed capillary development in fibroblastic foci in IPF [32]. Work by Strieter et al. [33] has shown that the CXC chemokine family may be important mediators of aberrant neoangiogenesis, and they may achieve some of their effects via increasing expression of MMPs. Most research on the role of MMPs in angiogenesis comes from the cancer field. MMPs shown to be proangiogenic include MMPs-1, -2, -7, -9, and MT1-MMP [34]. MMP-9 activity, for example, results in the release of VEGF from ECM [35]. Other MMPs have angiostatic actions, such as MMPs-7 and -12, which can block angiogenesis by conversion of plasminogen to angiostatin [36]. TIMP3 has opposing effects to MMP-9 in inhibiting binding of VEGF to VEGFR2, thus once more MMP/TIMP balance seems likely to influence cellular behavior in this setting [37].

2.12 Activation of Growth Factors

TGF-β_1 has been shown to be a central mediator in fibrosis, where it has diverse pro-fibrotic effects. Certain MMPs, including MMPs-2,-9,-13, and -14 activate latent *TGF*-β_1 by proteolysis [22] and MMPs can release *TGF*-β_1 from latent *TGF*-β_1-binding protein-1 (LTBP1) [38]. In turn, *TGF*-β_1 has been shown to stimulate expression of MMPs-1, -13, and -10 [39] by epithelial cells, which is likely to influence processes such as EMT. In *TGF*-β_1-induced lung fibrosis there was significant increased expression of MMP-12 and TIMP-1 with reduced expression of MMP-9. MMP-12-deficient animals also showed reduced fibrosis following treatment with either *TGF*-β_1 or bleomycin suggesting that MMP-12 plays an important role in *TGF*-β_1 effector pathways in generating fibrotic responses, whereas MMP-9 may be more important in the control of such responses.

TNF-α has also been shown to be a critical mediator in fibrosis of many tissues and an important stimulant of *TGF*-β_1 production by fibroblasts and macrophages. In models of cardiac fibrosis, treatment with a broad spectrum MMP inhibitor prevented increased levels of myocardial *TNF*-α, suggesting that MMP-mediated cleavage of latent extracellular membrane-bound *TNF*-α protein is a primary source of bioactive *TNF*-α in the myocardium of the volume overload heart [40]. Levels of *TNF*-α are increased in IPF, with greatest expression in alveolar epithelial cells [41]. In the bleomycin model of pulmonary fibrosis *TNF*-$\alpha^{-/-}$ animals are protected from inflammation and subsequent fibrosis [42], and it appears that it is soluble TNF that is required for lymphocyte recruitment and transition from the inflammatory to fibrotic stages [43]. TIMP3 is unique in its ability to inhibit *TNF*-α-converting enzyme and in TIMP3$^{-/-}$ animals exposed to liver injury there is uncontrolled *TNF*-α release and inflammation [44]. *TNF*-α is likely to achieve some of its effector functions via induction of MMP expression, and it has been shown to induce MMP-9 expression in human tracheal smooth muscle cells [45].

IGF-I has been postulated to play a role in fibrosis via its mitogenic, antiapoptotic, and profibrotic effects. The superfamily of IGF-binding proteins (IGFBP) are known to be the substrate of many MMPs [46]. IGFBP expression is increased in human IPF BAL cells and fluid, and in lung tissue of both patients with IPF and bleomycin-treated mice and IGF-1 expression also increases in IPF. Whilst IGF-1 itself has not been shown to be clearly profibrotic there is some evidence that IGFBP-3 and -5 induce a profibrotic phenotype in normal primary adult lung fibroblasts [47]. Changes in MMP levels during fibrosis are likely to modulate IGFBP levels but precise mechanisms have not yet been clarified.

2.13 Translational Work

It is clear that MMPs and their inhibitors are likely to play multiple roles in the pathogenesis of fibrotic disorders. Nevertheless, dissecting the precise roles of

MMPs and their inhibitors in fibrosis is challenging due to the apparent overlapping roles and functional redundancy of different MMP inhibitor systems. A further obstacle is that microlocalization of these proteases is likely to be of great importance in determining their effects on disease processes, yet is inaccessible to study. Identification of drug targets is complicated by the fact that there is often a shift in MMP/TIMP expression by cells depending on the disease stage. For example, in cirrhosis, HSCs express MMP-3 and MMP-13 in the early inflammatory stages of disease, exhibiting a matrix degrading phenotype which permits migration of inflammatory cells, but in later stages expression is shifted towards pro-MMP-2 and MT1-MMP and TIMP1 expression protecting degradation of fibrillar collagen which is causing fibrosis [48].

An area with potential translational benefits is in using MMPs as biomarkers of fibrosis. In IPF, serum levels of MMP-1 and MMP-7 [17] have been shown to be increased in plasma, serum, and BALF, which may help in distinguishing IPF from interstitial lung disease, or in following disease progression. However, normal levels do not exclude disease due to the extent of overlap with healthy patients and other chronic lung disease. More recently, it has been suggested that levels of the activation peptides of MMP-7 and MMP-9 may be more specific markers [49].

The ultimate goal of drug therapy might lie in attempting to shift MMP/TIMP balance to drive regression of fibrosis. A precedent has been set for this in animal models of liver fibrosis where by administration of mutant MMP-9 scavenging of TIMP1was achieved with consequent stimulation of ECM degradation and reversal of fibrosis [50]. To date trials with broad spectrum MMP inhibitors such as marimastat in cancer have failed due to adverse effects, and when this drug was tested in a mouse model of liver fibrosis it attenuated inflammation but aggravated fibrosis [51]. It seems likely that more specific inhibitors will be required to safely modulate MMP activity. Of interest doxycycline has some MMP inhibiting activity and has the advantage of being a drug with a proven safety profile. It is currently being trialed in cystic fibrosis with the aim of reducing sputum MMP-9 activity (clinicaltrials.gov trial number NCT01112059) and there is interest to test its use in IPF. The antifibrotic agent pirfenidone recently trialed in IPF has MMP inhibiting effects amongst other actions [52].

2.14 Summary

Matrix metalloproteinases are well placed to influence many of the processes believed to be important in fibrosis pathogenesis. The number of members of this class of proteases together with their multiplicity of targets, both matrix and biological, mean that fully elucidating their action in health and disease is challenging, but great progress is being made in this regard. Manipulation of MMP activity pharmacologically seems likely to be an important component of effective drug therapy of fibrotic disease, but achieving specificity of effect remains a challenge.

3 Cysteine Proteases and Fibrosis

Cysteine proteases have a common catalytic mechanism based on a nucleophilic attack of the cysteine residue, leading to a tetrahedral intermediate and then an acyl enzyme. Functions of cysteine proteases are diverse and include regulation of apoptosis, pro-hormone processing, and extracellular matrix remodeling. As for other protease classes, their expression and activity is highly regulated and dysregulation of expression contributes to a variety of diseases including fibrotic disorders. The enzymatic activity of cysteine proteases is controlled by their endogenous inhibitors, the cystatins. Proteinase/antiproteinase imbalance may further shift the balance of synthesis and degradation of extracellular matrix proteins and contribute to fibrosis progression. Prevalent mechanisms of how cysteine proteases impact fibrosis are discussed in this chapter.

3.1 Caspases and Fibrosis

Caspases belong to a 12-member family and serve as effector molecules for apoptosis. Their activation can be carried out by either receptor-mediated extrinsic or mitochondrial intrinsic pathways. The first pathway is activated by members of the tumor necrosis factor family, such as Fas ligand binding to its receptor, which results in activation of caspase-8. The second pathway is triggered in response to chemical and physical stress, which stimulates the cytoplasmic release of pro-apoptotic mitochondrial proteins leading to activation of caspase-9. These initiator caspases activate a set of effector caspases, notably caspases 3, 6, and 7, which then synchronously cleave proteins in many cell compartments. Several studies suggest that fibroblasts derived from fibrotic tissues are more resistant to apoptosis [53–55]. The mechanisms involved in fibroblast resistance to apoptosis remains obscure, although some experimental evidence suggest that IL-6 and PGE_2 pathways may contribute to down-regulation of death-receptor-mediated apoptosis in fibrotic fibroblasts [54, 55]. Increased survival of fibroblasts may tip the balance toward extracellular matrix accumulation and contribute to fibrosis development (Fig. 5). In contrast to fibroblasts, wide-spread apoptosis is observed in fibrotic epithelium [53] that is increasingly viewed as a nexus between tissue injury and fibrosis (Fig. 5). In agreement with that blocking caspase activity attenuates excessive collagen deposition in rodent models of pulmonary and liver fibrosis. For example, a broad-spectrum caspase inhibitor decreased the number of apoptotic cells, the pathological grade of lung inflammation and collagen content in a mouse model of lung fibrosis [56]. Administration of pan-caspase inhibitor to db/db mice-fed methionine/choline-deficient diets resulted in attenuation of liver fibrosis suggesting that caspase-dependent apoptosis may contribute to development of nonalcoholic steatohepatitis [57]. Liver injury measured by bile infarcts and serum alanine aminotransferase (ALT) values were reduced in caspase inhibitor-treated

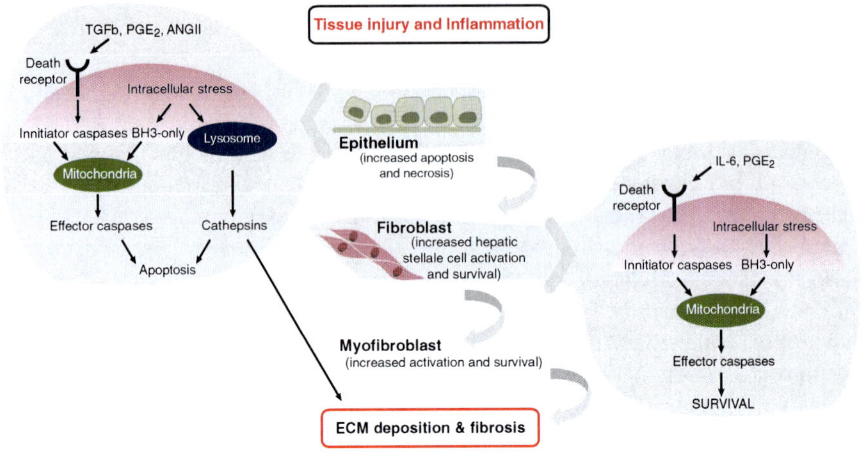

Fig. 5 Cysteine proteases contribute to tissue fibrosis through a variety of different mechanisms

versus saline-treated 3-day bile duct-ligated mice [58]. Interestingly, the same inhibitor (IDN-6556) decreased ALT activity in patients with chronic HCV infection and in patients with steatohepatitis during a 2-week dosing period [59]. Longer studies are required to assess whether caspase inhibitor may affect fibrosis-relevant endpoints and ultimately clinical outcomes of chronic liver disease.

Excess of epithelial cell apoptosis combined with fibroblast resistance to apoptosis appears to be important pathological processes underlying fibrosis progression. Extrinsic (death receptor mediated) and intrinsic (mitochondria mediated) caspase-dependent apoptotic pathways along with release lysosomal cathepsins can lead to epithelial cell death. Apoptosis and necrosis of epithelial cells propagates tissue damage and inflammation that in turn results in activation of fibroblasts or HSCs and their subsequent transdifferentiation to myofibroblasts with exaggerated ability to produce and deposit collagen. In addition, cathepsins may directly affect collagen degradation.

3.2 Cathepsins and Fibrosis

The cathepsins are a subclass of proteases with both intracellular and extracellular functions. Members of this family (11 in humans) are located primarily within the endosomal/lysosomal compartments of cells, but are also found as active forms in the pericellular environment as soluble enzymes or bound to cell surface receptors at the plasma membrane [60]. Indeed, some cathepsins are believed to take part in the extra cellular degradation of collagen [61] (Fig. 5); the resulting collagen fragments are phagocytosed by macrophages and fibroblasts. The collagen-containing phagosomes within these phagocytic cells fuse with lysosomes to generate a phagolysosome, in which intracellular cathepsins complete the degradation of collagen fragments [62]. Other cathepsins are important for cell survival as they mediate caspase-independent cell apoptosis [63] (Fig. 5). Lysosomal membrane permeabilization that may be triggered by both extrinsic and intrinsic stimulus is critical step in cathepsin-mediated apoptosis.

3.3 *Cathepsin K and Lung Fibrosis*

Cathepsin K expressed in epithelial cells and macrophages [64, 65] is a particularly interesting cysteine cathepsin, as it has a unique collagenolytic activity within cross-linked triple helices collagens I and II [61], that depends on its specificity for the Pro residue at P2 and its capacity to form a complex with chondroitin 4-sulfate. The alveolar walls of cathepsin K knockout mice are thicker after bleomycin treatment and have more deposited collagen than control mice, and their fibroblasts have less collagenolytic activities [61, 66]. Likewise, over-expression of cathepsin K in the lungs of mice is protective in bleomycin-induced pulmonary fibrosis despite similar inflammatory leukocyte mobilization in response to bleomycin challenge [67]. Interestingly, control mice in response to bleomycin revealed an increased expression of cathepsin K in fibrotic lung regions suggesting a protective role of cathepsin K to counter the excessive deposition of collagen matrix in the diseased lung. Similarly, in lung specimens obtained from patients with lung fibrosis fibroblasts expressed larger amounts of cathepsin K than those obtained from normal lungs [61]. Due to its potent collagenolytic activity, cathepsin K is an interesting target with therapeutic potential to attenuate pulmonary fibrosis.

3.4 *Cathepsin B and Liver Fibrosis*

There is compelling data indicating that liver fibrogenesis is reduced by cathepsin B inhibition. Genetic or pharmacologic inactivation of cathepsin B blocks fibrosis in a variety of animal models of liver injury including models of cholestasis, cold-induced perfusion injury and carbon tetrachloride-induced injury [68–70]. Cathepsin B inactivation and consequent attenuation of hepatocyte apoptosis proposed as the mechanism linking this proapoptotic protease to liver fibrogenesis. It has been further suggested that reduction in intrahepatic apoptosis leads to alleviation of inflammation and liver damage specifically when the extent of apoptosis overrides the phagocytic clearance. Decreased inflammation suppresses hepatic stellate cell activation and their subsequent transdifferentiation to myofibroblasts with exaggerated ability to produce and deposit collagen into the perisinusoidal space. In human non-alcoholic fatty liver disease, cathepsin B is redistributed into the cytosol and is correlated with disease severity suggesting activation of caspase B pathway in human fibrosis tissues [68]. These findings support an important role for cathepsin B in hepatic apoptosis and matrix deposition, suggesting that the antagonism of cathepsin B may be of relevance for the treatment of liver fibrosis.

3.5 *Summary*

Taken together, data reviewed here suggest that cysteine proteases play a key role in development and progression of fibrosis by affecting extracellular matrix accumulation directly via regulating collagen degradation and indirectly via controlling apoptosis and tissue injury with subsequent activation of collagen-producing lung cells (Fig. 5). Several cysteine proteases are dysregulated in multiple fibrotic human tissues such as cathepsin D in fibrotic lung and liver [71, 72] and caspase 3 in fibrotic lung, kidney, and skin [73–76] likely reflecting the common mechanisms contributing to fibrosis development in different organs. Although much work remains to be done in delineating the role of individual cysteine proteases in fibrotic process, the findings outlined here suggest a potential therapeutic benefit of inhibiting fibrosis by targeting cysteine proteases. Currently there is no clinical trial exploring the efficacy of cysteine protease inhibitors in patients with fibrosis. At least partially that can be explained by inability to produce small molecule inhibitors with the desired selectivity and bioavailability [77]. Recent interest in development of biologic inhibitors such as antibodies may stimulate the progress in the therapeutic targeting of protease species [77], although caution is warranted because anti-apoptotic drugs might potentially promote the excessive cell growth seen in malignancies.

4 Serine Proteases and Fibrosis

This diverse and abundant family of proteases is found throughout the circulation and body tissues. They can exist in monomeric or multimeric forms, alone or associated with key cofactors and stabilizers, active or zymogen forms and, soluble or membrane bound. All members of this family have a serine residue within the catalytic site that is essential for activity. They can alter cellular function though direct catalytic activation of surface receptors or through interaction with cells or molecules via intermolecular interactions unrelated to the catalytic site. This great diversity in structure and function of the serine protease family makes them likely participants, either directly or indirectly, in tissue repair and fibrosis. In this review we focus on coagulation proteases, mast cell proteases, neutrophil serine proteases as well as a less well-known serine protease on fibroblasts.

In the case of coagulation proteases these proteases are found in the circulation and in injured tissues. Inflammatory cells can also serve as an abundant source of serine proteases in injured and inflamed tissues. Serine proteases can activate cellular receptors, namely PAR's, which can have direct cellular affects such as induction of proliferation and increased matrix deposition. Either directly or indirectly many of these serine proteases are known to ultimately promote the activation of latent TGF-β leading to fibrotic effects.

4.1 Coagulation Proteases

Coagulation proteases represent a class of serine proteases that are produced in an inactive proenzyme (or zymogen) predominantly by the liver, although local production in tissues has recently been recognized [78]. The proenzymes are abundant in the circulation, hence this class of serine proteases represents the most widely distributed in the body, and thus has the greatest exposure to tissues, especially under conditions of local vascular injury, following trauma, for example. Local activation of coagulation proteases occurs with tissue injury and microvascular leakage that sets in motion a cascade of events, ultimately leading to the activation of the downstream zymogens factor X (FX) and pro-thrombin to FXa and thrombin, respectively. Beyond the classical coagulation activity of FXa and thrombin there now appears to be mounting evidence that these proteases play a role in fibrosis. Direct inhibition of FXa in a bleomycin-induced experimental model of pulmonary fibrosis in mice reduced fibrosis [78] (Table 1). Moreover, FXa and thrombin are found at increased levels in bronchoalveolar lavage fluid from patients with pulmonary fibrosis leading to alveolar fibrin accumulation in the lungs [82, 89]. Whether or not tissue deposition of fibrin is essential for the induction or maintenance of tissue fibrosis is still debatable. Kubo et al. showed, albeit in a small unblinded study, that anticoagulant therapy with warfarin or low-molecular-weight heparin provided benefit for patients with lung fibrosis [79] (Table 1). Although physicians may be hesitant to treat IPF patients with anticoagulants, this type of therapy is currently being evaluated in larger trials and thus far this has not been identified as a prohibitive problem.

Table 1 Serine proteases implicated in fibrosis. Table summarizes the evidence from both in vivo animal models and human association studies implicating various members of the serine protease family in fibrosis

Serine protease	Altered expression in human fibrotic tissues	Neutralization or overexpression in animal models impacted fibrosis
Thrombin	Lung [79]	Lung [80]
Factor Xa	Lung [81, 82]	Lung [78]
		Lung [85–89]
		Kidney [90]
Mast cell chymase	Lung [83, 84]	Cardiac [91–93]
Mast cell tryptase	Lung [83, 84]	Lung [85–87]
Neutrophil elastase	Lung [94]	Lung [95, 96]
	Liver [97, 98]	
Fibroblast activation protein-α	Lung [99, 100]	Lung [101]
Protease-activated receptor (PAR)-1	Liver [102]	Liver [102, 103]
	Lung [80]	Lung [104]

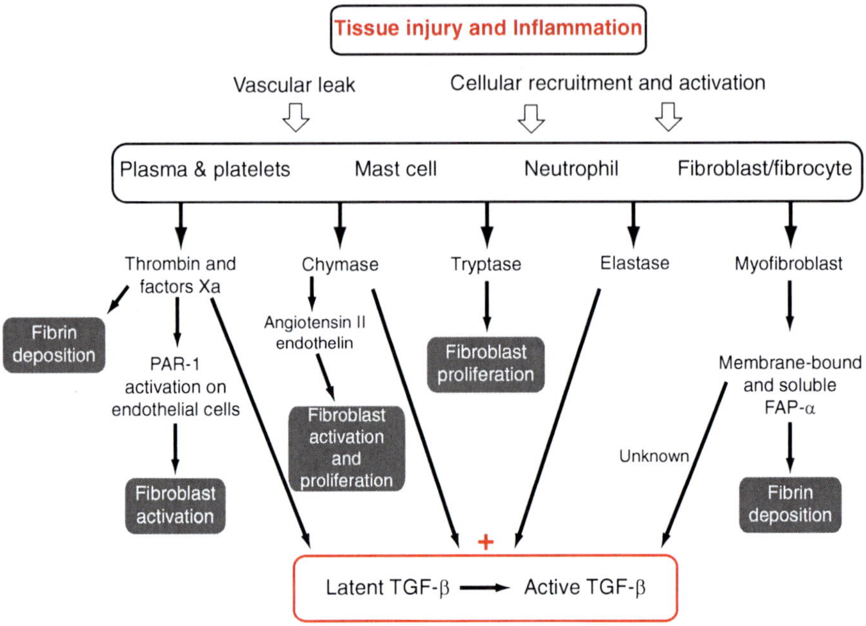

Fig. 6 Serine proteases are an abundant family of proteases that play a role in fibrosis

Coagulation proteases affect cellular responses through activation of the novel class of seven transmembrane G-protein coupled receptors referred to as protease-activated receptors (PARs), which exist in a family of four receptor subtypes PAR1-4; for example, thrombin can activate PAR-1, -3, and -4. PARs appear to be playing a role in the pathogenesis of fibrotic lung diseases [76]. For example, when thrombin, long known to be a potent mitogen for fibroblasts, activates PAR-1 localized on fibroblasts, it elicits a complex fibrotic response characterized by increased fibroblast proliferation and matrix deposition and transition to myofibroblasts [105–107] (Fig. 6). PAR-1 is upregulated in the liver during chronic injury, typically preceding fibroproliferative responses in rat and human HSCs [99]. Reduced PAR-1 activation, either through gene knock down or receptor antagonism, protected the lungs [104] and liver [103] from experimental fibrosis in animals. Thrombin can also contribute to the fibrotic process by increasing the expression [108] and activation [109] of MMPs.

4.2 Mast Cell Proteases

Mast cells contain in granular storage vesicles abundant amounts of the two serine proteases tryptase and chymase, tryptic and chymotryptic serine proteases, respectively. The mast cell, typically associated with allergic or anaphylactic conditions, has more recently been shown to be associated with fibrotic tissue responses [110]. Mast cells appear in greater numbers in the lungs of patients with interstitial pulmonary fibrosis and cystic fibrosis [83], and are colocalized with pulmonary fibroblasts in fibrotic lung tissue [84]. Reducing mast cell involvement by using mast cell-deficient mice [85] or treating with the mast cell stabilizer cromolyn [86] or ketotifen [87] significantly attenuated experimental lung fibrosis. The catalytic activities of tryptase and chymase can either directly or indirectly activate pathways that are known to be associated with fibrotic responses. Chymase, for example, converts angiotensin I to II [111], pro form of endothelin to mature endothelin [112], latent TGFb to its active form [113], and activate pro MMP2 and -9 [114], each of these products can elicit a fibrotic response (Fig. 6). Gene deletion of mouse mast cell protease-4, the mouse homologue of human chymase, has been shown to reduce experimental fibrosis in the kidneys [Scandiuzzi et al.] (Table 1). Moreover, small molecule inhibitors of chymase have been shown to prevent bleomycin-induced pulmonary fibrosis in rodents [88, 89]. Chymase appears to also have a prominent effect on the development of cardiac fibrosis associated with myocardial infarction in rats [91], and heart failure in dogs [92] and hamsters [93]. Accordingly, a small molecule chymase inhibitor has been advanced into clinical trials to blunt the fibroproliferative responses in the heart associated with myocardial infarction and heart failure (Teijin, company web site).

Tryptase is the most abundant human mast cell protease. Levi-Schaffer and Piliponsky propose that tryptase is a likely link between classically mast cell-mediated allergic inflammatory diseases and fibrosis [115]. Tryptase has been shown to be associated with fibrotic diseases in the lung [116] (Fig. 6), liver [117], kidney [118], and testes [119]. Tryptase is a mitogen for fibroblasts [120], airway smooth muscle [121], and epithelial cells [122]. Aker et al. demonstrated that tryptase could activate PAR-2 on fibroblasts resulting in increased cell proliferation [123] (Fig. 6). Frungieri et al. went on to show that via activation of PAR-2 by tryptase could indirectly modulate nuclear peroxisome proliferator-activated receptor (PPAR)-gamma resulting in increased fibroblast proliferation [124]. Interestingly, unlike other serine proteases, tryptase is not subject to tight control by endogenous inhibitors. Thus, through its abundant nature and general lack of control, the effect of tryptase on fibroblast proliferation suggests it likely plays a role in tissue fibrosis.

4.3 Neutrophil Elastase

Neutrophils have been implicated in the pathogenesis of interstitial pulmonary fibrosis in humans and experimental bleomycin-induced pulmonary fibrosis. This is evidenced by the presence of this cell in bronchoalveolar lavage fluid and lung tissue from patients with interstitial pulmonary fibrosis [94]. Consequently, the neutrophil-derived serine protease elastase can be found elevated in the lungs of patients with interstitial pulmonary fibrosis [94]. Both IPF and bleomycin-induced fibrosis are characterized by increased epithelial cell apoptosis and increased collagen production by fibroblasts and myofibroblasts. It has been shown that neutrophil elastase can induce pulmonary epithelial cell apoptosis [125, 126]. Song and coworkers showed that the small molecule elastase inhibitor ONO-5046 could reduce bleomycin-induced airway epithelial cell apoptosis in rats by blocking the elastase-induced activation of the mitochondrial apoptotic pathway [95]. Although TGF-β can be activated by a number of different mechanisms, it appears that neutrophil elastase can also actively participate in the activation of latent TGF-β to its active form [127] (Fig. 6). Dunsmore and coworkers initially showed that neutrophil elastase-deficient mice were resistant to bleomycin-induced pulmonary fibrosis [96]. Subsequently, Chua et al. confirmed and extended these findings by demonstrating that the resistance to the development of lung fibrosis was attributable to impairment in the activation of latent TGF-β [128].

4.4 Fibroblast Activation Protein-α

Fibroblast activation protein (FAP)-α, also known as seprase, is a dimeric type II membrane-bound serine protease on the cell surface of activated fibroblasts. In addition to dipeptidylpeptidase (DPP)-IV like activity, FAP-α also exhibits collagenase and gelatinase activity [129]. FAP-α was originally discovered as a protease that promoted tumor progression [130, 131]. More recently FAP-α has been shown to be associated with fibrotic diseases. Unlike other gelatinases that exist as zymogens, FAP-α has constitutive activity, therefore, the control of FAP-α activity lies in its expression pattern. Whereas FAP-α is not expressed in normal adult tissues, it can be found expressed in diseased human tissues. For example, FAP-α is strongly expressed on activated HSCs in close proximity to sites of cirrhosis-like injury and liver remodeling [97], and its expression is strongly correlated with the severity of disease [98] (Table 1). Although the precise regulatory mechanisms controlling FAP-α expression are unknown, Escobar et al. showed that cultured fibroblasts from a small sample of IPF patients exhibited high levels of FAP-α expression as compared to normal human lung fibroblasts (NHLF) [99]. Interestingly, TGF-β was shown to increase expression of FAP-α message in these cell cultures. In human lung tissue from patients with IPF, Achayra and coworkers showed that FAP-α was highly expressed on pulmonary fibroblasts located on the leading edge of fibrotic lesions within fibroblast foci and fibrotic interstitium, whereas it was absent in lung tissue from normal individuals [100]. Immunohistochemically FAP-α was localized in areas of fibrotic lesions in lungs from IPF patients but not in adjacent histologically normal lung tissue. FAP-α-deficient strains of mice appear to be fertile and do not exhibit overt developmental defects [101, 132]. In models of radiation and bleomycin-induced pulmonary fibrosis FAP-α mRNA and protein levels were both increased in the lungs of wild-type mice [101]. Compared with wild-type mice, FAP-α-deficient mice showed greater mortality and increased collagen deposition in each of these models of lung fibrosis.

Based on the postulated role that FAP-α may play in fibrosis, these findings were unexpected and show that more research, beyond that done in transgenic animals, is required. To that end, small molecule potent inhibitors of FAP-α have been identified [133] and should prove to be useful pharmacological tools to further assess the role of FAP-α in fibrotic diseases. Although the precise role of FAP-α in fibrotic disease remains to be elucidated, it is intriguing to consider that FAP-α, a protease with gelatinase and collagenase activity, may represent an alternate therapeutic approach to tissue remodeling and fibrosis that warrants further investigation.

4.5 Summary

Serine proteases represent a widely expressed diverse set of proteases that appear to play key roles in a various diseases including fibrotic disorders. Many of these proteases play key roles in critical physiological functions, including blood coagulation, wound healing, growth, and repair, wherein a fine balance is required between enzyme activity and inhibition. Research tools to investigate the role serine proteases play in physiology and fibrosis exist or are beginning to emerge for newer proteins such as FAP-α and should prove to be invaluable aides. Serine proteases appear to be involved in profibrotic responses through a variety of mediators. These include activation of PAR's by coagulation and mast cell proteases as well angiotensin II and endothelin formation, each of which have profibrotic properties. In addition, TGF-β, a key molecule in the stimulation of pro-fibrotic responses, can be enzymatically activated from its latent form by many serine proteases (Fig. 6). As such, TGF-β activation may represent a common pathway representing the involvement of serine proteases in fibrosis.

References

1. Schoenheimer R (1942) The dynamic state of body constituents. Harvard University Press, Cambridge, MA
2. Asokananthan N, Graham PT, Stewart DJ, Bakker AJ, Eidne KA, Thompson PJ, Stewart GA (2002) House dust mite allergens induce proinflammatory cytokines from respiratory epithelial cells: The cysteine protease allergen, Der p 1, activates protease-activated receptor (PAR)-2 and inactivates PAR-1. J Immunol 169:4572–4578
3. Maher TM, Wells AU, Laurent GJ (2007) Idiopathic pulmonary fibrosis: multiple causes and multiple mechanisms? Eur Respir J 30(5):835–839
4. Laurent GJ (1987) Dynamic State of collagen: pathways of collagen degradation in vivo and their possible role in regulation of collagen mass. Am J Physiol 252(1 Pt 1):C1–C9
5. Limb GA, Matter K, Murphy G, Cambrey AD, Bishop PN, Morris GE, Khaw PT (2005) Matrix metalloproteinase-1 associates with intracellular organelles and confers resistance to lamin A/C degradation during apoptosis. Am J Pathol 166(5):1555–1563

6. Merkle M, Ribeiro A, Sauter M, Ladurner R, Mussack T, Sitter T, Wörnle M (2010) Effect of activation of viral receptors on the gelatinases MMP-2 and MMP-9 in human mesothelial cells. Matrix Biol 29(3):202–208
7. Kim JH, Suk MH, Yoon DW, Lee SH, Hur GY, Jung KH et al (2006) Inhibition of matrix metalloproteinase-9 prevents neutrophilic inflammation in ventilator-induced lung injury. Am J Physiol Lung Cell Mol Physiol 291(4):L580–L587
8. Desai LP, Aryal AM, Ceacareanu B, Hassid A, Waters CM (2004) Rhoa and rac1 are both required for efficient wound closure of airway epithelial cells. Am J Physiol Lung Cell Mol Physiol 287(6):L1134–L1144
9. Planus E, Galiacy S, Matthay M, Laurent V, Gavrilovic J, Murphy G et al (1999) Role of collagenase in mediating in vitro alveolar epithelial wound repair. J Cell Sci 112(Pt 2): 243–252
10. McGuire JK, Li Q, Parks WC (2003) Matrilysin (matrix metalloproteinase-7) mediates e-cadherin ectodomain shedding in injured lung epithelium. Am J Pathol 162(6):1831–1843
11. Selman M, Pardo A, Barrera L, Estrada A, Watson SR, Wilson K et al (2006) Gene expression profiles distinguish idiopathic pulmonary fibrosis from hypersensitivity pneumonitis. Am J Respir Crit Care Med 173(2):188–198
12. Iredale JP, Murphy G, Hembry RM, Friedman SL, Arthur MJ (1992) Human hepatic lipocytes synthesize tissue inhibitor of metalloproteinases-1.Implications for regulation of matrix degradation in the liver. J Clin Invest 90(1):282–287
13. Ramachandran P, Iredale JP (2009) Reversibility of liver fibrosis. Ann Hepatol 8(4):283–291
14. Wang H, Zhang Y, Heuckeroth RO (2007) PAI-1 deficiency reduces liver fibrosis after bile duct ligation in mice through activation of tpa. FEBS Lett 581(16):3098–3104
15. Manoury B, Caulet-Maugendre S, Guénon I, Lagente V, Boichot E (2006) TIMP-1 is a key factor of fibrogenic response to bleomycin in mouse lung. Int J Immunopathol Pharmacol 19(3):471–487
16. Selman M, Ruiz V, Cabrera S, Segura L, Ramírez R, Barrios R, Pardo A (2000) TIMP-1, -2, -3, and −4 in idiopathic pulmonary fibrosis. A prevailing nondegradative lung microenvironment? Am J Physiol Lung Cell Mol Physiol 279(3):L562–L574
17. Rosas IO, Richards TJ, Konishi K, Zhang Y, Gibson K, Lokshin AE et al (2008) MMP1 and MMP7 as potential peripheral blood biomarkers in idiopathic pulmonary fibrosis. PLoS Med 5(4):e93
18. Zuo F, Kaminski N, Eugui E, Allard J, Yakhini Z, Ben-Dor A et al (2002) Gene expression analysis reveals matrilysin as a key regulator of pulmonary fibrosis in mice and humans. Proc Natl Acad Sci USA 99(9):6292–6297
19. Yu WH, Woessner JF, McNeish JD, Stamenkovic I (2002) CD44 anchors the assembly of matrilysin/MMP-7 with heparin-binding epidermal growth factor precursor and erbb4 and regulates female reproductive organ remodeling. Genes Dev 16(3):307–323
20. Arendt E, Ueberham U, Bittner R, Gebhardt R, Ueberham E (2005) Enhanced matrix degradation after withdrawal of tgf-beta1 triggers hepatocytes from apoptosis to proliferation and regeneration. Cell Prolif 38(5):287–299
21. Radisky DC, Levy DD, Littlepage LE, Liu H, Nelson CM, Fata JE et al (2005) Rac1B and reactive oxygen species mediate mmp-3-induced EMT and genomic instability. Nature 436(7047):123–127
22. Karsdal MA, Larsen L, Engsig MT, Lou H, Ferreras M, Lochter A et al (2002) Matrix metalloproteinase-dependent activation of latent transforming growth factor-beta controls the conversion of osteoblasts into osteocytes by blocking osteoblast apoptosis. J Biol Chem 277(46):44061–44067
23. Kasai H, Allen JT, Mason RM, Kamimura T, Zhang Z (2005) Tgf-Beta1 induces human alveolar epithelial to mesenchymal cell transition (EMT). Respir Res 6:56
24. Cheng S, Lovett DH (2003) Gelatinase A (MMP-2) is necessary and sufficient for renal tubular cell epithelial-mesenchymal transformation. Am J Pathol 162(6):1937–1949

25. Weathington NM, van Houwelingen AH, Noerager BD, Jackson PL, Kraneveld AD, Galin FS et al (2006) A novel peptide CXCR ligand derived from extracellular matrix degradation during airway inflammation. Nat Med 12(3):317–323
26. Nunnari G, Vancheri C, Gilli E, Migliore S, Palermo F, La Rosa C et al (2010) Circulating fibrocytes as a marker of liver fibrosis in chronic hepatitis C. Front Biosci (Elite Ed) 2:1241–1245
27. Sakai N, Furuichi K, Shinozaki Y, Yamauchi H, Toyama T, Kitajima S et al (2010) Fibrocytes are involved in the pathogenesis of human chronic kidney disease. Hum Pathol 41(5):672–678
28. Keeley EC, Mehrad B, Strieter RM (2010) Fibrocytes: Bringing new insights into mechanisms of inflammation and fibrosis. Int J Biochem Cell Biol 42(4):535–542
29. Garcia de Alba C, Becerril C, Reyes S, Selman M, Pardo A (2010) MMP-8 is a central player in fibrocytes transmigration. Am J Respir Crit Care Med 181(1_MeetingAbstracts):A5270
30. Deli G, Jin CH, Mu R, Yang S, Liang Y, Chen D, Makuuchi M (2005) Immunohistochemical assessment of angiogenesis in hepatocellular carcinoma and surrounding cirrhotic liver tissues. World J Gastroenterol 11(7):960–963
31. Turner-Warwick M (1963) Precapillary systemic-pulmonary anastomoses. Thorax 18:225–237
32. Yamashita M, Yamauchi K, Chiba R, Iwama N, Date F, Shibata N et al (2009) The definition of fibrogenic processes in fibroblastic foci of idiopathic pulmonary fibrosis based on morphometric quantification of extracellular matrices. Hum Pathol 40(9):1278–1287
33. Strieter RM, Gomperts BN, Keane MP (2007) The role of CXC chemokines in pulmonary fibrosis. J Clin Invest 117(3):549–556
34. Huo N, Ichikawa Y, Kamiyama M, Ishikawa T, Hamaguchi Y, Hasegawa S et al (2002) MMP-7 (matrilysin) accelerated growth of human umbilical vein endothelial cells. Cancer Lett 177(1):95–100
35. Keane MP, Strieter RM, Lynch JP, Belperio JA (2006) Inflammation and angiogenesis in fibrotic lung disease. Semin Respir Crit Care Med 27(6):589–599
36. Nishizuka I, Ichikawa Y, Ishikawa T, Kamiyama M, Hasegawa S, Momiyama N et al (2001) Matrilysin stimulates DNA synthesis of cultured vascular endothelial cells and induces angiogenesis in vivo. Cancer Lett 173(2):175–182
37. Bergers G, Brekken R, McMahon G, Vu TH, Itoh T, Tamaki K et al (2000) Matrix metalloproteinase-9 triggers the angiogenic switch during carcinogenesis. Nat Cell Biol 2(10):737–744
38. Dallas SL, Rosser JL, Mundy GR, Bonewald LF (2002) Proteolysis of latent transforming growth factor-beta (tgf-beta)-binding protein-1 by osteoclasts. A cellular mechanism for release of tgf-beta from bone matrix. J Biol Chem 277(24):21352–21360
39. Ishikawa F, Miyoshi H, Nose K, Shibanuma M (2009) Transcriptional induction of MMP-10 by TGF-[beta], mediated by activation of MEF2A and downregulation of class iia hdacs. Oncogene 29(6):909–919
40. Murray DB, Levick SP, Brower GL, Janicki JS (2010) Inhibition of matrix metalloproteinase activity prevents increases in myocardial tumor necrosis factor-alpha. J Mol Cell Cardiol 49(2):245–250
41. Zhao MQ, Stoler MH, Liu AN, Wei B, Soguero C, Hahn YS, Enelow RI (2000) Alveolar epithelial cell chemokine expression triggered by antigen-specific cytolytic CD8(+) T cell recognition. J Clin Invest 106(6):R49–R58
42. Ortiz LA, Lasky J, Hamilton RF, Holian A, Hoyle GW, Banks W et al (1998) Expression of TNF and the necessity of TNF receptors in bleomycin-induced lung injury in mice. Exp Lung Res 24(6):721–743
43. Oikonomou N, Harokopos V, Zalevsky J, Valavanis C, Kotanidou A, Szymkowski DE et al (2006) Soluble TNF mediates the transition from pulmonary inflammation to fibrosis. PLoS One 1:e108
44. Black RA (2004) TIMP3 checks inflammation. Nat Genet 36(9):934–935

45. Lee CW, Lin CC, Lin WN, Liang KC, Luo SF, Wu CB et al (2007) Tnf-Alpha induces MMP-9 expression via activation of src/EGFR, PDGFR/PI3K/akt cascade and promotion of nf-kappab/p300 binding in human tracheal smooth muscle cells. Am J Physiol Lung Cell Mol Physiol 292(3):L799–L812
46. Coppock HA, White A, Aplin JD, Westwood M (2004) Matrix metalloprotease-3 and -9 proteolyze insulin-like growth factor-binding protein-1. Biol Reprod 71(2):438–443
47. Pilewski JM, Liu L, Henry AC, Knauer AV, Feghali-Bostwick CA (2005) Insulin-Like growth factor binding proteins 3 and 5 are overexpressed in idiopathic pulmonary fibrosis and contribute to extracellular matrix deposition. Am J Pathol 166(2):399–407
48. Kisseleva T, Brenner DA (2007) Role of hepatic stellate cells in fibrogenesis and the reversal of fibrosis. J Gastroenterol Hepatol 22(Suppl 1):S73–S78
49. Voeghtly LM, Kaminski N, Oury TD (2009) MMP activation peptide detection in biological samples as a diagnostic marker of idiopathic pulmonary fibrosis. FASEB J 23(1_MeetingAbstracts):572.11
50. Roderfeld M, Weiskirchen R, Wagner S, Berres ML, Henkel C, Grötzinger J et al (2006) Inhibition of hepatic fibrogenesis by matrix metalloproteinase-9 mutants in mice. FASEB J 20(3):444–454
51. de Meijer VE, Sverdlov DY, Popov Y, Le HD, Meisel JA, Nosé V et al (2010) Broad-spectrum matrix metalloproteinase inhibition curbs inflammation and liver injury but aggravates experimental liver fibrosis in mice. PLoS One 5(6):e11256
52. Maher TM (2019) Pirfenidone in idiopathic pulmonary fibrosis. Drugs Today 46(7):432–482
53. Li X, Filippatos G, Uhal BD (2004) Apoptosis in lung injury and remodelling. J Appl Physiol 97(4):1535–1542
54. Moodley YP, Scaffidi AK, Fogel-Petrovic M, McAnulty RJ, Laurent GJ, Thompson PJ, Knight DA (2003) Inverse effects of interleukin-6 on apoptosis of fibroblasts from pulmonary fibrosis and normal lung. Am J Respir Cell Mol Biol 29(4):490–498
55. Maher TM, Bottoms SE, Mercer PF, Thorley AJ, Nicholson AG, Laurent GJ, Tetley TD, Chambers RC, McAnulty RJ (2010) Diminished prostaglandin E2 contributes to the apoptosis paradox in idiopathic pulmonary fibrosis. Am J Respir Crit Care Med 182(1):73–82
56. Kuwano K, Maeyama T, Hagimoto N, Kawasaki M, Matsuba T, Yoshimi M, Inoshima I, Yoshida K, Hara N (2001) Attenuation of bleomycin-induced pneumopathy in mice by a caspase inhibitor. Am J Physiol Lung Cell Mol Physiol 280(2):L316–L325
57. Witek RP, Karaca FG, Syn WK, Pereira TA, Agboola KM, Omenetti A, Jung Y, Teaberry V, Choi SS, Guy CD, Pollard J, Charlton P, Diehl AM (2009) Pan-caspase inhibitor VX-166 reduces fibrosis in an animal model of nonalcoholic steatohepatitis. Hepatology 50(5):1421–1430
58. Canbay A, Baskin-Bey E, Bronk SF, Gores GJ (2004) The caspase inhibitor IDN-6556 attenuates hepatic injury and fibrosis in the bile duct ligated mouse. J Pharmacol Exp Ther 308(3):1191–1196
59. Pockros PJ, Shiffman ML, McHutchison JG, Gish RG, Afdhal NH, Makhviladze M, Huyghe M, Hecht D, Oltersdorf T, Shapiro DA (2007) Oral IDN-6556, an antiapoptotic caspase inhibitor, may lower aminotransferase activity in patients with chronic hepatitis C. Hepathology 46(2):324–329
60. Brix K, Mayer K, Jordans S (2008) Cysteine cathepsins: cellular roadmap to different functions. Biochimie 90(2):194–207
61. Bühling F, Brasch F, Hartig R, Yasuda Y, Saftig P, Brömme D, Welte T (2004) Pivotal role of cathepsin K in lung fibrosis. Am J Pathol 164(6):2203–2216
62. Lecaille F, Brömme D (2002) Human and parasitic papain-like cysteine proteases: their role in physiology and pathology and recent developments in inhibitor design. Chem Rev 102(12):4459–4488
63. Tang PS, Seth R, Liu M (2008) Acute lung injury and cell death: how many ways can cells die? Am J Physiol Lung Cell Mol Physiol 294(4):L632–L641
64. Lecaille F, Lalmanach G (2008) Biochemical properties and regulation of cathepsin K activity. Biochimie 90(2):208–226

65. Bühling F, Gerber A, Krüger S, Weber E, Brömme D, Roessner A, Ansorge S, Welte T, Röcken C (2001) Cathepsin K – a marker of macrophage differentiation? J Pathol 195(3):375–382
66. van den Brûle S, Bühling F, Lison D, Huaux F (2005) Overexpression of cathepsin K during silica-induced lung fibrosis and control by TGF-beta. Respir Res 6:84
67. Srivastava M, Kiviranta R, Morko J, Hoymann HG, Länger F, Buhling F, Welte T, Maus UA (2008) Overexpression of cathepsin K in mice decreases collagen deposition and lung resistance in response to bleomycin-induced pulmonary fibrosis. Respir Res 9:54
68. Canbay A, Higuchi H, Feldstein A, Bronk SF, Rydzewski R, Taniai M, Gores GJ (2003) Cathepsin B inactivation attenuates hepatic injury and fibrosis during cholestasis. J Clin Invest 112(2):152–159
69. Guicciardi ME, Bronk SF, Gores GJ (2001) Cathepsin B knockout mice are resistant to tumor necrosis factor-alpha-mediated hepatocyte apoptosis and liver injury: implications for therapeutic applications. Am J Pathol 159(6):2045–2054
70. Roberts LR, Bronk SF, Fesmier PJ, Agellon LB, Leung WY, Mao F, Gores GJ (1997) Cathepsin B contributes to bile salt-induced apoptosis of rat hepatocytes. Gastroenterology 113(5):1714–1726
71. Kasper M, Haase M, Schuh D, Müller M (1996) Immunolocalization of cathepsin D in pneumocytes of normal human lung and in pulmonary fibrosis. Virchows Arch 428 (4–5):207–215
72. Leto G, Pizzolanti G, Montalto G, Soresi M, Gebbia N (1997) Lysosomal cathepsins B and L and Stefin A blood levels in patients with hepatocellular carcinoma and/or liver cirrhosis: potential clinical implications. Oncology 54(1):79–83
73. Okazaki S, Iwata Y, Hara T, Muroi E, Komura K, Takenaka M, Shimizu K, Hasegawa M, Fujimoto M, Sato S (2010) Autoantibody against caspase-3, an executioner of apoptosis, in patients with systemic sclerosis. Rheumatol Int 30(7):871–878
74. Plataki M, Darivianaki K, Delides G, Siafakas NM, Bouros D (2005) Expression of apoptotic and antiapoptotic markers in epithelial cells in idiopathic pulmonary fibrosis. Chest 127(1):266–274
75. Piekarska A, Omulecka A, Szymczak W, Piekarski J (2007) Expression of tumour necrosis factor-related apoptosis-inducing ligand and caspase-3 in relation to grade of inflammation and stage of fibrosis in chronic hepatitis C. Histopathology 51(5):597–604
76. Akasaka Y, Ono I, Fujita K, Masuda T, Asuwa N, Inuzuka K, Kiguchi H, Ishii T (2000) Enhanced expression of caspase-3 in hypertrophic scars and keloid: induction of caspase-3 and apoptosis in keloid fibroblasts in vitro. Lab Invest 80(3):345–357
77. Scott CJ (2010) Biologic protease inhibitors as novel therapeutic agents. Biochimie 1:8
78. Scotton CJ, Krupiczojc MA, Konigshoff M, Mercer PF, Lee YC, Kaminski N, Morser J, Post JM, Maher TM, Nicholson AG, Moffatt JD, Laurent GJ, Derian CK, Eickelberg O, Chambers RC (2009) Increased local expression of coagulation factor X contributes to the fibrotic response in human and murine lung injury. J Clin Invest 119:2550–2563
79. Kubo H, Nakayama K, Yanai M, Suzuki T, Yamaya M, Watanabe M, Sasaki H (2005) Anticoagulant therapy for idiopathic pulmonary fibrosis. Chest 128:1475–1482
80. Chambers RC (2003) Proteinase-activated receptors and the pathophysiology of pulmonary fibrosis. Drug Dev Res 60:29–35
81. Hernandez-Rodriguez NA, Cambrey AD, Harrison NK, Chambers RC, Gray AJ, Southcott AM, duBois RM, Black CM, Scully MF, McAnulty RJ et al (1995) Role of thrombin in pulmonary fibrosis. Lancet 346:1071–1073
82. Imokawa S, Sato A, Hayakawa H, Kotani M, Urano T, Takada A (1997) Tissue factor expression and fibrin deposition in the lungs of patients with idiopathic pulmonary fibrosis and systemic sclerosis. Am J Respir Crit Care Med 156:631–636
83. Andersson C, Sjoland AA, Mori M, Pardo A, Eriksson L, Bjermer L, Lofdahl CG, Selman M, Westergren-Thorsson G, Erjefalt J (2010) Altered lung mast cell populations infiltrate areas of inflammation and remodeling in patients with cystic fibrosis and idiopathic pulmonary fibrosis. Am J Respir Crit Care Med 181:A1426

84. Veerappan A, O'Connor N, Jung A, Brazin J, Reid AC, Kaner RJ, Silver RB (2010) Mast cells and fibroblasts work in tandem to promote pulmonary fibrosis. Am J Respir Crit Care Med 181:A1432
85. O'Brien-Ladner AR, Wesselius LJ, Stechschulte GJ (1993) Bleomycin injury of the lung in a mast-cell-deficient model. Agents Actions 39:20–24
86. Hemmati A, Nazari AZ, Motlagh ME, Goldasteh S (2002) The role of sodium cromolyn in treatment of paraquat-induced pulmonary fibrosis in rat. Pharmacol Res 46:229–234
87. Walker M, Harley R, LeRoy EC (1990) Ketotifen prevents skin fibrosis in the tight skin mouse. J Rheumatol 17:57–59
88. Sakaguchi M, Takai S, Jin D, Okamoto Y, Muramatsu M, Kim S, Miyazaki M (2004) A specific chymase inhibitor, NK3201, suppresses bleomycin-induced pulmonary fibrosis in hamsters. Eur J Pharmacol 493:173–176
89. Tomimori Y, Muto T, Saito K, Tanaka T, Maruoka H, Sumida M, Fukami H, Fukuda Y (2003) Involvement of mast cell chymase in bleomycin-induced pulmonary fibrosis in mice. Eur J Pharmacol 478(1):79–85
90. Scandiuzzi L, Beghdadi W, Daugas E, Abrink M, Tiwari N, Brochetta C, Claver J, Arouche N, Zang X, Pretolani M, Monteiro RC, Pejler G, Blank U (2003) Mouse mast cell protease-4 deteriorates renal function by contributing to inflammation and fibrosis in immune complex-mediated glomerulonephritis. J Immunol 185:624–633
91. Kanemitsu H, Takai S, Tsuneyoshi H, Nishina T, Yoshikawa K, Miyazaki M, Ikeda T, Komeda M (2006) Chymase inhibition prevents cardiac fibrosis and dysfunction after myocardial infarction in rats. Hypertens Res 29:57–64
92. Matsumoto T, Wada A, Tsutamoto T, Ohnishi M, Isono T, Kinoshita M (2003) Chymase inhibition prevents cardiac fibrosis and improves diastolic dysfunction in the progression of heart failure. Circulation 107:2555–2558
93. Takai S, Jin D, Sakaguchi M, Katayama S, Muramatsu M, Matsumura E, Kim S, Miyazaki M (2003) A novel chymase inhibitor, 4-[1-([bis-(4-methyl-phenyl)-methyl]-carbamoyl)3-(2-ethoxy-benzyl)-4-oxo-a zetidine-2-yloxy]-benzoic acid (BCEAB), suppressed cardiac fibrosis in cardiomyopathic hamsters. J Pharmacol Exp Ther 305:17–23
94. Yamanouchi H, Fujita J, Hojo S, Yoshinouchi T, Kamei T, Yamadori I, Ohtsuki Y, Ueda N, Takahara J (1998) Neutrophil elastase: alpha-1-proteinase inhibitor complex in serum and bronchoalveolar lavage fluid in patients with pulmonary fibrosis. Eur Respir J 11:120–125
95. Song JS, Kang CM, Rhee CK, Yoon HK, Kim YK, Moon HS, Park SH (2009) Effects of elastase inhibitor on the epithelial cell apoptosis in bleomycin-induced pulmonary fibrosis. Exp Lung Res 35:817–829
96. Dunsmore SE, Roes J, Chua FJ, Segal AW, Mutsaers SE, Laurent GJ (2001) Evidence that neutrophil elastase-deficient mice are resistant to bleomycin-induced fibrosis. Chest 120:35S–36S
97. Levy MT, McCaughan GW, Abbott CA, Park JE, Cunningham AM, Muller E, Rettig WJ, Gorrell MD (1999) Fibroblast activation protein: a cell surface dipeptidyl peptidase and gelatinase expressed by stellate cells at the tissue remodelling interface in human cirrhosis. Hepatology 29:1768–1778
98. Levy MT, McCaughan GW, Marinos G, Gorrell MD (2002) Intrahepatic expression of the hepatic stellate cell marker fibroblast activation protein correlates with the degree of fibrosis in hepatitis C virus infection. Liver 22:93–101
99. Escobar NRD, Lopez R, de Lara LGV (2010) Fibroblast activation protein-a (FAP) is differentailly expressed in fibroblast cultures from normal human and idiopathic pulmonary fibrosis lungs. Am J Respir Crit Care Med 181:A3513
100. Acharya PS, Zukas A, Chandan V, Katzenstein AL, Pure E (2006) Fibroblast activation protein: a serine protease expressed at the remodelling interface in idiopathic pulmonary fibrosis. Hum Pathol 37:352–360

101. Fan MH, Majumdar RS, Kinder M, Razvi M, Christofidou-Solomiduo M (2010) A protective role for fibroblast activation protein (FAP) in pulmonary fibrosis. Am J Respir Crit Care Med 181:A1974
102. Duplantier J, Dubuisson L, Senant N, Freyburger G, Laurendeau I, Herbert JM, Desmouliere A, Rosenbaum J (2004) A role for thrombin in liver fibrosis. Gut 53:1682–1687
103. Fiorucci S, Antonelli E, Distrutti E, Severino B, Fiorentina R, Baldoni M, Caliendo G, Santagada V, Morelli A, Cirino G (2004) PAR1 antagonism protects against experimental liver fibrosis. Role of proteinase receptors in stellate cell activation. Hepatology 39:365–375
104. Howell DC, Johns RH, Lasky LA, Shan B, Scotton CJ, Laurent GJ, Chambers RC (2005) Absence of proteinase-activated receptor-1 signaling affords protection from bleomycin-induced lung inflammation and fibrosis. Am J Pathol 166:1353–1365
105. Chambers RC, Dabbagh K, McAnulty RJ, Gray AJ, Blanc-Brude OP, Laurent GJ (1998) Thrombin stimulates fibroblast procollagen production via proteolytic activation of protease-activated receptor 1. Biochem J 333(Pt 1):121–127
106. Bogatkevich GS, Tourkina E, Silver RM, Ludwicka-Bradley A (2001) Thrombin differentiates normal lung fibroblasts to a myofibroblast phenotype via the proteolytically activated receptor-1 and a protein kinase C-dependent pathway. J Biol Chem 276: 45184–45192
107. Chen LB, Buchanan JM (1975) Mitogenic activity of blood components I. Thrombin and prothrombin. Proc Natl Acad Sci USA 72:131–135
108. Duhamel-Clerin E, Orvain C, Lanza F, Cazenave JP, Klein-Soyer C (1997) Thrombin receptor-mediated increase of two matrix metalloproteinases, MMP-1 and MMP-3, in human endothelial cells. Arterioscler Thromb Vasc Biol 17:1931–1938
109. Lafleur MA, Hollenberg MD, Atkinson SJ, Knauper V, Murphy G, Edwards DR (2001) Activation of pro-(matrix metalloproteinase-2) (pro-MMP-2) by thrombin is membrane-type-MMP-dependent in human umbilical vein endothelial cells and generates a distinct 63 kDa active species. Biochem J 357:107–115
110. Gruber BL (2003) Mast cells in the pathogenesis of fibrosis. Curr Rheumatol Rep 5:147–153
111. Lindberg BF, Gyllstedt E, Andersson KE (1997) Conversion of angiotensin I to angiotensin II by chymase activity in human pulmonary membranes. Peptides 18:847–853
112. Wypij DM, Nichols JS, Novak PJ, Stacy DL, Berman J, Wiseman JS (1992) Role of mast cell chymase in the extracellular processing of big-endothelin-1 to endothelin-1 in the perfused rat lung. Biochem Pharmacol 43:845–853
113. Lindstedt K, Wang AY, Shiota N, Saarinen J, Hyytiainen M, Kokkonen JO, Keski-Oja J, Kovanen PT (2001) Activation of paracrine TGF-beta1 signaling upon stimulation and degranulation of rat serosal mast cells: a novel function for chymase. FASEB J 15:1377–1388
114. Tchougounova E, Lundequist A, Fajardo I, Winberg JO, Abrink M, Pejler G (2005) A key role for mast cell chymase in the activation of pro-matrix metalloprotease-9 and pro-matrix metalloprotease-2. J Biol Chem 280:9291–9296
115. Levi-Schaffer F, Piliponsky AM (2003) Tryptase, a novel link between allergic inflammation and fibrosis. Trends Immunol 24:158–161
116. Pesci A, Majori M, Piccoli ML, Casalini A, Curti A, Franchini D, Gabrielli M (1996) Mast cells in bronchiolitis obliterans organizing pneumonia. Mast cell hyperplasia and evidence for extracellular release of tryptase. Chest 110:383–391
117. Matsunaga Y, Terada T (2000) Mast cell subpopulations in chronic inflammatory hepatobiliary diseases. Liver 20:152–156
118. Kondo S, Kagami S, Kido H, Strutz F, Muller GA, Kuroda Y (2001) Role of mast cell tryptase in renal interstitial fibrosis. J Am Soc Nephrol 12:1668–1676
119. Apa DD, Cayan S, Polat A, Akbay E (2002) Mast cells and fibrosis on testicular biopsies in male infertility. Arch Androl 48:337–344
120. Ruoss SJ, Hartmann T, Caughey GH (1991) Mast cell tryptase is a mitogen for cultured fibroblasts. J Clin Invest 88:493–499

121. Brown JK, Jones CA, Rooney LA, Caughey GH, Hall IP (2002) Tryptase's potent mitogenic effects in human airway smooth muscle cells are via nonproteolytic actions. Am J Physiol Lung Cell Mol Physiol 282:L197–L206
122. Cairns JA, Walls AF (1996) Mast cell tryptase is a mitogen for epithelial cells. Stimulation of IL-8 production and intercellular adhesion molecule-1 expression. J Immunol 156:275–283
123. Akers IA, Parsons M, Hill MR, Hollenberg MD, Sanjar S, Laurent GJ, McAnulty RJ (2000) Mast cell tryptase stimulates human lung fibroblast proliferation via protease-activated receptor-2. Am J Physiol Lung Cell Mol Physiol 278:L193–L201
124. Frungieri MB, Weidinger S, Meineke V, Kohn FM, Mayerhofer A (2002) Proliferative action of mast-cell tryptase is mediated by PAR2, COX2, prostaglandins, and PPARgamma: Possible relevance to human fibrotic disorders. Proc Natl Acad Sci USA 99:15072–15077
125. Ginzberg HH, Shannon PT, Suzuki T, Hong O, Vachon E, Moraes T, Abreu MT, Cherepanov V, Wang X, Chow CW, Downey GP (2004) Leukocyte elastase induces epithelial apoptosis: role of mitochondial permeability changes and Akt. Am J Physiol Gastrointest Liver Physiol 287:G286–G298
126. Wallach-Dayan SB, Izbicki G, Cohen PY, Gerstl-Golan R, Fine A, Breuer R (2006) Bleomycin initiates apoptosis of lung epithelial cells by ROS but not by Fas/FasL pathway. Am J Physiol Lung Cell Mol Physiol 290:L790–L796
127. Taipale J, Lohi J, Saarinen J, Kovanen PT, Keski-Oja J (1995) Human mast cell chymase and leukocyte elastase release latent transforming growth factor-beta 1 from the extracellular matrix of cultured human epithelial and endothelial cells. J Biol Chem 270:4689–4696
128. Chua F, Dunsmore E, Clingen PH, Mutsaers SE, Shapiro SD, Segal AW, Roes J, Laurent GJ (2007) Mice lacking neutrophil elastase are resistant to bleomycin-induced pulmonary fibrosis. Am J Pathol 170:65–74
129. Park JE, Lenter MC, Zimmermann RN, Garin-Chesa P, Old LJ, Rettig WJ (1999) Fibroblast activation protein, a dual specificity serine protease expressed in reactive human tumor stromal fibroblasts. J Biol Chem 274:36505–36512
130. Goodman JD, Rozypal TL, Kelly T (2003) Seprase, a membrane-bound protease, alleviates the serum growth requirement of human breast cancer cells. Clin Exp Metastasis 20:459–470
131. Huang Y, Wang S, Kelly T (2004) Seprase promotes rapid tumor growth and increased microvessel density in a mouse model of human breast cancer. Cancer Res 64:2712–2716
132. Niedermeyer J, Kriz M, Hilberg F, Garin-Chesa P, Bamberger U, Lenter MC, Park J, Viertel B, Puschner H, Mauz M, Rettig WJ, Schnapp A (2000) Targeted disruption of mouse fibroblast activation protein. Mol Cell Biol 20:1089–1094
133. Juillerat-Jeanneret L, Gerber-Lemaire S (2009) The prolyl-aminodipeptidases and their inhibitors as therapeutic targets for fibrogenic disorders. Mini Rev Med Chem 9:215–226

Proteases/Antiproteases in Inflammatory Bowel Diseases

Jean-Paul Motta, Laurence Martin, and Nathalie Vergnolle

Abstract Protein turnover is orchestrated by a large array of proteases, and the gastrointestinal tract is the organ the most exposed to proteases, whether they originate from the pancreas for digestive purpose, from resident cells, from infiltrated inflammatory cells, or from microorganisms present in the intestinal lumen. To maintain tissue homeostasis, the gastrointestinal tract has developed mechanisms that tightly regulate the proteolytic balance in the tissues. Those mechanisms range from physical barriers (mucus layer, tight control of intestinal epithelial cell para- and transcellular passages) to molecular control of proteolytic activity through endogenous protease inhibitors. In the setting of inflammation, gastrointestinal tissues are overflowed with massive proteolytic activity that contributes to the generation of inflammatory symptoms. The present chapter analyses the disruption of the protease–antiprotease balance in the context of inflammatory bowel disease. It proposes to review the type of proteases that are present in the gastrointestinal tract, their known effects on the generation and/or maintenance of inflammatory symptoms, and their mechanisms of action. Further, the role of endogenous protease inhibitors is discussed, as well as the potential use of protease inhibition, as new possible therapeutic approach to treat chronic inflammatory disorders of the gut.

Keywords Alarmins • Antiproteases • Colitis • Colon • Crohn's disease • Gut • Inflammation • Inflammatory bowel disease • Intestinal mucosa • Intestine • Pain • Proteases • Serpins • Ulcerative colitis

---

The authors Jean-Paul Motta and Laurence Martin contributed equally.

J.-P. Motta • L. Martin • N. Vergnolle (✉)
Inserm, U1043, Toulouse 31300, France

CNRS, U5282, Toulouse 31300, France

Université de Toulouse, UPS, Centre de Physiopathologie de Toulouse Purpan (CPTP), Toulouse 31300, France
e-mail: nathalie.vergnolle@inserm.fr

N. Vergnolle and M. Chignard (eds.), *Proteases and Their Receptors in Inflammation*, Progress in Inflammation Research, DOI 10.1007/978-3-0348-0157-7_8,

Abbreviations

ACE	Angiotensin-converting enzyme
APN	Aminopeptidase N
CathG	Cathepsin G
CD	Crohn's disease
CoPP	Cobalt protoporphyrin
CTL	Cytotoxic T lymphocyte
DNBS	Dinitrobenzene sulfonic acid
DPP	Dipeptidyl peptidase
DSS	Dextran sulfate sodium
Gzm	Granzyme
HO-1	Heme-oxygenase-1
IBD	Inflammatory bowel disease
ICE	IL-1β converting enzyme
IEC	Intestinal epithelial cell
IL	Interleukin
LPMC	Lamina propria mononuclear cells
LPS	Lipopolysaccharide
MIF	Macrophage migration inhibitory factor
MMP	Matrix metalloproteinase
NF-κB	Nuclear factor-kappa B
NK	Natural killer
NLR	Nucleotide-binding domain and leucine-rich repeat containing
PAR	Protease-activated receptor
PBMC	Peripheral blood mononuclear cells
PPAR-γ	Peroxisome proliferator-activated receptor-gamma
PR3	Proteinase 3
RAS	Renin angiotensin system
SLPI	Secretory leukocyte protease inhibitor
SPATE	Serine protease autotransporters
TACE	TNF-α converting enzyme
TIMP	Tissue inhibitor of metalloprotease
TLR	Toll-like receptor
TNBS	2,4,6-Trinitrobenzene sulfonic acid
TNF	Tumor necrosis factor
UC	Ulcerative colitis

1 Introduction

Inflammatory bowel diseases (IBD) are chronic inflammatory disorders of the gastrointestinal tract characterized by alternating phases of relapse and remission periods [1]. The two major clinical forms of IBD are Crohn's Disease (CD) and

Ulcerative Colitis (UC). CD primarily involves the small and large bowel with transmural lesions, whereas UC is restricted to the colon and confined to the mucosa [2]. The main symptoms of these diseases are fever, rectal bleeding, diarrhea, and weight loss [2]. IBD are a major source of morbidity in children and adults. Their incidence is rising, particularly in young people. Moreover, IBD carry a lifelong risk of cancer, which is proportional to disease duration. Drug and surgical treatments cannot be considered as cures, as they solely relieve from some, but not all, symptoms. In addition, those treatments often induce a number of side effects. The etiology of IBD has not been fully elucidated but genetic and environmental factors seem to be involved in the pathogenesis of IBD [3]. In addition, it has been shown that germ-free mice did not develop colitis, suggesting that gut bacteria would be involved in IBD pathogenesis [4]. Proteases are heavily present in the gastrointestinal tract, where they are known to be involved in several biological processes, such as digestion of food proteins, extracellular matrix remodeling, protein processing, blood coagulation, and apoptosis. Many proteases secreted by resident cells (intestinal epithelial cells, smooth muscle cells, leukocytes) or commensal bacteria are found in the gut under physiological conditions. Antiproteases are also naturally present under physiological conditions to inhibit the prospective excessive activity of these proteases. During inflammation, the production and the activity of proteases are increased because of infiltration of activated leukocytes, apoptosis of intestinal epithelial cells, inflammatory insults to cells, and infection by pathogenic bacteria [5, 6]. Several studies, including ours, have demonstrated a crucial role for proteases in the maintenance of chronic inflammatory response in the gastrointestinal tract. Increased proteolytic activity that could be explained both by the increased production of proteases and a decreased expression of endogenous antiproteases highlights the dysregulation of the protease/antiprotease balance during the course of IBD [7–11].

This chapter will provide an overview of the involvement of proteases and their inhibitors in the context of IBD, suggesting that protease/antiprotease imbalance could lead to chronic gut inflammation. We will also discuss potential therapeutic strategies targeting these proteases and their inhibitors for the treatment of IBD.

2 Proteases and Their Inhibitors in IBD

2.1 Serine Proteases

All serine proteases are named after their conserved amino acid serine residue present at the active enzymatic site. The family of serine proteases is by far the most represented among proteases [12]. It includes neutrophil and mast cell proteases, leukocyte granzymes, kallikreins, trypsins, and coagulation cascade proteases. In rodent animal models of IBD, colitis is associated with an increase of serine protease activity by up to several fold ([13] and unpublished data from Vergnolle et al.). Intestinal samples from IBD patients also exhibit increased levels of serine protease activity in particular trypsin, chymotrypsin, and elastase [7, 8, 11]. These data revealed a state of disequilibrium between serine proteases and their endogenous inhibitors during gut inflammation. Considering the pro-inflammatory effects of a number of serine proteases, it is this tempting to consider protease inhibitors as valuable pharmacological agents to treat IBD patients.

2.1.1 Neutrophil Serine Proteases

Neutrophils are the first cells to reach the site of inflammation, thus providing a primary line of defense against pathogens. Neutrophil elastase, proteinase 3 (PR3) and cathepsin G (cathG) are serine proteases belonging to the chymotrypsin family. They are stored in primary (azurophilic) granules in active form until they are released with an array of other pro-inflammatory molecules following exposure to inflammatory stimuli. They can also act intracellularly to digest phagocytozed particles in combination with antimicrobial peptides and the NADPH-oxidase system that produces oxygen metabolites. They have long been studied for their role in the regulation of innate immunity, inflammation, and tissue destruction. This paradoxal protective (particule digestion) and destructive (tissue degradation) functions of neutrophil serine proteases emphasized the great interest to understand their respective roles and to design specific drugs that would target extra- or intracellular expression of serine proteases in neutrophils.

Neutrophil activation and prolonged survival is a hallmark of persistent intestinal inflammation such as IBD [3]. Increased levels of neutrophil elastase in plasma and feces may reflect the local status of inflammation in active stage of CD and UC. Elastase levels can thus be used to monitor disease progression [14, 15]. In a similar manner, animal models of IBD harbor increased elastase activity both in colonic lumenal washes and tissues [16]. Neutrophil elastase could act on several populations of cells to induce the release of pro-inflammatory cytokines [17–19]. Increased elastase activity could also lead to the disruption of epithelial paracellular permeability [20] indeed participating to intestinal inflammation, although this concept is still debated concerning ion permeability [21].

Interestingly, antineutrophilic cytoplasmic antibodies against PR3, named pANCA (antineutrophilic complex antibodies), develop during the disease and could be used as therapeutic marker of intestinal inflammation particularly with UC [22]. However, there is no further link yet between PR3 and IBD. In a recent study, we have demonstrated that mice deficient for both elastase and PR3 are protected against the development of colitis in a model of IBD [16]. However, because both PR3 and elastase genes were disrupted in those mice, it was impossible to conclude on the role of PR3 versus elastase in colitis pathogenesis.

Cathepsin G has long been involved in inflammation [17] but its specific role in IBD is still poorly documented. For example, it has been proposed to modify intestinal permeability in animal models of IBD [23], but the intramucosal presence of cathG activity still has to be demonstrated. The use of animal models invalidated for the different neutrophil serine proteases will be valuable for future research to define the respective roles of those proteases in the context of IBD.

2.1.2 Mast Cell Proteases

Mast cells have been extensively studied in the context of asthma, allergy, and innate immunity [24, 25]. Upon degranulation, various inflammatory mediators are released, including histamine, cytokines, proteoglycans, and the serine proteases tryptase and chymase. Tryptase and chymase constitute the major protein content of mast cells, but other serine proteases such as cathG and granzyme B are also present to a lower extent in human mast cells [26, 27]. Different studies have suggested that mast cells may play an important role in the pathogenesis of IBD [28] partly due to their serine proteases content. Chymase was reported to alter intestinal epithelial cell permeability [29, 30], but also to simultaneously activate matrix metalloproteases and inhibit TIMP (tissue inhibitor of metalloproteases) activation [31, 32]. Cultured mast cells from UC patients secreted increased levels of tryptase compared to non-IBD patients [33]. Both chymase and tryptase were detected in the mucosa of patients with IBD (both CD and UC) [34]. In non-IBD colonic mucosa, a small number of chymase immunopositive mast cells were detected contrasting with significantly higher level in active CD mucosa [35]. Although mast cell proteases are clearly pro-inflammatory in the gut, the specific role of mast cell proteases, and in particular tryptase and chymase in the pathogenesis of IBD still need to be addressed.

2.1.3 Serine Proteases of the Coagulation Cascade

Coagulation processes involve a large number of proteases that are activated sequentially. Several mediators of the coagulation cascade are altered in IBD patients, including increased fibrinogen [36], increased prothrombin fragments$_{1+2}$ and thrombin–antithrombin complexes [37], and increased factors V and VIII [36]. This increase in coagulation factors favors clots formation, and therefore predisposes IBD patients to thrombosis. Impaired fibrinolytic capacity could contribute to this hypercoagulable state in IBD patients. The plasma level of tissue plasminogen activator (tPA), the principal activator of plasminogen, the main fibrinolytic enzyme, is significantly lower in IBD patients than in the general population [38]. In addition, increases in two proteins that inhibit fibrinolysis, plasminogen activator inhibitor, and thrombin-activable fibrinolysis inhibitor have been reported [39]. Thus, both decreased activation and increased inhibition of the fibrinolytic system can contribute to hypofibrinolysis in IBD patients. Several studies report that the prevalence of factor V Leiden mutation (conferring resistance to the protein C) is more frequent in IBD patients than in the general population, although other works refute this theory [40, 41].

2.1.4 Trypsins

Trypsin is usually merely considered as a digestive enzyme released in the upper gastrointestinal tract. Three isoforms of trypsin have been cloned from the human pancreas: *PRSS1* for cationic trypsin, *PRSS2* for anionic trypsin, and *PRSS3* for mesotrypsin [42, 43]. Mesotrypsin is resistant to natural serine protease inhibitors such as soybean trypsin inhibitor [44], but also to all the endogenous protease inhibitors tested so far. Further, mesotrypsin is also capable to cleave and inactivate these inhibitors [45], thereby appearing as a form of trypsin whose activity is not endogenously controlled. Although traditionally considered as pancreatic enzymes, the different forms of trypsin can also be released in other tissues, synthesized by a number of different cell types, such as neurons, epithelial and endothelial cells [46]. In particular, mesotrypsin is expressed in epithelial cell lines derived from colon adenocarcinoma (Caco-2, T84, HT-29) and from normal colonic mucosa (NCM-460) ([47] and unpublished data from J.P. Motta). A significant increase in trypsin-like activity was detected in colonic lumenal washes and colonic tissues from animal models of IBD [16]. Data collected from IBD patients tissue biopsies showed an increased serine protease activity similar to trypsin-like enzymes, compared to the activity detected in healthy individuals [11]. These results strongly suggest that uncontrolled increase of intestinal trypsin activity takes place in chronically inflamed gut and might play a role in the pathogenesis of IBD. The cellular origin, function, and regulation of this unique form of trypsin still remain to be elucidated.

2.1.5 Kallikreins

Kallikreins are specialized serine proteases that cleave kininogens, their physiological substrates, into active kinins. The kallikrein–kinin system is involved in many inflammatory diseases and evidence has emerged indicating the role of activation of the kinin system in IBD [48]. In experimental models of IBD, up-regulation of the kallikrein–kinin system has been observed, leading the authors to use a kallikrein inhibitor strategy to reduce intestinal inflammation [49]. In preliminary data, activation of intestinal kallikreins in a small series of patients with UC was observed, compared to healthy individuals. More recently, this result was confirmed in a larger study including both CD and UC patients [50–52]. Intestinal tissue kallikrein, the key kallikrein protease in the gut could also play a key role in IBD by releasing bradykinin, a potent mediator of inflammation. Bradykinin has multiple pathophysiological effects in the intestine that correspond to clinical features of IBD: it increases intestinal capillary permeability [53], induces pain [54], contributes to diarrhea [55], and alters intestinal motility [56]. Bradykinin can also prompt macrophages to secrete interleukin-1β (IL-1β) and tumor necrosis factor-α (TNF-α), two cytokines that play a key role in experimental and human IBD [57]. Bradykinin receptors are present in the human gut and are up-regulated during inflammation [58]. Indeed, therapeutic benefits of bradykinin receptor antagonism have been demonstrated in an experimental model of colitis [59]. Nevertheless, there is a lack of clinical studies to demonstrate the involvement of the kallikrein–bradykinin system in colonic tissues taken from IBD patients.

2.1.6 Granzymes

Leukocytes like natural killer (NK) cells, cytotoxic T lymphocytes (CTL), mast cells, and basophils employ serine proteases called granzymes. Those serine proteases participate in the clearance of infected cells via intracellular caspase-mediated cell death [60]. The role of cell-mediated cytotoxicity in the pathogenesis of IBD has been controversial as reports indicate either a decreased or an increased activity of cytotoxic T cells in active stages of the disease [61]. However, in the context of IBD, data reported an increased number of activated cytotoxic cells close to the epithelial mucosa. In the same study, the authors report that intestinal epithelial cells themselves induced activation of cytotoxic T cells [62, 63]. Close proximity of epithelial cells to cytotoxic T cells naturally lead to intestinal epithelial cell death. There is also evidence of intestinal epithelial cell apoptosis up-regulation not only in UC [64, 65], but also in CD [66, 67]. Granzyme A (GzmA), also named leukocyte tryptase, clearly participates in the pathogenesis of intestinal inflammation and epithelial cell loss in animal models of IBD [68]. Using an infectious colitis model, researchers reported increased specific GzmA activity in the large intestine, and this was associated with increased NK and CTL leukocyte infiltration [69]. In a study using both UC and CD biopsies, GzmA mRNA-expressing cells were significantly higher in the lamina propria and the epithelial layer [62]. Another member of granzyme family, the GzmB is also suspected to be involved in IBD [67]. In this study, authors have shown that GzmB mRNA was increased in mucosal lesion areas of CD, supporting a role for this protease in properties of lymphocytes to induce cell death during IBD. Despite their intracellular localization and functions during apoptosis, granzymes have alternative extracellular functions in the context of inflammation such as extracellular matrix remodeling [60]. During inflammatory disorders, levels of granzymes like GzmA and GzmB is elevated in the plasma of patients [70, 71]. However, further experiments are needed to clarify the intracellular and extracellular roles of granzymes in IBD.

2.1.7 Serine Protease Inhibitors in IBD

Under physiological conditions, a balance exists between serine proteases and their endogenous inhibitors. Serine antiproteases have co-evolved with proteases in order to control their destructive nature to the surrounding intestinal tissue. The crucial role of protease inhibitors in intestinal mucosal immunity has also led researchers to investigate their therapeutic potential [72, 73]. The activities of serine proteases are mainly regulated by the serpin and chelonianin families and to a lesser extent by other classes of antiproteases.

Serpins

The serpins superfamilly is composed of multiple members classified in nine clades. Many of them have antiproteolytic capacities via a suicide substrate-like inhibitor [74]. Clade A, also called antitrypsin-like, is the largest clade of serpin and contains extracellular inhibitors.

SerpinA1 or α1-antitrypsin is the most expressed antiprotease, produced constitutively by the liver, and inhibits neutrophil and mast cell serine proteases, thrombin, trypsins, and kallikreins [75]. Some authors suggest that genetic deficiency of SerpinA1 might be associated with IBD [76–80], whereas others did not find correlation between SerpinA1 expression and clinical UC activity [80].

SerpinA3 or α1-antichymotrypsin is constitutively produced in the liver and inhibits digestive chymotrypsin and mast cell chymases and cathepsin G [74]. This antiprotease was highly detected in human IBD feces. This expression was such that it was proposed to be used as a marker [81].

Kallistatin or SerpinA4 is a specific regulator of tissue kallikrein in vivo and like tissue kallikrein, it is widely distributed in tissues [74]. Studies have reported a decreased plasma level and mRNA of kallistatin in UC and CD patients [51, 52, 82, 83]. However, more studies are needed to evaluate a possible defect in kallikrein–kallistatin balance in IBD. Hypofibrinolysis process that prevails in IBD is a consequence of the reduction in fibrinolysis proteases activators (see previous paragraph), and could be explained by an increase in fibrinolysis proteases inhibitors, such as SerpinE1 or PAI-1 (plasminogen activator inhibitor) [84]. In that context, the level of SerpinC1 (antithrombin or antithrombin III), which is a potent endogenous thrombin inhibitor, is significantly decreased in the plasma of IBD patients [85, 86].

Chelonianin Inhibitors

Secretory leukocyte protease inhibitor (SLPI) and Skin-derived antileukocyte protease (SKALP)/elafin are the main chelonianin endogenous inhibitors of serine proteases. Those antiproteases are locally secreted by resident cells or inflammatory cells at the site of inflammation [75]. Their spectrum of inhibition is different since elafin inhibits neutrophil elastase and proteinase 3, while SLPI inhibits neutrophil elastase, cathepsin G, trypsin, chymotrypsin, and mast cell tryptase and chymase [75, 87]. In the colon, elafin and SLPI have been found in normal tissue [88, 89]. In UC tissues, mRNA and immunostaining for elafin and

SLPI are increased [10, 90–92] whereas in CD patient tissues, level of elafin was not changed between inflamed and noninflamed tissues [10, 90]. In a recent study, we have investigated the effects of elafin delivery through the intralumenal administration of adenoviral vectors expressing elafin, in mouse models of IBD. Our results have demonstrated that elafin transient delivery exerted a strong anti-inflammatory effect, against all the parameters of inflammation in those animal models of IBD. The protective effects of elafin delivery were further demonstrated in mice transgenic for the expression of elafin, which were protected against the development of inflammatory parameters, in two models of IBD [16]. Interestingly, elafin delivery was not further beneficial against the development of colitis, in mice deficient for the expression of PR3 and elastase, suggesting that elafin exerts its anti-inflammatory properties through the inhibition of those enzymes [16].

Other Inhibitors

Serine protease could be inhibited by the polyvalent protease inhibitor α2-macroglobulin. It is a large plasma protein produced by the liver that plays an important role in the regulation of the coagulation cascade by inhibiting various classes of proteases (serine-, cystein-, aspartic-, and metalloproteases). Some data propose to graduate inflammatory state in IBD by dosing this protein in human stool because its quantity may reflect the disease activity [80]. Under normal conditions, activated protein C and its cofactor protein S inhibit activated factors V and VIII, and so prevent thrombosis processes [93]. Although reports on the levels of protein C and protein S in IBD patients are conflicting, few reports show a deficiency of protein C and of protein S during active bowel disease [94–97].

In order to establish the concept that antiproteases might be beneficial in intestinal inflammation, authors have used pharmacological inhibitors. For instance, inhibition of tryptase and chymase was shown to inhibit leukocyte peritoneum infiltration [98]. Nafamostat mesilate (or FUT-175) a broad-spectrum serine protease inhibitor and neutrophil elastase specific inhibitor has been used and efficiently reduced parameters of inflammation in a model of IBD [28, 99, 100]. Altogether, these findings underline that defects in serine antiprotease functions and/or expression can have serious clinical consequences in the context of IBD. And thus the serine protease inhibitors may be considered as therapeutic agents for IBD.

2.2 *Aspartate Proteases and Their Inhibitors*

Aspartate proteases (also called aspartic or aspartyl proteases) are a subfamily of proteolytic enzymes which belong to endonucleases family. These enzymes are called "aspartate proteases" because they use an aspartate residue for catalysis of their peptide substrates.

2.2.1 Cathepsin D

Cathepsin D belongs to this protease family. Its major function is the digestion of proteins and peptides within the acidic compartment of lysosomes [101]. It has been shown that cathepsin D protein is expressed in intestinal mucosa of mice with acute or chronic dextran sulfate sodium (DSS)-induced colitis, while this lysosomal enzyme is absent in noninflamed mucosa [102]. Moreover, two studies have demonstrated that cathepsin D, not detected in control patients, is expressed at the protein level in lamina propria mononuclear cells (LPMC) from IBD patients [103], and more precisely in intestinal macrophages [102]. Such expression profile of cathepsin D might only reflect the active metabolic state of infiltrated leukocytes in inflamed tissues, and not necessarily the implication of this protease directly in the pathogenesis of IBD. However, the inhibition of cathepsin D by pepstatin A ameliorated the DSS-induced colitis in mice [104]. The inhibition of cathepsin D could thus prevent mucosal damage in the context of IBD, most likely by inhibiting the activated status of infiltrated leukocytes and could, from that point of view, represent an interesting therapeutic approach in IBD.

2.2.2 Renin

Renin, also known as angiotensinogenase, belongs to the aspartate protease family too. Renin is responsible for the production of angiotensin I from angiotensinogen. Angiotensin I is then processed into angiotensin II thanks to the angiotensin converting enzyme (ACE). It was demonstrated that colonic mucosal levels of angiotensin I and II were greater in patients with IBD than in normal subjects. Mucosal levels of angiotensin I and II were also higher in CDs than in UC, suggesting that angiotensin I and II may have a role in IBD [105]. More recent studies have demonstrated that the renin angiotensin system (RAS) is involved in the pathogenesis of IBD. In comparison with wild-type mice, the trinitrobenzene sulfonic acid (TNBS)-induced colitis is less severe in angiotensinogen-deficient ($Atg^{-/-}$) mice and the colonic production of IL-1β and interferon-γ (IFN-γ) is impaired in $Atg^{-/-}$ mice [106]. Interestingly, administration of an angiotensin II receptor antagonist inhibited the induction of colitis by TNBS [106]. Moreover, DSS-induced colitis is improved in mice deficient for the angiotensin II type 1 receptor. Thus, interference with angiotensin receptors, or the formation of angiotensins through renin inhibition for example, might constitute a novel therapeutic outcome in IBD. Nevertheless, the RAS is also involved in physiological processes such as arterial blood pressure control, which might be severely impaired, provoking heavy side-effects, if RAS control is proposed as a therapeutic option.

2.3 *Cysteine Proteases and Their Inhibitors*

2.3.1 Caspases

Caspases are an evolutionary conserved family of aspartate-specific cysteine proteases. These enzymes always cleave proteins following an aspartic residue. Most caspases are involved in the enzymatic pathways leading to apoptosis [107]. To prevent undesired cell death as a consequence of unscheduled caspase activity, these proteases are produced as latent zymogens. CARD (Caspase activation recruitment domain) motifs are present on a number of proteins that promote apoptosis, including caspases. Mutations in the CARD15 gene have been associated to susceptibility to CD [108, 109].

In contrast to most caspases, the main function of caspase-1 is not related to apoptosis. Caspase-1, also known as IL-1β converting enzyme (ICE), is required for the processing of IL-1β and IL-18, proinflammatory cytokines involved in the pathogenesis of IBD, into their bioactive mature forms [110]. ICE-deficient mice have a defective production and release of mature, bioactive IL-1β and IL-18, whereas the precursor forms are normally synthesized. Caspase-1 itself exists as an inactive precursor and requires two internal cleavages to become active [111, 112]. Several studies demonstrated that the active form of ICE was expressed in LMPC from IBD patients [113], mainly in intestinal macrophages [114, 115] but not in colonic mucosa of non-IBD controls, suggesting a role for this cysteine protease in IBD. Moreover, it has been demonstrated that ICE-deficient mice were less susceptible to chronic DSS-induced colitis [116]. Caspase-1 can also form an inflammasome complex with the adaptor molecule ASC and the NLR (nucleotide-binding domain and leucine-rich repeat containing) protein NLRP3 in response to various stimuli [117]. It was demonstrated that DSS-induced colitis provoked the activation of caspase-1 via the NLRP3 inflammasome [118]. The pharmacological inhibition of caspase-1 with pralnacasan ameliorated DSS-induced colitis both in Balb/c [119] and in C57BL/6 mice [120] and led to a mucosal protection comparable to NLRP3 deficiency [118]. These results are counterbalanced by the findings of Zaki and collaborators, who have demonstrated that NLRP3-deficient mice were highly susceptible to DSS-induced colitis [121].

Neutrophil-mediated damage has been shown to be associated with the development of IBD. Neutrophils isolated from patients with IBD showed decreased spontaneous apoptosis and a reduced expression of pro-caspase-3, which may lead to the persistence of the inflammation associated with IBD [122].

Defective apoptosis of lamina propria T lymphocytes has also been proposed as a key pathogenic mechanism in IBD. Under normal conditions, activated T lymphocytes are rapidly eliminated by apoptosis. However, increased expression of antiapoptotic molecules has been observed in lamina propria lymphocytes from CD patients, in comparison with lamina propria lymphocytes from noninflammatory intestinal mucosa. Moreover, it has been shown that lamina propria lymphocytes from CD and UC patients displayed a reduced sensitivity to apoptosis in response to different pro-apoptotic stimuli [123]. In CD, decreased levels of caspase-3 and caspase-8 were described in activated memory T cells [124]. The induction of T cell apoptosis in the intestine could lead to a rapid elimination of activated T cells in the gut mucosa and thus represents an interesting option for IBD therapy. As a matter of fact, several strategies leading to caspase activation were tested. Vitamin D derivatives activated caspase-3 in peripheral blood mononuclear cells (PBMC) [125]. Sulphasalazine, the conjugate of 5-amino salicylic acid (5-ASA), and sulphapyridine decreased the expression of antiapoptotic molecules and increased the activation of caspase-3 and caspase-9, thus inducing lamina propria T lymphocytes apoptosis [123]. L-Cysteine and L-tryptophan independently restored susceptibility of activated T cells to apoptosis by increasing expression of the apoptosis initiator caspase-8. Moreover, the administration of L-cysteine or L-tryptophan improved DSS-induced colitis in piglets [126, 127]. Visilizumab, a humanized low Fc receptor binding anti-CD3 antibody, induced dose- and time-dependent apoptosis of lamina propria T cells from control and IBD patients. This apoptosis was dependent on caspase-3 and caspase-8 but not caspase-9 activation [128]. Moreover, the targeted depletion of T cells expressing the IL-2 receptor with the fusion protein IL-2-caspase-3 ameliorated the outcome of experimental inflammatory colitis [129–131].

The balance between apoptosis and regeneration of epithelial cells seems to be a key factor to maintain intestinal homeostasis [132]. In contrast to T lymphocytes, an increase in pro-apoptotic molecules including caspases has been described in intestinal epithelial cells of IBD patients. An apoptosis-specific gene array expression profiling of intestinal epithelial cells from UC patients demonstrated an increased expression of caspase-1, caspase-5, and the inhibitor of apoptosis protein-2 (c-IAP2), whereas caspase-14 expression was unchanged in comparison to controls [133]. Moreover, caspase-3 activity was increased in colonic tissues in an animal model of colitis [134, 135]. Villin, an epithelial cell-specific antiapoptotic protein, has been shown to inhibit the activation of caspase-3 and caspase-9. Absence of villin predisposed mice to DSS-induced colitis by promoting apoptosis [136]. The inhibition of caspase expression and/or activity could decrease intestinal epithelial cells apoptosis and improve intestinal inflammation.

Oxidative stress takes place in the pathogenesis of UC. It has been shown that deoxyschisandrin inhibited H_2O_2-induced caspase-3 activation in intestinal epithelial cells (HCT116) in vitro by blocking cleavage of pro-caspase-3 [127]. In vitro, induction of Heme-oxygenase-1 (HO-1) by the HO-1 inducer cobalt protoporphyrin (CoPP) down-regulated caspase-3 activation in HT-29 cells [137]. In vivo, preventive HO-1 induction by CoPP in acute DSS-induced colitis decreased the number of apoptotic epithelial cells in the colon and led to a significant down-regulation of colonic inflammation. However, the induction of HO-1 may not be a promising approach for the treatment of IBD as no beneficial effects were observed if HO-1 was induced by CoPP after the onset of acute colitis or in chronic DSS-induced colitis [138].

In vitro, TNF-α/butyrate-induced apoptosis led to caspase-3 activation in Caco-2 cells. Inhibitors of caspase-8 (z-IETD.fmk) and caspase-10 (z-AEVD.fmk) reduced TNF-α/butyrate-induced apoptosis, thus reducing cell death. The inhibitor of caspase-8, but not that of caspase-10, prevented the decrease of transmembrane resistance induced by TNF-α/butyrate [138].

In animal models of colitis, melatonin decreased the caspase-3 activity in colonic tissues of rats treated with TNBS and reduced mucosal damage [138, 139]. Glutamine treatment also reduced colonic damage in TNBS-induced colitis by increasing HO-1 expression and decreasing caspase-3 activity [135]. Cyclosporine, a potent immunomodulator, has been shown to decrease epithelial apoptosis in DSS-induced colitis through inhibition of caspase-8 activity [132]. This mechanism could explain the beneficial effect of cyclosporine in the treatment of UC.

Anti-TGF-β antibodies up-regulated caspase-8 activity and reduced the expression of c-FLIP (an inhibitor of caspase-8 activity) in intestinal epithelial cells in vivo, thus increasing their apoptosis and leading to an amplification of the inflammation in gut mucosa. This effect could explain, at least in part, the protective role of TGF-β shown against intestinal injury in vivo [140].

2.3.2 Cathepsins B and L

In contrast to cathepsin D (see paragraph "Aspartate proteases and their inhibitors"), cathepsin B and cathepsin L belong to the cysteine protease family. The expression of these cathepsins was increased in intestinal macrophages from IBD patients [104]. Moreover, the combined inhibition of cathepsin B and cathepsin L by CA-074 and Z-Phe-Tyr-aldehyde was followed by an amelioration of DSS-induced colitis in mice [104] and IL-1β secretion was significantly reduced in cathepsin B- and, to a lesser extent, in cathepsin L-deficient macrophages treated with DSS in vitro [118].

2.3.3 Calpains

Calpains are intracellular nonlysosomal proteins which belong to the cysteine protease family too. Calpain inhibitor I reduces the degradation of the inhibitor IκB in the proteasome and hence prevents the translocation of nuclear factor-κB (NF-κB) from the cytosol into the nucleus. The administration of calpain inhibitor I to rats reduced the severity of the DNBS-induced colitis (weight loss, hemorrhagic diarrhea, colon injury, MPO activity) [141], but no further studies have demonstrated the involvement of calpains in IBD. Here again, the use of mice genetically modified for the expression of calpains would be of great interest to understand the role of these proteases in the setting of IBD.

2.4 Metalloproteases and Their Inhibitors

2.4.1 Matrix Metalloproteases and Their Inhibitors

Matrix metalloproteases (MMPs) are a family of metal-dependent enzymes responsible for the degradation and remodeling of extracellular matrix and basement membrane proteins that occurs during both normal physiologic activity and disease [142]. These proteases cleave components of the extracellular matrix and their dysregulation leads to damage to the mucosa [143]. Several studies have reported an up-regulation of MMPs in IBD, suggesting an important role for these enzymes in the process of tissue remodeling and destruction in these inflammatory diseases. MMPs are now considered the predominant proteases involved in the pathogenesis of mucosal ulcerations associated with IBD [144].

The study of gene expression patterns of MMPs and their endogenous tissue inhibitors (tissue inhibitors of metalloprotease, TIMPs) in single endoscopic biopsies of patients with IBD revealed an increase in MMP-1 (interstitial collagenase) and MMP-3 (stromelysin-1) mRNA expression in inflamed versus normal colon samples from patients with CD and UC. MMP-2, MMP-14, and TIMP-1 mRNAs were increased in ulcerated colonic mucosa of IBD patients, whereas TIMP-2 mRNA expression remained unchanged [90, 145]. MMP-12 mRNA expression was also increased in UC patients [90]. MMP-3 and its endogenous tissue inhibitor TIMP-1 were undetectable at the protein level in noninflamed samples from either healthy subjects or IBD patients. However, it has been demonstrated that the production of these mediators was significantly increased in inflamed mucosa, both in CD and UC [146]. A more recent study also showed a spontaneous expression of several MMPs transcripts in human colonic epithelial cells (CEC): MMP-1, -3, -7, -9, -10, and -12. All of these, except MMP-12, were expressed at significantly increased levels in cells from inflamed IBD mucosa. Importantly, proteolytic MMP activity has been detected in CEC supernatants. This proteolytic activity, increased in inflamed IBD epithelium, was strongly inhibited by GM 6001, a specific MMP inhibitor [147]. MMP-3 and TIMP-1 seem to have an important role in IBD as allelic composition at the examined single nucleotide polymorphism (SNP) in genes coding for these two proteins affects CD susceptibility and/or phenotype, i.e., fistulizing disease, stricture pathogenesis, and first disease localization [148]. In almost all the studies, no difference in MMPs expression was found between intestinal tissues from CD or UC patients. A recent study has demonstrated that in the colon, expression of MMP-7 in epithelium was significantly greater in CD samples compared to UC. From this work, it was suggested that expression of this MMP could help in the diagnosis of CD versus UC [145].

In human colonic cell lines (HT-29 and DLD-1 cells), a spontaneous expression of several MMPs has also been shown at the transcript level (MMP-1, -3, -7, 10, -11, -13). The expression of MMP-3, -10, and -13 was increased after TNF-α exposure. It is noteworthy that MMP-12 transcripts were only detected after

TNF-α treatment in these cells, demonstrating that the expression pattern of MMPs in in vitro cell models was different from that of ex vivo CEC [149].

Although the expression of several MMPs has been shown to increase in IBD, all MMPs are not involved in intestinal inflammation. For example, MMP-28 (epilysin) expression has not been associated with inflammatory and destructive changes seen in IBD [149].

MMPs can be produced by several cell types in the intestinal mucosa such as myofibroblasts, smooth muscle cells, intestinal epithelial cells, endothelial cells, activated T cells, and macrophages [146, 147, 150–152]. It appears that MMP-3 can also be produced by mucosal IgG positive plasma cells during IBD [153].

The expression of MMPs can be modulated in vitro by several pro-inflammatory cytokines released during IBD, such as TNF-α, IL-1β, and IL-21. For example, MMP-9 expression and secretion were up-regulated by IL-1β and TNF-α in Caco-2 cells, but not by lipopolysaccharide (LPS) [154]. However, it has been demonstrated that LPS, as well as phytohemagglutinin (PHA) or TNF-α, increased MMPs activity of PBMC from IBD patients [155]. IL-21, a cytokine highly produced by T cells in IBD, enhanced fibroblast production of MMP-1, MMP-2, MMP-3, and MMP-9, but not tissue inhibitors of MMP-1 and MMP-2, without affecting gene transcription and protein synthesis. IL-21 synergized with TNF-α to increase MMPs synthesis. It has also been demonstrated that supernatants of cultured LPMC harvested from IBD patients stimulated MMP secretion by intestinal fibroblasts, an effect partly inhibited by antibodies directed against IL-21 receptor [156].

As MMPs expression and activity were shown to be up-regulated in IBD, the use of MMPs inhibitors may be useful for IBD therapy. Several studies have investigated the effect of direct or indirect inhibition of MMPs expression or activity. The concept of beneficial effects resulting from MMP inhibition in IBD was proven with the use of siRNA in a mouse model of colitis. Indeed, siRNA that targeted MMP-3 and MMP-10 reduced both the transcription of these MMPs and the severity of DSS-induced colitis [157].

MMP-2 and MMP-9 are two known gelatinases with opposite actions and which expression and activity are up-regulated during IBD. Epithelial-derived MMP-9 is an important mediator of tissue injury in colitis, whereas MMP-2 protects against tissue damage and maintains gut barrier function [158]. These two proteins are structurally similar, so it is possible that inhibition of MMP-9 would also likely inhibit MMP-2. Experiments carried out in double knockout mice (dKO) lacking both MMP-2 and MMP-9 demonstrated that dKO mice were resistant to the development of colitis, whereas wild-type mice had extensive inflammation and tissue damage after administration of DSS, *Salmonella typhimurium*, or TNBS. Those studies suggest that MMP-9 could have an overriding role in mediating tissue injury, compared with the protective role of MMP-2 in the development of colitis. Thus, inhibition of MMP-9 may be beneficial to treat gut inflammation, even if it also results in the inhibition of MMP-2 [160]. Several companies have developed inhibitors of MMP-9 with the aim of having a potential benefit for IBD therapy. Rosiglitazone, a peroxisome proliferator-activated receptor-gamma (PPAR-γ) agonist, reduced MMP-9 secretion in Caco-2 cells, indicating that agents that activate PPAR-γ may have therapeutic use in patients with IBD [154]. ZK 156979, a vitamin D analog, also inhibited MMP-9 activity in PBMC from IBD patients [155]. MMP-9 activity was also down-regulated in human intestinal epithelial cells in inflammatory conditions by three N-un-substituted curcuminoid pyrazoles synthesized from the corresponding β-diketones (including curcumin), making them original candidates for the treatment of IBD [159]. Curcumin, a component of the spice turmeric, and an anti-inflammatory and anticancer agent, was shown to inhibit MMP-3 expression in colonic myofibroblasts (CMF) from children and adults with active IBD [160]. Although all those drugs act, at least in part, through the inhibition of MMPs, this might not be their only mechanism of action, and most of them act also on other inflammatory mediators. Therefore, MMP inhibition cannot be considered as the unique target for the treatment of IBD.

Plasma MIF (macrophage migration inhibitory factor) has been shown to be elevated in patients with CD or UC compared with healthy individuals. In animal models of colitis, an anti-MIF antibody prevented MMP-13 up-regulation by DSS and ameliorated TNBS colitis [161].

Mucosal inflammation leads to accumulation of ceramide in the intestinal epithelium. Ceramides are second messengers produced after hydrolysis or de novo synthesis of plasma membrane sphingomyelin via the action of sphingomyelinases In vitro, it was shown that generation of ceramide by exogenous sphingomyelinases increased MMP-1-protein production by Caco-2 cells. Inhibition of acid sphingomyelinase by imipramine completely blocked induction of MMP-1 by Caco-2 cells under inflammatory conditions [162].

As we discussed earlier, several studies have shown the involvement of MMPs in the pathogenesis of IBD. It was recently demonstrated for the first time that MMP levels were elevated in urine of children and young adults with known or suspected IBD. Thus, in addition of being interesting molecular targets for the treatment of IBD, MMPs might also serve as biomarkers of IBD activity, at least in young people [144].

2.4.2 TNF-α Converting Enzyme

TNF-α plays a central role in the pathogenesis of IBD. TNF-α production is regulated by a pleiotropic metalloprotease: TNF-α converting enzyme (TACE), also known as ADAM-17. TACE protein is widely expressed in colonic LPMC and in human colonic epithelium [163] and is the major enzyme responsible for the release of TNF-α in human colonic mucosa [164]. TACE thus appears as an attractive target in IBD. Nevertheless, studies using inhibitors of TACE led to conflicting results. In a rat model of colitis, TNBS increased TACE expression and activity in rat colonic samples. B1110, a TACE inhibitor, blocked TNBS-induced increase in TACE activity and ameliorated TNBS-induced damage and inflammation [165]. In humans, functional TACE activity has been demonstrated in HT-29 and DLD-1 CEC lines and in CEC of healthy individuals or IBD patients, irrespective of disease activity. CH4474, another synthetic TACE inhibitor, almost completely abolished the release of soluble TNF-α protein by HT-29 and DLD-1 cells [163]. These studies show that the inhibition of TACE may be beneficial in the treatment against IBD. In contrast, a recent study has demonstrated that TACE inhibition, either by its endogenous inhibitor TIMP-3 or by Tapi-2, a pharmacological inhibitor of ADAM17, amplified the TNF-α-mediated hyper-permeability of an in vitro intestinal epithelial cell monolayer. The authors suggest that TIMP-3 could exert an autocrine effect, leading to amplified epithelial barrier hyper-permeability in inflammatory conditions [166]. These conflicting results may be explained by the specificity of the inhibitors used, as CH4474 has been shown to be a broad spectrum MMP inhibitor that also inhibits TACE [167], whereas Tapi-2 seems to be a selective TACE inhibitor [167, 169]. Besides, it was demonstrated that inhibition of TNF-α, especially with the anti-TNF-α antibody infliximab, improved symptoms in IBD patients. Infliximab was shown to down-regulate MMP-1, -3, and -9 expression and to decrease MMP-1 and -3 activities. The action of infliximab on MMPs may thus contribute to a large extent, but not solely, to the observed therapeutic efficacy of this drug in IBD [168].

2.5 *Exopeptidases*

2.5.1 Dipeptidyl Peptidase

The dipeptidyl peptidase (DPP) family has rare substrate specificity by cleaving N-terminal dipeptides in the penultimate position at a proline or alanine residue [169]. The most studied member of this family in the context of IBD is DPP4, a ubiquitously expressed transmembrane protein, also known as CD26 which exert co-stimulatory functions on the immune response [170]. While resting B cells expressed low amounts of DPP4 at their surface, its expression is up-regulated upon B cell activation [171]. DPP4 has multiple substrates, which includes chemokines, cytokines, peptide hormones, and neuropeptides [172]. In animal models of IBD, increased DPP4 and DPP2 mRNA expression and increased DPP activity have been shown [173], contrasting with another study performed in tissues from CD patients [174]. Yet, another study has also reported that CD patients harbor an increased number of DPP4 expressing T cell [175], suggesting that DPP4 might play a role in perpetuating inflammatory response with IBD.

2.5.2 Aminopeptidase

Aminopeptidase N (APN), or CD13, is a zinc-dependent exopeptidase. Resting T lymphocytes lack immunohistochemically detectable APN, while elevated APN mRNA and surface expression were reported on activated T cells derived from local sites of inflammation [176]. Only few studies mentioned APN in the context of

IBD, and in a large clinical analysis authors did not find any correlation between IBD and APN expression [90]. Although this proteolytic target might be relevant in the context of visceral pain originating from the colon (see Chapter 11 by N. Cenac in this book), no evidence points yet to CD13, as a potential therapeutic target to treat gut inflammation per se.

2.5.3 Carboxypeptidases

Mast cells expressing both chymase and tryptase also largely express an exopeptidase called carboxypeptidase A [177]. Although numerous studies point to a role for mast cell proteases in IBD pathogenesis (see earlier paragraph in the present chapter), the involvement of the mast cell carboxypeptidase A in IBD has been barely investigated. The role of carboxypeptidase B2 in IBD is even more obscure. Conflicting reports showed either low plasma levels of carboxypeptidase B2 or thrombin-activatable fibrinolysis inhibitor (TAFI) that could contribute to the prothrombotic state associated with IBD [39], or increased levels of TAFI in IBD [84].

2.5.4 Angiotensin-Converting Enzyme

The RAS is strictly related to the kallikrein–kinin system and both are suspected to be involved in the pathogenesis of IBD. Angiotensin-converting enzyme (ACE) is the key exopeptidase of the RAS and by the way the main degradative enzyme of bradykinin. In accordance to the putative role of bradykinin in IBD, it has been reported that ACE level in both CD and UC patients was lower compared to non-IBD biopsies [178] and was not due to a genetic defect [179].

2.5.5 Exopeptidase Inhibitors

The protease/antiprotease concept has led researchers to investigate whether DPP inhibitors had anti-inflammatory properties. This was proven by a combined therapy against DPP4 and APN in animal models of IBD [180]. Targeting of DPP4 or APN on activated T lymphocytes from mouse and humans results in a suppression of cell proliferation and a reduced synthesis of pro-inflammatory cytokines [172]. In animal models of IBD, others have tested pharmacological inhibitors of DPP to treat the colitis and showed reduced disease activity in those treated mice [181]. These are further evidences that DPP4 inhibition could be essential for a reduced inflammatory response. A possible mechanism by which DPP4 could be deleterious for the intestinal recovery after burden could be in part by its hydrolysis of intestinal growth factors like Glucagon-like peptide 2 (GLP-2). Using DPP4-deficient mice, authors argued a possible protection in those mice against colitis. Although mice developed a colitis as severe as control mice, authors detected as much DPP activity as in control mice, which thus could be explained in part by other exopeptidases redundancy [182]. Therapeutic usefulness of strategy using inhibitor of ACE exopeptidase in experimental models of IBD is still highly controversial [183–185]. Further studies are yet required to better characterize the specific role of these enzymes in the course of colitis.

2.6 Microbial Proteases and Their Inhibitors

2.6.1 Bacteria

Several studies have suggested that intestinal lumenal content, including commensal bacteria, may be involved in the development of IBD. This theory is mostly supported by the fact that germ-free mice do not develop colitis [4], and the fact that in IBD patients, the microbiota composition seems to differ from healthy subjects [186]. However, the specific role of intestinal bacteria in IBD remains elusive. A dysregulation of the balance between species having putative "protective" versus "harmful" properties has been suggested in the context of IBD. The origin of this "dysbiosis" is not well defined and whether dysbiosis is the cause of IBD development or a secondary phenomenon due to the chronic intestinal inflammation remains a key question [187]. Analyses of intestinal microflora have shown increased concentrations of *Bacteroides* species [187] and *Enterobacteriaceae* (especially *Escherichia coli*) in CD and UC patients [188] but no "pathogenic" bacteria strain has been clearly associated with IBD development so far.

The high prevalence of *E. coli* belonging to the B2 + D phylogenetic group, described in IBD patients, has been associated with the expression of serine protease autotransporters (SPATE) and adherence factors [188]. SPATE constitute a large family of extracellular proteases secreted by *Enterobacteriaceae*, which contain a conserved serine-protease motif and exhibit two distinct proteolytic activities [189]. Most SPATE are proven or putative virulence factors; these proteases could thus be involved in the pathogenesis of IBD.

Enterotoxigenic *Bacteroides fragilis* (ETBF), a molecular subgroup of *B. fragilis*, has been identified to cause diarrhea in animals and humans. ETBF has also been suggested to be associated with active IBD [190, 191]. The only known virulence factor of ETBF is the *B. fragilis* toxin (BFT), also called fragilysin. This protease was shown to decrease transepithelial electrical resistance of intestinal epithelial cells in vitro and to increase human colonic permeability ex vivo [192, 193]. Moreover, ETBF induced acute and persistent colitis in specific pathogen-free mice, whereas a nontoxicogenic strain of *B. fragilis* overexpressing a biologically inactive mutated fragilysin caused no colonic histopathology [194]. Analysis of several *B. fragilis* strains demonstrated that fragilysin proteolytic activity was only detected in strains isolated from patients with diarrhea [195]. Fragilysin could thus be responsible of the pathogenicity of *B. fragilis* in IBD. Another study suggested that MMPs produced by *B. fragilis* could be implicated in the induction of transmural inflammation by this bacteria strain in mice [196].

Clostridium difficile is an intestinal pathogen responsible for severe diarrheas and colitis. It produces toxins or virulence factors (toxin A and toxin B) that have been identified as proteases. Indeed, protease inhibitors delayed intoxication of cells and blocked *C. difficile* toxin B autocatalytic cleavage [197, 198]. Moreover, it has been shown that auto-catalytic cleavage of *C. difficile* toxins A and B was dependent on cysteine protease activity [199]. Although both toxins A and B are potent cytotoxins, only toxin A possesses enterotoxic effects in the rodent intestine [200]. Toxin B, and to a lesser extent toxin A were shown to cause tissue damage and electrophysiological changes in normal human colon in vitro [201], suggesting that both *C. difficile* toxins are involved in the pathophysiology of human colitis. Although *C. difficile* is an intestinal pathogen that induces diarrhea and colitis, its specific association with IBD has never been demonstrated.

2.6.2 Yeasts

Saccharomyces boulardii is a nonpathogenic yeast, which seems to exert protective effects in different inflammatory pathologies of the gut, including IBD and bacterially mediated or enterotoxin-mediated diarrhea. In human, treatment with *S. boulardii* has demonstrated beneficial effects both in CD and UC patients [202]. *S. boulardii* treatment was shown to improve intestinal permeability in CD patients in remission, when added to other more stringent therapies [203]. Several mechanisms have been described to explain the anti-inflammatory effects of *S. boulardii*: interference with NF-κB, ERK1/2, and p38 signaling, suppression of "bacteria overgrowth" and host cell adherence, and trapping of T cells in mesenteric lymph nodes [204]. Interestingly, *S. boulardii* was shown to release a 54-kDa protease that cleaved both *C. difficile* toxin A and its brush border receptor and inhibited toxin A-induced ileal permeability in rats [205]. Moreover, *S. boulardii* protease prevented the reduction of human colonic mucosa tissue resistance mediated by *C. difficile* toxin A and toxin B [206]. The beneficial effects carried on by *S. boulardii* in IBD patients could thus be at least in part related to its proteolytic activity.

2.6.3 Helminths

The filarial nematode *Acanthocheilonema viteae* secretes cystatins which are cysteine protease inhibitors. Cystatin from this nematode modulated macrophage-mediated inflammation in a murine model of DSS-induced colitis, leading to the reduction of inflammatory cell infiltration and epithelial damage [207]. This effect could be explained by the up-regulation of IL-10 production by macrophages. Interestingly, recombinant cystatins of the free-living nematode *Caenorhabditis elegans* failed to modulate macrophage functions in vitro [208]. This suggests that the anti-inflammatory effects of cystatin on macrophages could be specific for cystatins of filarial nematodes (Tables 1–3).

Table 1 Endogenous proteases in the gastrointestinal tract: sources and expression during IB

	Protease	Source BioGPS: biogps.gnf/org	Expression in IBD
Serine proteases	Elastase	Neutrophil/Mast cell/Monocyte/Eosino	↑ in CD/UC
	Proteinase 3	*Neutrophil/Monocytes*	↑ in CD/UC
	Chymase	Mast cell/Basophils	↑ in CD/UC
	Tryptase	*Mast cell/Basophils*	↑ in CD/UC
	Thrombin	Hepatocyte	↑ in CD/UC
	Factor V and VIII	*Various (Hepatocyte/ Placenta/Leukocytes...)*	↑ in UC
	Plasminogen activator	Various (small intestine/ colon/myocytes...)	↓ in CD/UC
	Protein C	*Hepatocyte*	↓ in CD/UC
	Trypsins	Various (Pancreas/Epithelial cells/ Endothelial cells/Neuronal cells)	↑ in CD/UC
	Kallikreins	*Various (pancreas/salivary glands/ prostate/leukocytes...)*	↑ in CD/UC
	Granzymes	Leukocyte ($CD8^+$/$CD4^+$)	↑ in CD/UC
Metalloproteases	MMP-1	Various (leukocyte/fibroblasts/muscle cells/epithelial cells...)	↑ in CD/UC
	MMP-2	Various (muscle cells/adipocytes/lung/ liver)	↑ in UC
	MMP-3	Ubiquitous (largely by smooth muscle cells)	↑ in UC/CD
	MMP-7	Pancreas/skin/B cells/salivary gland/lung	in CD
	MMP-12	Ubiquitous	↑ in UC/CD
	MMP-14	Ubiquitous (largely by myocytes)	↑ in UC
	MMP-28	Ubiquitous (largely by lung)	= in CD/UC
	TACE/ADAM17	Ubiquitous	= in CD/UC
Cysteine proteases	Caspase-1	LMPC/macrophages/Epithelial cells	↑ in UC
	Caspase-3	*Ubiquitous*	↓ in memory T cells from CD
	Caspase-5	Ubiquitous	↑ in UC
	Caspase-8	*Ubiquitous*	↓ in memory T cells from CD
	Caspase-14	Keratinocytes/placenta	= in UC
	Cathepsin B	*Various (thyroid/liver/kidney/leukocytes)*	↑
	Cathepsin L	Various (placenta/intestine/pancreas/ macrophages)	↑
Aspartate proteases	Cathepsin D	Lamina propria mononuclear cells/ macrophages	↑
Exopeptidase	DPP4	Leukocytes/smooth muscle cells/salivary gland	= or ↑ (human) and ↑ (mouse)
	Aminopeptidase N	Multiorgans (kidney/prostase/intestine/ liver/leukocyte/myocytes)	= in UC/CD
	Carboxypeptidase B2	Hepatocyte	↓ or ↑
	Angiotensinogen-converting enzyme (ACE)	Ubiquitous	↓ in CD/UC

Table 2 Endogenous antiproteases in the gastrointestinal tract: sources and expression in IBD

	Antiproteases	Proteases targets	Source BioGPS: biogps.gnf/org	Expression in IBD
Serpin family	SerpinA1	Trypsins/chymase/trypase/elastase/proteinase3/cathG/thrombin/kallikreins	Hepatocyte	= in UC or ↓ in CD/UC
	SerpinA3	*Chymotrypsin/chymase/cathepsinG*	*Hepatocyte*	↑ in CD/UC
	SerpinA4	Kallikreins	Hepatocyte	↓ in CD/UC
	SerpinE1	*Plasminogen activator*	*Myocytes/Intestine/Placenta*	↑ in CD/UC
	Serpin C1	Thrombin	Hepatocyte	↓ in CD/UC
Chelonianin family	SLPI	Elastase/cathepsin G/trypsin/chymotrypsin/tryptase/chymase	Leukocyte/Epithelial cells/Lung	↑ in UC and ↓ in CD
	Elafin	*Elastase and Proteinase3*	*Thymus/tonsils/leukocytes/epithelial cells*	= in CD/UC or ↑ in UC
TIMP family	TIMP-1	MMP-1/ MMP-2/ MMP-3/MMP-4/MMP-6/MMP-19/ ADAM10/ADAM17	Ubiquitous	↑ in UC/CD
	TIMP-2	*MMP-1/MMP-2/MMP-2/MMP-14*	Leukocytes/placenta	=
	c-IAP2	Caspase-9	B cells	↓ in UC

3 Mechanism of Action of Proteases During IBD

Proteases present at the site of intestinal inflammation dispose of several mechanisms of action to participate in inflammatory processes. They could act by proteolytic processing of other molecules, by inducing intracellular signals, but also by extracellular pathways, in particular through the activation of membrane-associated receptors. Better understanding of molecular functions of proteases and of their mechanisms of action is necessary if we want to consider them as potential targets for therapeutic development.

3.1 Activation of Receptors

In the gastrointestinal tract, the receptors of exogenous and endogenous serine proteases that have been mainly studied are the members of the protease-activated receptor (PAR) family. PARs are G-protein coupled receptors that undergo N-terminal cleavage, leading to the exposure of a tethered ligand. There are several experimental and human evidences that PAR receptors are key players of the intestinal mucosal immunity. For instance, in vivo administration of serine proteases (trypsin and tryptase) lead to the development of colitis resembling human IBD

Table 3 Pharmocological inhibitors of proteases to control intestinal inflammation

	Protease inhibitor	Targets	Action/effect	References
Serine proteases	Z-Ile-Glu-Pro-Phe-CO2Me (ZIGPPF)	Chymase inhibition	Inhibited chymase-induced nucleated cell accumulation in vivo	[28, 98]
	Soy bean trypsin inhibitor (SBTI)	Chymase inhibition	Inhibited chymase-induced nucleated cell accumulation in vivo	[28, 98]
	Chymostatin	Chymase inhibition	Inhibited chymase-induced nucleated cell accumulation in vivo	[28, 98]
	Alpha-1 antitrypsin	*Broad spectrum inhibition*	Inhibited chymase-induced nucleated cell accumulation in vivo	[28, 98]
	Leupeptin	Tryptase inhibition	Inhibited chymase-induced nucleated cell accumulation in vivo	[28, 98]
	APC366	*Tryptase inhibition*	Inhibited chymase-induced nucleated cell accumulation in vivo	[28, 98]
	Aprotinin	Tryptase inhibition	Inhibited chymase-induced nucleated cell accumulation in vivo	[28, 98]
	Nafamostat mesilate	*Broad spectrum inhibitor*	*Reduced the severity of TNBS-induced colitis in rats*	[99]
	Silvelestat sodium hydrate (ONO-5046)	Elastase inhibition	ONO-5046 showed therapeutic effects in DSS-treated mice	[100]
	Z-Ile-Glu-Pro-Phe-CO2Me (ZIGPFM)	*Chymase inhibition*	*Inhibited tryptase released from dispersed colon mast cells*	[28, 98]
	Tosyl-L-Phenylalanyl-chloromethylketone (TPCK)	Chymase inhibition	*Inhibited tryptase released from dispersed colon mast cells*	[28, 98]
	Tosyl-L-lysinechloromethyl ketone (TLCK)	*Tryptase inhibition*	*Inhibited tryptase released from dispersed colon mast cells*	[28, 98]
Metalloproteases	GM 6001	Broad spectrum inhibitor	Inhibition of MMP release of IEC from IBD patients	[147]
	ZK 156979	*Vitamin D analog*	*Inhibition of MMP-9 activity of PBMC from IBD patients*	[155]
	Curcumin		Inhibition of MMP-3 activity of CMF from IBD patients	[160]
	Curcuminoid pyrazoles		*Inhibition of MMP-9 activity of IEC under inflammatory conditions*	[159]
	Rosiglitazone	PPAR-gamma agonist	Inhibition of MMP-9 activity of IEC under inflammatory conditions	[154]

(continued)

Table 3 (continued)

	Protease inhibitor	Targets	Action/effect	References
Metalloproteases (continued)	*Infliximab*	*Anti-TNF-α antibody*	*Downregulation of MMP-1, -3, and -9 relative to TIMP-1 and -2 and decrease of MMP-1 and -3 activities*	[168]
		Anti-MIF (macrophage migration inhibitory factor) antibody	Inhibition of MMP-13 up-regulation by DSS and reduction of severity of TNBS colitis	[161]
		Anti-IL-21 antibody	*Partial inhibition of intestinal fibroblasts MMP secretion induced by LPMC supernatants from IBD patients*	[156]
	Imipramine	Spingomyelinase inbibition	Indirect reduction of MMP-1 expression and tissue destruction under inflammatory conditions	[162]
	CH4474	*Broad spectrum*	*Inhibition of TNFα release by IEC under inflammatory conditions*	[163]
	B1110	TACE inhibition	Reduced the severity of TNBS-induced colitis	[165]
	Tapi-2	*TACE inhibition*	*Increased TNFα-induced loss of IEC permeability*	[166]
Cysteine proteases	CA-074 + Z-Phe-Tyr-aldehyde	Cathepsin B and L inhibition	Reduced the severity of DSS-induced colitis in mice	[104]
	Pralnacasan	*Caspase-1 inhibition*	*Reduced the severity of DSS-induced colitis*	[118–120]
	Calpain I inhibitor	Calpain I inhibition	Reduced the severity of DNBS-induced colitis in rats	[141]
	Deoxyschisandrin	*Calpain inhibition*	*Inhibition of H_2O_2-induced caspase-3 activation in IEC*	[127]
	z-IETD.fmk	Caspase-8 inhibition	Reduction of TNF α/butyrate-induced apoptosis and loss of IEC permeability	[138]
	z-AEVD.fmk	*Caspase-10 inhibition*	*Reduction of TNF α/butyrate-induced IEC apoptosis*	[138]
	Glutamine	Caspase-3 inhibition	Reduced the severity of in TNBS-induced colitis	[135]
	Cyclosporine	*Caspase-8 inhibition*	*Reduced epithelial apoptosis in DSS-induced colitis*	[132]

(continued)

Table 3 (continued)

	Protease inhibitor	Targets	Action/effect	References
Cysteine proteases (continued)	Anti-TGF-β antibodies	Caspase 8 up-regulation	Reduction of c-FLIP (caspase-8 inhibitor) expression in IEC	[140]
	Vitamin D derivatives	*Caspase-3 activation*	*Activation of caspase-3 in PBMC*	[125]
	Sulphasalazine	Caspase-3 and Caspase-9	Increase activation of caspase-3 and caspase-9, induction of lamina propria T lymphocytes apoptosis	[123]
	L-Cysteine	*Caspase-8 up-regulation*	*Up-regulation of caspase-8, restoration of activated T cell susceptibility to apoptosis*	[126]
	L-Tryptophan	Caspase-8 up-regulation	Reduced the severity of DSS-induced colitis in piglets	[126]
	Visilizumab	*Caspase-3 and Caspase-8 activation*	*Induction of apoptosis of lamina propria T cells"*	[128]
Aspartate proteases	Pepstatin A	Cathepsin D inhibition	Reduced the severity of DSS-induced colitis in mice	[104]
	Losartan	*Angiotensin II receptor antagonist*	*Reduced the severity of TNBS-induced colitis*	[106]
Exopeptidases	Isoleucyl-cyano-pyrrolidine (P59/99)	DPP inhibition	Reduced the severity of DSS-induced colitis in mice	[181]
	Isoleucyl-thiazolidine (P32/98)	*DPP inhibition*	Reduced the severity of DSS-induced colitis in mice	[181]

[209] and this seems to be mainly mediated by the activation of PAR_1, PAR_2, and PAR_4 ([209–211] and unpublished data from N. Vergnolle). Those three receptors have been shown to be heavily involved in the pathogenesis of animal models of colitis ([212, 213], and unpublished data). As per their mode of action, PAR activation is able to modify all the physiological functions in the gut: ion exchange, motility, nociception, permeability, etc., as they are present and functional in a number of different cell types (intestinal epithelial cells, fibroblasts, neurons, smooth muscle cells, infiltrated immune cells, or resident cells, etc.) [214]. PAR activation is able to induce chemokine and cytokine release from epithelial or immune cells [215], it does increase the permeability of intestinal epithelial cells monolayers (at least PAR_1 and PAR_2) [216, 217], it modifies gut motility, and also signals to enteric neurons, modifying nociceptive functions [11, 218, 219]. From a gut innate immune response point of view, the activation of the three PARs: PAR_1,

PAR_2, and PAR_4, clearly leads to a pro-inflammatory phenotype. However, from a nociception point of view, PAR_2 seems to have opposite effects compared to PAR_1 and PAR_4 activation: PAR_2 being involved as a causative agent in inflammatory pain, while PAR_1 and PAR_4 activation relieve from pain (please refer to the Chapter 11 of N. Cenac in this book). Therefore, it appears important to define the types of PAR-activating proteases that are released in IBD tissues, as well as the relative contribution of each of the PAR family members to the generation of inflammatory symptoms.

Constant sampling and recognition of lumenal microbial content occurs in the gastrointestinal tract, and this recognition process occurs, at least in part, via the activation of members of the family of the toll-like receptors (TLR). Interestingly, it has been reported that neutrophil elastase could up-regulate the release of CXCL-8 through TLR4 receptor, the main receptor of Gram-negative bacteria LPS [220]. Similarly, specific inhibitor of neutrophil elastase also inhibited CXCL-2 release by LPS-stimulated macrophages [221]. Proteolytic activity of the protease was found to be necessary to induce this process, but the exact mechanism is still unknown. Some authors proposed a MD-2 (TLR4 co-receptor) mimicking effect [222]. Some insights into this concept might come from the observation that trypsin reduced TLR4 signaling through cleavage of MD2 co-receptor on intestinal epithelial cells [223] which thus could be a physiologic mechanism to LPS tolerance in the intestine. Receptor cooperativity between PAR_2 and TLR4 receptors both in vitro and in vivo in the intestinal mucosal tissue has been proposed. In intestinal epithelial cells, PAR_2 and TLR4 crosstalk in intracellular NF-κB-dependant cytokine induction signaling (CXCL-8/IL-1β) and thus may explain the mechanism of action of proteases-induced inflammation in the gut [224, 225].

Another serine protease, cathG, has been shown to interact with a G-protein coupled formyl peptide receptor (G-FPR) leading to MAP kinase pathways activation [226]. This receptor for intestinal bacterial chemotactic formylated peptide could be implicated in the pathogenesis of IBD [227, 228], however further studies would have to demonstrate such role.

3.2 *Induction of Apoptosis*

Even if the concept of increased mortality of the epithelium and its role in the pathogenesis of IBD is debated, proteases constitute active mediators of this process. Granzymes are efficient inducers of cell death due to their ability to activate multiple mediators of apoptosis, including caspase-3, caspase-8, and BID [229]. Blockade of caspase activity does not prevent GzmB-induced cell death [230], while inhibition of mitochondrial pathway via Bcl-2 overexpression prevents apoptosis [231]. GzmA has been reported to induce reactive oxygen species (ROS) and loss of mitochondrial potential after entering target cells [232]. The serine protease thrombin could also induce apoptosis in vitro and in vivo via extracellular activation of the receptor PAR_1 leading to caspase-mediated intestinal epithelial cell death [216]. Therefore, it appears that proteases could participate to IBD-associated tissue damage, through their pro-apoptotic properties on intestinal epithelial cells.

3.3 Cytokines and Chemokines Processing

N-terminal cleavage of CXCL-8 by PR3 and CXCL-5 by cathG releases truncated forms of chemokines that have higher chemotactic activity towards neutrophils than full-length molecules [233, 234]. Proteinase 3 could also activate IL-18, members of IL-1 family [235], IL-1β [236] and TNF-α [237], all cytokines suspected to be involved in IBD pathogenesis [3]. From those data, it could be considered that proteases actively participate in inflammatory processes, by releasing and generating pro-inflammatory metabolites. However, some proteases released at the site of inflammation could also have opposite effects on chemokines. For instance, elastase and cathepsin G both degrade mature TNF [238] and elastase, proteinase-3, and cathepsin G can inactivate IL-6 [239]. This regulation of pleiotropic pro-inflammatory molecules by proteases could either limit the activation of leukocytes or exacerbate inflammatory processes. The net effects of proteolytic chemokines/cytokines modification during IBD would need to be elucidated in vivo (Figs. 1 and 2).

4 Relevance of Antiprotease-Based Treatments for IBD

Overall, increased tissue proteolytic activity associated with IBD exerts rather pro-inflammatory properties, which naturally led to think that inhibition of this activity could have therapeutic benefits to treat inflammatory symptoms. However, considering the large number of proteases present, their diverse activities and functions, it appears rather difficult to identify single molecular targets among all those proteases. In addition, it has been difficult to evaluate the relative contribution of individual proteases, or even families of proteases, to intestinal inflammatory disorders. One important question to solve if protease inhibition is considered for IBD therapy is the spectrum of protease inhibition that would be optimal to achieve sufficient anti-inflammatory properties, and that would avoid potential deleterious side effects. Should we consider large-spectrum protease inhibitors that would inhibit both serine proteases from inflammatory cells and MMPs for example? Or should we consider more specific inhibitors of some subclasses of proteases? From all the families of proteases that are present and up-regulated in the setting of IBD, metalloproteases have raised so far, the most interest, and several MMP inhibitors have been developed by the industry. They exert good anti-inflammatory properties in animal models of IBD (cf previous paragraph), but in human, they seem to be more efficient at helping mucosal repair than reducing acute inflammation bouts.

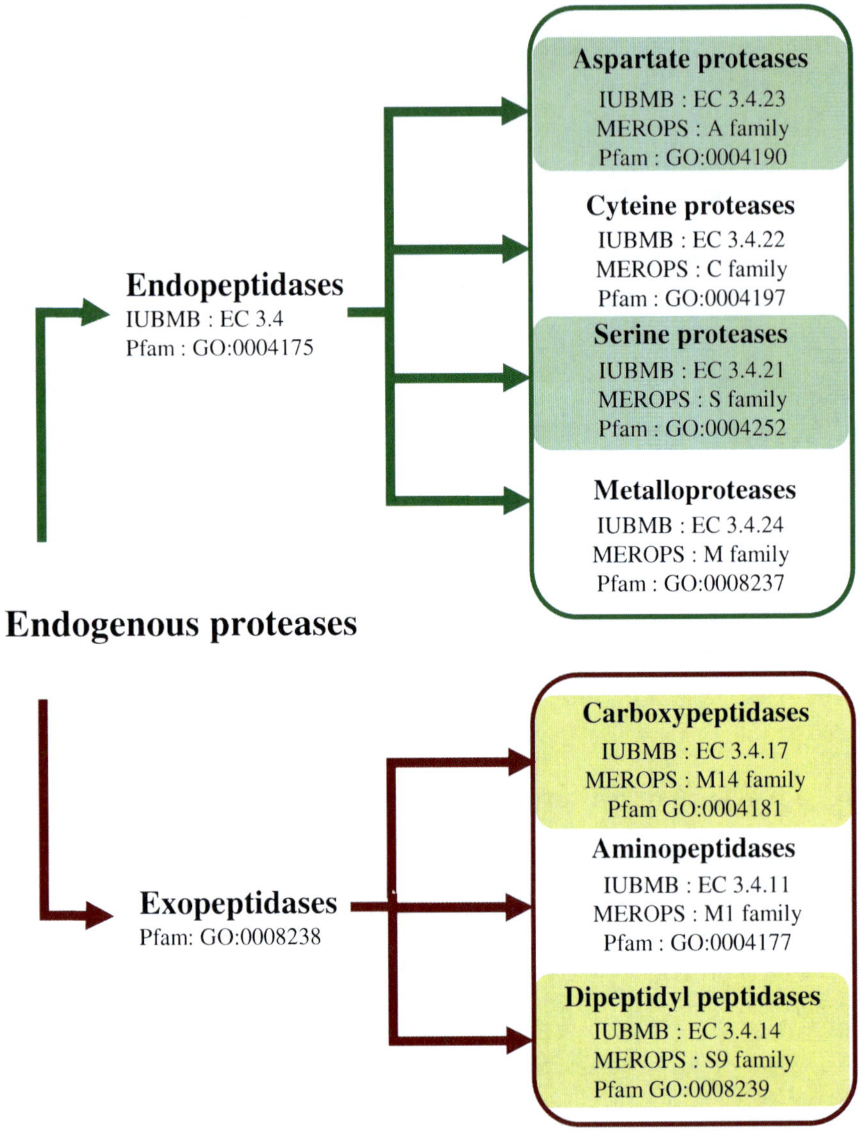

Fig. 1 Endogenous proteases present in the human gastrointestinal tract during IBD. During the course of colitis, the human gastrointestinal tract is flooded with several classes of proteases that belong to the large family of peptide hydrolases. Their classification is difficult to define, depending on the nature of amino acid residues around the peptide bond and their catalytic mechanism. A classification involving both criteria is therefore used based on the Enzyme Commission (EC) code by the International Union of Biochemistry and Molecular Biology (IUBMB), the MEROPS peptidase database [240], and the Pfam protein family database [241]. Endopeptidases act internally in polypeptide chains and are divided into the subclasses of Aspartate-, Cysteine-, Serine-, and Metallo-proteases. In contrast, exopeptidases act near the ends of polypeptide chains. Those proteases act at a free C-terminus and liberate a single residue, for this reason, they are called Carboxypeptidases. Those acting at the N-terminus can either liberate a single amino-acid residue (Aminopeptidases), or a dipeptide (Dipeptidyl-peptidases)

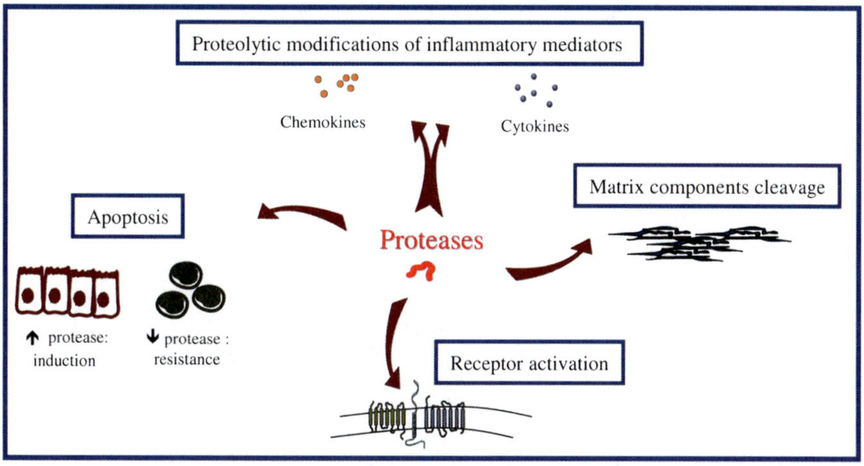

Fig. 2 Mechanisms of action of proteases in the gastrointestinal tract. An increased proteolytic activity in the gastrointestinal tract can lead to chronic inflammation through the cleavage of extracellular matrix components, the processing of inflammatory mediators, and the modulation of apoptosis (increased mucosal layer apotosis vs. leukocyte resistance to apoptosis). They can also activate intracellular inflammatory pathways via the activation of several transmembranous receptors

Therefore, the use of MMP inhibitors might be restricted to an association with other anti-inflammatory approaches, such as anti-TNF or antiadhesion molecules therapies. Clinical studies using other protease inhibitors have not yet been issued, although a number of synthetic serine protease inhibitors have demonstrated good anti-inflammatory properties in animal models of IBD.

Another option to develop protease inhibition-based therapy, and to re-equilibrate the protease–antiprotease balance in the inflamed gut, would be to favor the expression of endogenous protease inhibitors. A better knowledge of the mechanisms regulating the endogenous expression of antiproteases will be important, in order to develop such strategy. Even more challenging, and in order to avoid potential side effects, the expression of protease inhibitors could be, for the most part, restricted to the gut. To that aim, the use of genetically modified bacteria could be of major importance. This concept, recently developed by Steidler for interleukin-10 delivery to the gut, could be applied to antiprotease therapy. Commensal or probiotic bacteria that can colonize the gut (lactic bacteria for the most part) are genetically transformed to overexpress protease inhibitors naturally present in the intestine. Those bacteria could thus be used as carriers for antiprotease delivery to the gut. Preliminary experiments have demonstrated in animal models of IBD that oral treatment with *Lactococcus lactis* and *Lactobacillus casei* recombinant for the expression of elafin, or SLPI have strong anti-inflammatory effects (Motta, Martin et al. unpublished results). The use of such strategy in humans will have to consider the development of nondisseminating bacteria because of their genetically modified nature. However, the bacteria-mediated delivery of antiproteases to the gut appears as a relevant way to re-equilibrate locally the proteolytic balance, and to induce anti-inflammatory signals.

5 Conclusions

Taken together, data reviewed in this chapter present the proteolytic balance as a key player in the development and progression of IBD. A number of proteases are up-regulated in IBD and exert pro-inflammatory effects, but some also have opposite effects, reducing some aspects of the acute inflammatory response. Thus, protease inhibition appears as an interesting alternative to reduce inflammatory symptoms associated with CD or UC, but a clear characterization of the protease–antiprotease pattern in human pathologic intestinal tissues is necessary. Currently, no clinical trial explores the efficacy of protease inhibitors other than MMP inhibitors, in patients with IBD. This can be explained by the relatively recent concept that proteases by in large are active mediators of IBD, but also by the fact that a large array of proteases and not a single molecular target has been identified to play a role in IBD. It could also be explained by the general belief that inhibition of proteases in the gut obviously conveys the message that digestive functions will be significantly altered by such an approach. Options towards local protease inhibition, and in particular inhibition in the lower gastrointestinal tract, where digestion does not take place, could seriously be considered for future therapeutic developments.

References

1. Hanauer SB (2006) Inflammatory bowel disease: epidemiology, pathogenesis, and therapeutic opportunities. Inflamm Bowel Dis 12(Suppl 1):S3–S9
2. Podolsky D (2002) Inflammatory bowel disease. N Engl J Med 347:6
3. Xavier R, Podolsky D (2007) Unravelling the pathogenesis of inflammatory bowel disease. Nature 448:7152
4. Strober W, Fuss IJ, Blumberg RS (2002) The immunology of mucosal models of inflammation. Annu Rev Immunol 20:495–549
5. McHugh KJ, Svensjo E, Persson CG (1996) Exudative and absorptive permeability in different phases of an experimental colitis condition. Scand J Gastroenterol 31:9900–9905
6. Vergnolle N (2008) Proteinase-activated receptors (PARs) in infection and inflammation in the gut. Int J Biochem Cell Biol 40(6–7):1219–1227
7. Bustos D, Negri G, De Paula J, Di Carlo M, Yapur V, Facente A, De Paula A (1998) Colonic proteinases: increased activity in patients with ulcerative colitis. Medicina (B Aires) 58:3
8. KjeldsenJ, Lassen J, Brandslund I, Schaffalitzky de Muckadell O (1998) Markers of coagulation and fibrinolysis as measures of disease activity in inflammatory bowel disease. Scand J Gastroenterol 33:6

9. Schurmann G, Bruwer M, Klotz A, Schmid K, Senninger N, Zimmer K (1999) Transepithelial transport processes at the intestinal mucosa in inflammatory bowel disease. Int J Colorectal Dis 14:1
10. Schmid M, Fellermann K, Fritz P, Wiedow O, Stange E, Wehkamp J (2007) Attenuated induction of epithelial and leukocyte serine antiproteases elafin and secretory leukocyte protease inhibitor in Crohn's disease. J Leukoc Biol 81:4
11. Cenac N, Andrews C, Holzhausen M, Chapman K, Cottrell G, Andrade-Gordon P, Steinhoff M, Barbara G, Beck P, Bunnett N, Sharkey K, Ferraz J, Shaffer E, Vergnolle N (2007) Role for protease activity in visceral pain in irritable bowel syndrome. J Clin Invest 117:3
12. Di Cera E (2009) Serine proteases. IUBMB Life 61:5
13. Hawkins J, Emmel E, Feuer J, Nedelman M, Harvey C, Klein H, Rozmiarek H, Kennedy A, Lichtenstein G, Billings P (1997) Protease activity in a hapten-induced model of ulcerative colitis in rats. Dig Dis Sci 42:9
14. Gouni-Berthold I, Baumeister B, Wegel E, Berthold H, Vetter H, Schmidt C (1999) Neutrophil-elastase in chronic inflammatory bowel disease: a marker of disease activity? Hepatogastroenterology 46:28
15. Langhorst J, Elsenbruch S, Koelzer J, Rueffer A, Michalsen A, Dobos G (2008) Noninvasive markers in the assessment of intestinal inflammation in inflammatory bowel diseases: performance of fecal lactoferrin, calprotectin, and PMN-elastase, CRP, and clinical indices. Am J Gastroenterol 103:1
16. Motta JP, Magne L, Descamps D, Rolland C, Squarzoni-Dale C, Rousset P, Martin L, Cenac N, Balloy V, Huerre M, Frohlich LF, Jenne D, Wartelle J, Belaaouaj A, Mas E, Vinel JP, Alric L, Chignard M, Vergnolle N, Sallenave JM (2011) Modifying the protease, antiprotease pattern by elafin overexpression protects mice from colitis. Gastroenterology 140:1272–1282
17. Pham C (2006) Neutrophil serine proteases: specific regulators of inflammation. Nat Rev Immunol 6:7
18. Wiedow O, Meyer-Hoffert U (2005) Neutrophil serine proteases: potential key regulators of cell signalling during inflammation. J Intern Med 257:4
19. Fitch P, Roghanian A, Howie S, Sallenave J (2006) Human neutrophil elastase inhibitors in innate and adaptive immunity. Biochem Soc Trans 34(Pt 2):279–282
20. Chin A, Lee W, Nusrat A, Vergnolle N, Parkos C (2008) Neutrophil-mediated activation of epithelial protease-activated receptors-1 and −2 regulates barrier function and transepithelial migration. J Immunol 181:8
21. Swystun V, Renaux B, Moreau F, Wen S, Peplowski M, Hollenberg M, MacNaughton W (2009) Serine proteases decrease intestinal epithelial ion permeability by activation of protein kinase Czeta. Am J Physiol Gastrointest Liver Physiol 297:1
22. Patel R, Stokes R, Birch D, Ibbotson J, Keighley M (1994) Influence of total colectomy on serum antineutrophil cytoplasmic antibodies in inflammatory bowel disease. Br J Surg 81:5
23. Dabek M, Ferrier L, Roka R, Gecse K, Annahazi A, Moreau J, Escourrou J, Cartier C, Chaumaz G, Leveque M, Ait-Belgnaoui A, Wittmann T, Theodorou V, Bueno L (2009) Luminal cathepsin g and protease-activated receptor 4: a duet involved in alterations of the colonic epithelial barrier in ulcerative colitis. Am J Pathol 175:1
24. Bradding P, Walls A, Holgate S (2006) The role of the mast cell in the pathophysiology of asthma. J Allergy Clin Immunol 117:6
25. Mekori Y, Metcalfe D (2000) Mast cells in innate immunity. Immunol Rev 173:131–140
26. Meier H, Heck L, Schulman E, MacGlashan DJ (1985) Purified human mast cells and basophils release human elastase and cathepsin G by an IgE-mediated mechanism. Int Arch Allergy Appl Immunol 77:1–2
27. Strik M, de Koning P, Kleijmeer M, Bladergroen B, Wolbink A, Griffith J, Wouters D, Fukuoka Y, Schwartz L, Hack C, van Ham S, Kummer J (2007) Human mast cells produce and release the cytotoxic lymphocyte associated protease granzyme B upon activation. Mol Immunol 44:14

28. He S (2004) Key role of mast cells and their major secretory products in inflammatory bowel disease. World J Gastroenterol 10:3
29. Scudamore C, Jepson M, Hirst B, Miller H (1998) The rat mucosal mast cell chymase, RMCP-II, alters epithelial cell monolayer permeability in association with altered distribution of the tight junction proteins ZO-1 and occludin. Eur J Cell Biol 75:4
30. Groschwitz K, Ahrens R, Osterfeld H, Gurish M, Han X, Abrink M, Finkelman F, Pejler G, Hogan S (2009) Mast cells regulate homeostatic intestinal epithelial migration and barrier function by a chymase/Mcpt4-dependent mechanism. Proc Natl Acad Sci USA 106:52
31. Frank B, Rossall J, Caughey G, Fang K (2001) Mast cell tissue inhibitor of metalloproteinase-1 is cleaved and inactivated extracellularly by alpha-chymase. J Immunol 166:4
32. Caughey G (2007) Mast cell tryptases and chymases in inflammation and host defense. Immunol Rev 217:141–154
33. Raithel M, Winterkamp S, Pacurar A, Ulrich P, Hochberger J, Hahn E (2001) Release of mast cell tryptase from human colorectal mucosa in inflammatory bowel disease. Scand J Gastroenterol 36:2
34. Bischoff S, Wedemeyer J, Herrmann A, Meier P, Trautwein C, Cetin Y, Maschek H, Stolte M, Gebel M, Manns M (1996) Quantitative assessment of intestinal eosinophils and mast cells in inflammatory bowel disease. Histopathology 28:1
35. Andoh A, Deguchi Y, Inatomi O, Yagi Y, Bamba S, Tsujikawa T, Fujiyama Y (2006) Immunohistochemical study of chymase-positive mast cells in inflammatory bowel disease. Oncol Rep 16:1
36. Lee L, Spittell JJ, Sauer W, Owen CJ, Thompson JJ (1968) Hypercoagulability associated with chronic ulcerative colitis: changes in blood coagulation factors. Gastroenterology 54:1
37. Chamouard P, Grunebaum L, Wiesel M, Frey P, Wittersheim C, Sapin R, Baumann R, Cazenave J (1995) Prothrombin fragment 1 + 2 and thrombin-antithrombin III complex as markers of activation of blood coagulation in inflammatory bowel diseases. Eur J Gastroenterol Hepatol 7:12
38. Gris J, Schved J, Raffanel C, Dubois A, Aguilar-Martinez P, Arnaud A, Sanchez N, Sarlat C, Balmes J (1990) Impaired fibrinolytic capacity in patients with inflammatory bowel disease. Thromb Haemost 63:3
39. Saibeni S, Bottasso B, Spina L, Bajetta M, Danese S, Gasbarrini A, de Franchis R, Vecchi M (2004) Assessment of thrombin-activatable fibrinolysis inhibitor (TAFI) plasma levels in inflammatory bowel diseases. Am J Gastroenterol 99:10
40. Papa A, Danese S, Grillo A, Gasbarrini G, Gasbarrini A (2003) Review article: inherited thrombophilia in inflammatory bowel disease. Am J Gastroenterol 98:6
41. Danese S, Papa A, Saibeni S, Repici A, Malesci A, Vecchi M (2007) Inflammation and coagulation in inflammatory bowel disease: The clot thickens. Am J Gastroenterol 102:1
42. Emi M, Nakamura Y, Ogawa M, Yamamoto T, Nishide T, Mori T, Matsubara K (1986) Cloning, characterization and nucleotide sequences of two cDNAs encoding human pancreatic trypsinogens. Gene 41:2–3
43. Rinderknecht H, Renner I, Abramson S, Carmack C (1984) Mesotrypsin: a new inhibitor-resistant protease from a zymogen in human pancreatic tissue and fluid. Gastroenterology 86:4
44. Nyaruhucha C, Kito M, Fukuoka S (1997) Identification and expression of the cDNA-encoding human mesotrypsin(ogen), an isoform of trypsin with inhibitor resistance. J Biol Chem 272:16
45. Szmola R, Kukor Z, Sahin-Toth M (2003) Human mesotrypsin is a unique digestive protease specialized for the degradation of trypsin inhibitors. J Biol Chem 278:49
46. Koshikawa N, Hasegawa S, Nagashima Y, Mitsuhashi K, Tsubota Y, Miyata S, Miyagi Y, Yasumitsu H, Miyazaki K (1998) Expression of trypsin by epithelial cells of various tissues, leukocytes, and neurons in human and mouse. Am J Pathol 153:3

47. Cottrell G, Amadesi S, Grady E, Bunnett N (2004) Trypsin IV, a novel agonist of protease-activated receptors 2 and 4. J Biol Chem 279:14
48. Moreau M, Garbacki N, Molinaro G, Brown N, Marceau F, Adam A (2005) The kallikrein-kinin system: current and future pharmacological targets. J Pharmacol Sci 99:1
49. Stadnicki A, DeLa Cadena R, Sartor R, Bender D, Kettner C, Rath H, Adam A, Colman R (1996) Selective plasma kallikrein inhibitor attenuates acute intestinal inflammation in Lewis rat. Dig Dis Sci 41:5
50. Zeitlin I, Smith A (1973) Mobilization of tissue kallikrein in inflammatory disease of the colon. Gut 14:2
51. Devani M, Cugno M, Vecchi M, Ferrero S, Di Berardino F, Avesani E, de Franchis R, Colman R (2002) Kallikrein-kinin system activation in Crohn's disease: differences in intestinal and systemic markers. Am J Gastroenterol 97:8
52. Stadnicki A, Mazurek U, Plewka D, Wilczok T (2003) Intestinal tissue kallikrein-kallistatin profile in inflammatory bowel disease. Int Immunopharmacol 3:7
53. Barrowman J, Perry M, Kvietys P, Ulrich M, Granger D (1981) Effects of bradykinin on intestinal transcapillary fluid exchange. Can J Physiol Pharmacol 59:8
54. Julia V, Mezzasalma T, Bueno L (1995) Influence of bradykinin in gastrointestinal disorders and visceral pain induced by acute or chronic inflammation in rats. Dig Dis Sci 40:9
55. Baird A, Skelly M, O'Donoghue D, Barrett K, Keely S (2008) Bradykinin regulates human colonic ion transport in vitro. Br J Pharmacol 155:4
56. Murrell T, Deller D (1967) Intestinal motility in man: the effect of bradykinin on the motility of the distal colon. Am J Dig Dis 12:6
57. Tiffany C, Burch R (1989) Bradykinin stimulates tumor necrosis factor and interleukin-1 release from macrophages. FEBS Lett 247:2
58. Stadnicki A, Pastucha E, Nowaczyk G, Mazurek U, Plewka D, Machnik G, Wilczok T, Colman R (2005) Immunolocalization and expression of kinin B1R and B2R receptors in human inflammatory bowel disease. Am J Physiol Gastrointest Liver Physiol 289:2
59. Hara D, Leite D, Fernandes E, Passos G, Guimaraes A, Pesquero J, Campos M, Calixto J (2008) The relevance of kinin B1 receptor upregulation in a mouse model of colitis. Br J Pharmacol 154:6
60. Chowdhury D, Lieberman J (2008) Death by a thousand cuts: granzyme pathways of programmed cell death. Annu Rev Immunol 26:389–420
61. Kappeler A, Mueller C (2000) The role of activated cytotoxic T cells in inflammatory bowel disease. Histol Histopathol 15:1
62. Muller S, Lory J, Corazza N, Griffiths G, Z'graggen K, Mazzucchelli L, Kappeler A, Mueller C (1998) Activated CD4+ and CD8+ cytotoxic cells are present in increased numbers in the intestinal mucosa from patients with active inflammatory bowel disease. Am J Pathol 152:1
63. Bisping G, Lugering N, Lutke-Brintrup S, Pauels H, Schurmann G, Domschke W, Kucharzik T (2001) Patients with inflammatory bowel disease (IBD) reveal increased induction capacity of intracellular interferon-gamma (IFN-gamma) in peripheral CD8+ lymphocytes co-cultured with intestinal epithelial cells. Clin Exp Immunol 123:1
64. Souza H, Tortori C, Castelo-Branco M, Carvalho A, Margallo V, Delgado C, Dines I, Elia C (2005) Apoptosis in the intestinal mucosa of patients with inflammatory bowel disease: evidence of altered expression of FasL and perforin cytotoxic pathways. Int J Colorectal Dis 20:3
65. Seidelin J, Nielsen O (2009) Epithelial apoptosis: cause or consequence of ulcerative colitis? Scand J Gastroenterol 44:12
66. Di Sabatino A, Ciccocioppo R, Luinetti O, Ricevuti L, Morera R, Cifone M, Solcia E, Corazza G (2003) Increased enterocyte apoptosis in inflamed areas of Crohn's disease. Dis Colon Rectum 46:11
67. Jenkins D, Seth R, Kummer J, Scott B, Hawkey C, Robins R (2000) Differential levels of granzyme B, regulatory cytokines, and apoptosis in Crohn's disease and ulcerative colitis at first presentation. J Pathol 190:2

68. Hirayasu H, Yoshikawa Y, Tsuzuki S, Fushiki T (2007) A role of a lymphocyte tryptase, granzyme A, in experimental ulcerative colitis. Biosci Biotechnol Biochem 71:1
69. Hansen K, Sherman P, Cellars L, Andrade-Gordon P, Pan Z, Baruch A, Wallace J, Hollenberg M, Vergnolle N (2005) A major role for proteolytic activity and proteinase-activated receptor-2 in the pathogenesis of infectious colitis. Proc Natl Acad Sci USA 102:23
70. Kondo H, Hojo Y, Tsuru R, Nishimura Y, Shimizu H, Takahashi N, Hirose M, Ikemoto T, Ohya K, Katsuki T, Yashiro T, Shimada K (2009) Elevation of plasma granzyme B levels after acute myocardial infarction. Circ J 73:3
71. Accardo-Palumbo A, Ferrante A, Cadelo M, Ciccia F, Parrinello G, Lipari L, Giardina A, Riili M, Giardina E, Dieli F, Triolo G (2004) The level of soluble Granzyme A is elevated in the plasma and in the Vgamma9/Vdelta2 T cell culture supernatants of patients with active Behcet's disease. Clin Exp Rheumatol 22(4 Suppl):34
72. Virca G, Metz G, Schnebli H (1984) Similarities between human and rat leukocyte elastase and cathepsin G. Eur J Biochem 144:1
73. Sallenave J, Xing Z, Graham F, Gauldie J (1997) In vivo adenovirus-mediated expression of human pre-elafin, a potent neutrophil elastase inhibitor. Chest 111(6 Suppl):128S–129S
74. Silverman GA, Lomas DA (2007) Molecular and cellular aspects of the serpinopathies and disorders in serpin activity. World Scientific Pub, Hackensack, NJ
75. Sallenave J (2010) Secretory leukocyte protease inhibitor and elafin/trappin-2: versatile mucosal antimicrobials and regulators of immunity. Am J Respir Cell Mol Biol 42:6
76. Yang P, Tremaine W, Meyer R, Prakash U (2000) Alpha1-antitrypsin deficiency and inflammatory bowel diseases. Mayo Clin Proc 75:5
77. Elzouki A, Eriksson S, Lofberg R, Nassberger L, Wieslander J, Lindgren S (1999) The prevalence and clinical significance of alpha 1-antitrypsin deficiency (PiZ) and ANCA specificities (proteinase 3, BPI) in patients with ulcerative colitis. Inflamm Bowel Dis 5:4
78. Bohe M, Genell S, Ohlsson K (1986) Protease inhibitors in plasma and faecal extracts from patients with active inflammatory bowel disease. Scand J Gastroenterol 21:5
79. Bohe M (1987) Pancreatic and granulocytic endoproteases in faecal extracts from patients with active ulcerative colitis. Scand J Gastroenterol 22:1
80. Becker K, Niederau C, Frieling T (1999) Fecal excretion of alpha 2-macroglobulin: a novel marker for disease activity in patients with inflammatory bowel disease. Z Gastroenterol 37:7
81. Huet G, Balduyck M, Mizon C, Colombel J, Mizon J (1990) Fecal alpha 1-antichymotrypsin in inflammatory bowel diseases. Clin Chem 36:3
82. Devani M, Vecchi M, Ferrero S, Avesani E, Arizzi C, Chao L, Colman R, Cugno M (2005) Kallikrein-kinin system in inflammatory bowel diseases: Intestinal involvement and correlation with the degree of tissue inflammation. Dig Liver Dis 37:9
83. Stadnicki A, Mazurek U, Gonciarz M, Plewka D, Nowaczyk G, Orchel J, Pastucha E, Plewka A, Wilczok T, Colman R (2003) Immunolocalization and expression of kallistatin and tissue kallikrein in human inflammatory bowel disease. Dig Dis Sci 48:3
84. Koutroubakis I, Sfiridaki A, Tsiolakidou G, Coucoutsi C, Theodoropoulou A, Kouroumalis E (2008) Plasma thrombin-activatable fibrinolysis inhibitor and plasminogen activator inhibitor-1 levels in inflammatory bowel disease. Eur J Gastroenterol Hepatol 20:9
85. Lake A, Stauffer J, Stuart M (1978) Hemostatic alterations in inflammatory bowel disease: response to therapy. Am J Dig Dis 23:10
86. Lam A, Borda I, Inwood M, Thomson S (1975) Coagulation studies in ulcerative colitis and Crohn's disease. Gastroenterology 68:2
87. Moreau T, Baranger K, Dade S, Dallet-Choisy S, Guyot N, Zani M (2008) Multifaceted roles of human elafin and secretory leukocyte proteinase inhibitor (SLPI), two serine protease inhibitors of the chelonianin family. Biochimie 90:2
88. Suzuki Y, Furukawa M, Abe J, Kashiwagi M, Hirose S (2000) Localization of porcine trappin-2 (SKALP/elafin) in trachea and large intestine by in situ hybridization and immunohistochemistry. Histochem Cell Biol 114:1

89. Bergenfeldt M, Nystrom M, Bohe M, Lindstrom C, Polling A, Ohlsson K (1996) Localization of immunoreactive secretory leukocyte protease inhibitor (SLPI) in intestinal mucosa. J Gastroenterol 31:1
90. Lawrance I, Fiocchi C, Chakravarti S (2001) Ulcerative colitis and Crohn's disease: distinctive gene expression profiles and novel susceptibility candidate genes. Hum Mol Genet 10:5
91. Flach C, Eriksson A, Jennische E, Lange S, Gunnerek C, Lonnroth I (2006) Detection of elafin as a candidate biomarker for ulcerative colitis by whole-genome microarray screening. Inflamm Bowel Dis 12:9
92. Eriksson A, Flach C, Lindgren A, Kvifors E, Lange S (2008) Five mucosal transcripts of interest in ulcerative colitis identified by quantitative real-time PCR: a prospective study. BMC Gastroenterol 8:34
93. Esmon C (2003) The protein C pathway. Chest 124(3 Suppl):26S–32S
94. Aadland E, Odegaard O, Roseth A, Try K (1992) Free protein S deficiency in patients with chronic inflammatory bowel disease. Scand J Gastroenterol 27:11
95. Souto J, Martinez E, Roca M, Mateo J, Pujol J, Gonzalez D, Fontcuberta J (1995) Prothrombotic state and signs of endothelial lesion in plasma of patients with inflammatory bowel disease. Dig Dis Sci 40:9
96. Heneghan M, Cleary B, Murray M, O'Gorman T, McCarthy C (1998) Activated protein C resistance, thrombophilia, and inflammatory bowel disease. Dig Dis Sci 43:6
97. Saibeni S, Vecchi M, Valsecchi C, Faioni E, Razzari C, de Franchis R (2001) Reduced free protein S levels in patients with inflammatory bowel disease: prevalence, clinical relevance, and role of anti-protein S antibodies. Dig Dis Sci 46:3
98. He S, Chen H, Zheng J (2004) Inhibition of tryptase and chymase induced nucleated cell infiltration by proteinase inhibitors. Acta Pharmacol Sin 25:12
99. Isozaki Y, Yoshida N, Kuroda M, Handa O, Takagi T, Kokura S, Ichikawa H, Naito Y, Okanoue T, Yoshikawa T (2006) Anti-tryptase treatment using nafamostat mesilate has a therapeutic effect on experimental colitis. Scand J Gastroenterol 41:8
100. Morohoshi Y, Matsuoka K, Chinen H, Kamada N, Sato T, Hisamatsu T, Okamoto S, Inoue N, Takaishi H, Ogata H, Iwao Y, Hibi T (2006) Inhibition of neutrophil elastase prevents the development of murine dextran sulfate sodium-induced colitis. J Gastroenterol 41:4
101. Fusek M, Vetvicka V (2005) Dual role of cathepsin D: ligand and protease Biomed Pap Med Fac Univ Palacky Olomouc. Czech Repub 149:143–150
102. Hausmann M, Obermeier F, Schreiter K, Spottl T, Falk W, Scholmerich J, Herfarth H, Saftig P, Rogler G (2004) Cathepsin D is up-regulated in inflammatory bowel disease macrophages. Clin Exp Immunol 136:1157–1167
103. Lugering N, Kucharzik T, Stein H, Winde G, Lugering A, Hasilik A, Domschke W, Stoll R (1998) IL-10 synergizes with IL-4 and IL-13 in inhibiting lysosomal enzyme secretion by human monocytes and lamina propria mononuclear cells from patients with inflammatory bowel disease. Dig Dis Sci 43:4706–4714
104. Menzel K, Hausmann M, Obermeier F, Schreiter K, Dunger N, Bataille F, Falk W, Scholmerich J, Herfarth H, Rogler G (2006) Cathepsins B, L and D in inflammatory bowel disease macrophages and potential therapeutic effects of cathepsin inhibition in vivo. Clin Exp Immunol 146:1169–1180
105. Jaszewski R, Tolia V, Ehrinpreis MN, Bodzin JH, Peleman RR, Korlipara R, Weinstock JV (1990) Increased colonic mucosal angiotensin I and II concentrations in Crohn's colitis. Gastroenterology 98:61543–61548
106. Inokuchi Y, Morohashi T, Kawana I, Nagashima Y, Kihara M, Umemura S (2005) Amelioration of 2,4,6-trinitrobenzene sulphonic acid induced colitis in angiotensinogen gene knockout mice. Gut 54:349–356
107. Lamkanfi M, Declercq W, Kalai M, Saelens X, Vandenabeele P (2002) Alice in caspase land. A phylogenetic analysis of caspases from worm to man. Cell Death Differ 9:4358–4361

108. Hugot JP, Chamaillard M, Zouali H, Lesage S, Cezard JP, Belaiche J, Almer S, Tysk C, O'Morain CA, Gassull M, Binder V, Finkel Y, Cortot A, Modigliani R, Laurent-Puig P, Gower-Rousseau C, Macry J, Colombel JF, Sahbatou M, Thomas G (2001) Association of NOD2 leucine-rich repeat variants with susceptibility to Crohn's disease. Nature 411:6837599–6837603
109. Ogura Y, Bonen DK, Inohara N, Nicolae DL, Chen FF, Ramos R, Britton H, Moran T, Karaliuskas R, Duerr RH, Achkar JP, Brant SR, Bayless TM, Kirschner BS, Hanauer SB, Nunez G, Cho JH (2001) A frameshift mutation in NOD2 associated with susceptibility to Crohn's disease. Nature 411:6837603–6837606
110. Siegmund B (2002) Interleukin-1beta converting enzyme (caspase-1) in intestinal inflammation. Biochem Pharmacol 64:11–18
111. Gu Y, Kuida K, Tsutsui H, Ku G, Hsiao K, Fleming MA, Hayashi N, Higashino K, Okamura H, Nakanishi K, Kurimoto M, Tanimoto T, Flavell RA, Sato V, Harding MW, Livingston DJ, Su MS (1997) Activation of interferon-gamma inducing factor mediated by interleukin-1beta converting enzyme. Science 275:5297206–5297209
112. Wilson KP, Black JA, Thomson JA, Kim EE, Griffith JP, Navia MA, Murcko MA, Chambers SP, Aldape RA, Raybuck SA (1994) Structure and mechanism of interleukin-1 beta converting enzyme. Nature 370:6487270–6487275
113. Monteleone G, Trapasso F, Parrello T, Biancone L, Stella A, Iuliano R, Luzza F, Fusco A, Pallone F (1999) Bioactive IL-18 expression is up-regulated in Crohn's disease. J Immunol 163:1143–1147
114. McAlindon ME, Hawkey CJ, Mahida YR (1998) Expression of interleukin 1 beta and interleukin 1 beta converting enzyme by intestinal macrophages in health and inflammatory bowel disease. Gut 42:2214–2219
115. McAlindon ME, Galvin A, McKaig B, Gray T, Sewell HF, Mahida YR (1999) Investigation of the expression of IL-1beta converting enzyme and apoptosis in normal and inflammatory bowel disease (IBD) mucosal macrophages. Clin Exp Immunol 116:2251–2257
116. Siegmund B, Lehr HA, Fantuzzi G, Dinarello CA (2001) IL-1 beta -converting enzyme (caspase-1) in intestinal inflammation. Proc Natl Acad Sci USA 98:2313249–2313254
117. Martinon F, Mayor A, Tschopp J (2009) The inflammasomes: guardians of the body. Annu Rev Immunol 27:229–265
118. Bauer C, Duewell P, Mayer C, Lehr HA, Fitzgerald KA, Dauer M, Tschopp J, Endres S, Latz E, Schnurr M (2010) Colitis induced in mice with dextran sulfate sodium (DSS) is mediated by the NLRP3 inflammasome. Gut 59:1192–1199
119. Loher F, Bauer C, Landauer N, Schmall K, Siegmund B, Lehr HA, Dauer M, Schoenharting M, Endres S, Eigler A (2004) The interleukin-1 beta-converting enzyme inhibitor pralnacasan reduces dextran sulfate sodium-induced murine colitis and T helper 1 T-cell activation. J Pharmacol Exp Ther 308:2583–2590
120. Bauer C, Loher F, Dauer M, Mayer C, Lehr HA, Schonharting M, Hallwachs R, Endres S, Eigler A (2007) The ICE inhibitor pralnacasan prevents DSS-induced colitis in C57BL/6 mice and suppresses IP-10 mRNA but not TNF-alpha mRNA expression. Dig Dis Sci 52:71642–71652
121. Zaki MH, Boyd KL, Vogel P, Kastan MB, Lamkanfi M, Kanneganti TD (2010) The NLRP3 inflammasome protects against loss of epithelial integrity and mortality during experimental colitis. Immunity 32:3379–3391
122. Brannigan AE, O'Connell PR, Hurley H, O'Neill A, Brady HR, Fitzpatrick JM, Watson RW (2000) Neutrophil apoptosis is delayed in patients with inflammatory bowel disease. Shock 13:5361–5366
123. Doering J, Begue B, Lentze MJ, Rieux-Laucat F, Goulet O, Schmitz J, Cerf-Bensussan N, Ruemmele FM (2004) Induction of T lymphocyte apoptosis by sulphasalazine in patients with Crohn's disease. Gut 53:111632–111638
124. Berndt BE, Zhang M, Chen GH, Huffnagle GB, Kao JY (2007) The role of dendritic cells in the development of acute dextran sulfate sodium colitis. J Immunol 179:96255–96262

125. Martinesi M, Treves C, d'Albasio G, Bagnoli S, Bonanomi AG, Stio M (2008) Vitamin D derivatives induce apoptosis and downregulate ICAM-1 levels in peripheral blood mononuclear cells of inflammatory bowel disease patients. Inflamm Bowel Dis 14:5597–5604
126. Kim CJ, Kovacs-Nolan J, Yang C, Archbold T, Fan MZ, Mine Y (2009) L-cysteine supplementation attenuates local inflammation and restores gut homeostasis in a porcine model of colitis. Biochim Biophys Acta 1790:101161–101169
127. Gu BH, Minh NV, Lee SH, Lim SW, Lee YM, Lee KS, Kim DK (2010) Deoxyschisandrin inhibits H2O2-induced apoptotic cell death in intestinal epithelial cells through nuclear factor-kappaB. Int J Mol Med 26:3401–3406
128. Yu QT, Saruta M, Papadakis KA (2008) Visilizumab induces apoptosis of mucosal T lymphocytes in ulcerative colitis through activation of caspase 3 and 8 dependent pathways. Clin Immunol 127:3322–3329
129. Sagiv Y, Kaminitz A, Lorberboum-Galski H, Askenasy N, Yarkoni S (2009) A fusion protein composed of IL-2 and caspase-3 ameliorates the outcome of experimental inflammatory colitis. Ann N Y Acad Sci 1173:791–797
130. Shteingart S, Rapoport M, Grodzovski I, Sabag O, Lichtenstein M, Eavri R, Lorberboum-Galski H (2009) Therapeutic potency of IL2-caspase 3 targeted treatment in a murine experimental model of inflammatory bowel disease. Gut 58:6790–6798
131. Yarkoni S, Sagiv Y, Kaminitz A, Farkas DL, Askenasy N (2009) Targeted therapy to the IL-2R using diphtheria toxin and caspase-3 fusion proteins modulates Treg and ameliorates inflammatory colitis. Eur J Immunol 39:102850–102864
132. Satoh Y, Ishiguro Y, Sakuraba H, Kawaguchi S, Hiraga H, Fukuda S, Nakane A (2009) Cyclosporine regulates intestinal epithelial apoptosis via TGF-beta-related signaling. Am J Physiol Gastrointest Liver Physiol 297:3G514–G519
133. Seidelin JB, Nielsen OH (2006) Expression profiling of apoptosis-related genes in enterocytes isolated from patients with ulcerative colitis. APMIS 114:7-8508-517
134. Necefli A, Tulumoglu B, Giris M, Barbaros U, Gunduz M, Olgac V, Guloglu R, Toker G (2006) The effect of melatonin on TNBS-induced colitis. Dig Dis Sci 51:91538–91545
135. Giris M, Erbil Y, Dogru-Abbasoglu S, Yanik BT, Alis H, Olgac V, Toker GA (2007) The effect of heme oxygenase-1 induction by glutamine on TNBS-induced colitis. The effect of glutamine on TNBS colitis. Int J Colorectal Dis 22:6591–6599
136. Wang Y, Srinivasan K, Siddiqui MR, George SP, Tomar A, Khurana S (2008) A novel role for villin in intestinal epithelial cell survival and homeostasis. J Biol Chem 283: 149454–149464
137. Paul G, Bataille F, Obermeier F, Bock J, Klebl F, Strauch U, Lochbaum D, Rummele P, Farkas S, Scholmerich J, Fleck M, Rogler G, Herfarth H (2005) Analysis of intestinal haem-oxygenase-1 (HO-1) in clinical and experimental colitis. Clin Exp Immunol 140: 3547–3555
138. Jones SA, Butler RN, Sanderson IR, Wilson JW (2004) The effect of specific caspase inhibitors on TNF-alpha and butyrate-induced apoptosis of intestinal epithelial cells. Exp Cell Res 292:129–139
139. Akcan A, Kucuk C, Sozuer E, Esel D, Akyildiz H, Akgun H, Muhtaroglu S, Aritas Y (2008) Melatonin reduces bacterial translocation and apoptosis in trinitrobenzene sulphonic acid-induced colitis of rats. World J Gastroenterol 14:6918–6924
140. Sakuraba H, Ishiguro Y, Yamagata K, Munakata A, Nakane A (2007) Blockade of TGF-beta accelerates mucosal destruction through epithelial cell apoptosis. Biochem Biophys Res Commun 359:3406–3412
141. Cuzzocrea S, McDonald MC, Mazzon E, Mota-Filipe H, Centorrino T, Terranova ML, Ciccolo A, Britti D, Caputi AP, Thiemermann C (2001) Calpain inhibitor I reduces colon injury caused by dinitrobenzene sulphonic acid in the rat. Gut 48:4478–4488
142. Manfredi MA, Zurakowski D, Rufo PA, Walker TR, Fox VL, Moses MA (2008) Increased incidence of urinary matrix metalloproteinases as predictors of disease in pediatric patients with inflammatory bowel disease. Inflamm Bowel Dis 14:81091–81096

143. Makitalo L, Kolho KL, Karikoski R, Anthoni H, Saarialho-Kere U (2010) Expression profiles of matrix metalloproteinases and their inhibitors in colonic inflammation related to pediatric inflammatory bowel disease. Scand J Gastroenterol 45:7-8862-871
144. Rath T, Roderfeld M, Halwe JM, Tschuschner A, Roeb E, Graf J (2010) Cellular sources of MMP-7, MMP-13 and MMP-28 in ulcerative colitis. Scand J Gastroenterol 45:1186–1196
145. von Lampe B, Barthel B, Coupland SE, Riecken EO, Rosewicz S (2000) Differential expression of matrix metalloproteinases and their tissue inhibitors in colon mucosa of patients with inflammatory bowel disease. Gut 47:163–173
146. Louis E, Ribbens C, Godon A, Franchimont D, De Groote D, Hardy N, Boniver J, Belaiche J, Malaise M (2000) Increased production of matrix metalloproteinase-3 and tissue inhibitor of metalloproteinase-1 by inflamed mucosa in inflammatory bowel disease. Clin Exp Immunol 120:2241–2246
147. Pedersen G, Saermark T, Kirkegaard T, Brynskov J (2009) Spontaneous and cytokine induced expression and activity of matrix metalloproteinases in human colonic epithelium. Clin Exp Immunol 155:2257–2265
148. Meijer MJ, Mieremet-Ooms MA, van Hogezand RA, Lamers CB, Hommes DW, Verspaget HW (2007) Role of matrix metalloproteinase, tissue inhibitor of metalloproteinase and tumor necrosis factor-alpha single nucleotide gene polymorphisms in inflammatory bowel disease. World J Gastroenterol 13:212960–212966
149. Bister VO, Salmela MT, Karjalainen-Lindsberg ML, Uria J, Lohi J, Puolakkainen P, Lopez-Otin C, Saarialho-Kere U (2004) Differential expression of three matrix metalloproteinases, MMP-19, MMP-26, and MMP-28, in normal and inflamed intestine and colon cancer. Dig Dis Sci 49:4653–4661
150. Kruidenier L, MacDonald TT, Collins JE, Pender SL, Sanderson IR (2006) Myofibroblast matrix metalloproteinases activate the neutrophil chemoattractant CXCL7 from intestinal epithelial cells. Gastroenterology 130:1127–1136
151. Vaalamo M, Karjalainen-Lindsberg ML, Puolakkainen P, Kere J, Saarialho-Kere U (1998) Distinct expression profiles of stromelysin-2 (MMP-10), collagenase-3 (MMP-13), macrophage metalloelastase (MMP-12), and tissue inhibitor of metalloproteinases-3 (TIMP-3) in intestinal ulcerations. Am J Pathol 152:41005–41014
152. McKaig BC, McWilliams D, Watson SA, Mahida YR (2003) Expression and regulation of tissue inhibitor of metalloproteinase-1 and matrix metalloproteinases by intestinal myofibroblasts in inflammatory bowel disease. Am J Pathol 162:41355–41360
153. Gordon JN, Pickard KM, Di Sabatino A, Prothero JD, Pender SL, Goggin PM, MacDonald TT (2008) Matrix metalloproteinase-3 production by gut IgG plasma cells in chronic inflammatory bowel disease. Inflamm Bowel Dis 14:2195–2203
154. Gan X, Wong B, Wright SD, Cai TQ (2001) Production of matrix metalloproteinase-9 in CaCO-2 cells in response to inflammatory stimuli. J Interferon Cytokine Res 21:293–298
155. Martinesi M, Treves C, Bonanomi AG, Milla M, Bagnoli S, Zuegel U, Steinmeyer A, Stio M (2010) Down-regulation of adhesion molecules and matrix metalloproteinases by ZK 156979 in inflammatory bowel diseases. Clin Immunol 136:151–160
156. Monteleone G, Caruso R, Fina D, Peluso I, Gioia V, Stolfi C, Fantini MC, Caprioli F, Tersigni R, Alessandroni L, MacDonald TT, Pallone F (2006) Control of matrix metalloproteinase production in human intestinal fibroblasts by interleukin 21. Gut 55:121774–121780
157. Kobayashi K, Arimura Y, Goto A, Okahara S, Endo T, Shinomura Y, Imai K (2006) Therapeutic implications of the specific inhibition of causative matrix metalloproteinases in experimental colitis induced by dextran sulphate sodium. J Pathol 209:3376–3383
158. Garg P, Vijay-Kumar M, Wang L, Gewirtz AT, Merlin D, Sitaraman SV (2009) Matrix metalloproteinase-9-mediated tissue injury overrides the protective effect of matrix metalloproteinase-2 during colitis. Am J Physiol Gastrointest Liver Physiol 296:2G175–G184
159. Claramunt RM, Bouissane L, Cabildo MP, Cornago MP, Elguero J, Radziwon A, Medina C (2009) Synthesis and biological evaluation of curcuminoid pyrazoles as new therapeutic

agents in inflammatory bowel disease: effect on matrix metalloproteinases. Bioorg Med Chem 17:31290–31296

160. Epstein J, Docena G, MacDonald TT, Sanderson IR (2010) Curcumin suppresses p38 mitogen-activated protein kinase activation, reduces IL-1beta and matrix metalloproteinase-3 and enhances IL-10 in the mucosa of children and adults with inflammatory bowel disease. Br J Nutr 103:6824–6832
161. Ohkawara T, Nishihira J, Takeda H, Hige S, Kato M, Sugiyama T, Iwanaga T, Nakamura H, Mizue Y, Asaka M (2002) Amelioration of dextran sulfate sodium-induced colitis by anti-macrophage migration inhibitory factor antibody in mice. Gastroenterology 123:1256–1270
162. Bauer J, Liebisch G, Hofmann C, Huy C, Schmitz G, Obermeier F, Bock J (2009) Lipid alterations in experimental murine colitis: role of ceramide and imipramine for matrix metalloproteinase-1 expression. PLoS One 4:9e7197
163. Kirkegaard T, Pedersen G, Saermark T, Brynskov J (2004) Tumour necrosis factor-alpha converting enzyme (TACE) activity in human colonic epithelial cells. Clin Exp Immunol 135:1146–1153
164. Brynskov J, Foegh P, Pedersen G, Ellervik C, Kirkegaard T, Bingham A, Saermark T (2002) Tumour necrosis factor alpha converting enzyme (TACE) activity in the colonic mucosa of patients with inflammatory bowel disease. Gut 51:137–143
165. Colon AL, Menchen LA, Hurtado O, De Cristobal J, Lizasoain I, Leza JC, Lorenzo P, Moro MA (2001) Implication of TNF-alpha convertase (TACE/ADAM17) in inducible nitric oxide synthase expression and inflammation in an experimental model of colitis. Cytokine 16:6220–6226
166. Freour T, Jarry A, Bach-Ngohou K, Dejoie T, Bou-Hanna C, Denis MG, Mosnier JF, Laboisse CL, Masson D (2009) TACE inhibition amplifies TNF-alpha-mediated colonic epithelial barrier disruption. Int J Mol Med 23:141–148
167. Yamashita Y, Kawashima I, Yanai Y, Nishibori M, Richards JS, Shimada M (2007) Hormone-induced expression of tumor necrosis factor alpha-converting enzyme/A disintegrin and metalloprotease-17 impacts porcine cumulus cell oocyte complex expansion and meiotic maturation via ligand activation of the epidermal growth factor receptor. Endocrinology 148:126164–126175
168. Meijer MJ, Mieremet-Ooms MA, van Duijn W, van der Zon AM, Hanemaaijer R, Verheijen JH, van Hogezand RA, Lamers CB, Verspaget HW (2007) Effect of the anti-tumor necrosis factor-alpha antibody infliximab on the ex vivo mucosal matrix metalloproteinase-proteolytic phenotype in inflammatory bowel disease. Inflamm Bowel Dis 13:2200–2210
169. Abbott C, Yu D, Woollatt E, Sutherland G, McCaughan G, Gorrell M (2000) Cloning, expression and chromosomal localization of a novel human dipeptidyl peptidase (DPP) IV homolog, DPP8. Eur J Biochem 267:20
170. Ohnuma K, Dang N, Morimoto C (2008) Revisiting an old acquaintance: CD26 and its molecular mechanisms in T cell function. Trends Immunol 29:6
171. Bauvois B, De Meester I, Dumont J, Rouillard D, Zhao H, Bosmans E (1999) Constitutive expression of CD26/dipeptidylpeptidase IV on peripheral blood B lymphocytes of patients with B chronic lymphocytic leukaemia. Br J Cancer 79:7–8
172. Yazbeck R, Howarth G, Abbott C (2009) Dipeptidyl peptidase inhibitors, an emerging drug class for inflammatory disease? Trends Pharmacol Sci 30:11
173. Yazbeck R, Sulda M, Howarth G, Bleich A, Raber K, von Horsten S, Holst J, Abbott C (2010) Dipeptidyl peptidase expression during experimental colitis in mice. Inflamm Bowel Dis 16:8
174. Rybaczyk L, Rozmiarek A, Circle K, Grants I, Needleman B, Wunderlich J, Huang K, Christofi F (2009) New bioinformatics approach to analyze gene expressions and signaling pathways reveals unique purine gene dysregulation profiles that distinguish between CD and UC. Inflamm Bowel Dis 15:7
175. Hildebrandt M, Rose M, Ruter J, Salama A, Monnikes H, Klapp B (2001) Dipeptidyl peptidase IV (DP IV, CD26) in patients with inflammatory bowel disease. Scand J Gastroenterol 36:10

176. Lendeckel U, Arndt M, Bukowska A, Tadje J, Wolke C, Kahne T, Neubert K, Faust J, Ittenson A, Ansorge S, Reinhold D (2003) Synergistic action of DPIV and APN in the regulation of T cell function. Adv Exp Med Biol 524:123–131
177. Pejler G, Knight S, Henningsson F, Wernersson S (2009) Novel insights into the biological function of mast cell carboxypeptidase A. Trends Immunol 30:8
178. Matsuda T, Suzuki J, Furuya K, Masutani M, Kawakami Y (2001) Serum angiotensin I-converting enzyme is reduced in Crohn's disease and ulcerative colitis irrespective of genotype. Am J Gastroenterol 96:9
179. Saibeni S, Spina L, Virgilio T, Folcioni A, Borsi G, de Franchis R, Cugno M, Vecchi M (2007) Angiotensin-converting enzyme insertion/deletion gene polymorphism in inflammatory bowel diseases. Eur J Gastroenterol Hepatol 19:11
180. Bank U, Heimburg A, Helmuth M, Stefin S, Lendeckel U, Reinhold D, Faust J, Fuchs P, Sens B, Neubert K, Tager M, Ansorge S (2006) Triggering endogenous immunosuppressive mechanisms by combined targeting of Dipeptidyl peptidase IV (DPIV/CD26) and Aminopeptidase N (APN/CD13) – a novel approach for the treatment of inflammatory bowel disease. Int Immunopharmacol 6:13–14
181. Yazbeck R, Howarth G, Geier M, Demuth H, Abbott C (2008) Inhibiting dipeptidyl peptidase activity partially ameliorates colitis in mice. Front Biosci 13:6850–6858
182. Geier M, Tenikoff D, Yazbeck R, McCaughan G, Abbott C, Howarth G (2005) Development and resolution of experimental colitis in mice with targeted deletion of dipeptidyl peptidase IV. J Cell Physiol 204:2
183. Jahovic N, Ercan F, Gedik N, Yuksel M, Sener G, Alican I (2005) The effect of angiotensin-converting enzyme inhibitors on experimental colitis in rats. Regul Pept 130:1–2
184. Spencer A, Yang H, Haxhija E, Wildhaber B, Greenson J, Teitelbaum D (2007) Reduced severity of a mouse colitis model with angiotensin converting enzyme inhibition. Dig Dis Sci 52:4
185. Byrnes J, Gross S, Ellard C, Connolly K, Donahue S, Picarella D (2009) Effects of the ACE2 inhibitor GL1001 on acute dextran sodium sulfate-induced colitis in mice. Inflamm Res 58:11
186. Sokol H, Pigneur B, Watterlot L, Lakhdari O, Bermudez-Humaran L, Gratadoux J, Blugeon S, Bridonneau C, Furet J, Corthier G, Grangette C, Vasquez N, Pochart P, Trugnan G, Thomas G, Blottiere H, Dore J, Marteau P, Seksik P, Langella P (2008) Faecalibacterium prausnitzii is an anti-inflammatory commensal bacterium identified by gut microbiota analysis of Crohn disease patients. Proc Natl Acad Sci USA 105:43
187. Tamboli C, Neut C, Desreumaux P, Colombel J (2004) Dysbiosis in inflammatory bowel disease. Gut 53:1
188. Kotlowski R, Bernstein CN, Sepehri S, Krause DO (2007) High prevalence of *Escherichia coli* belonging to the B2 + D phylogenetic group in inflammatory bowel disease. Gut 56:5669–5675
189. Dautin N, Bernstein HD (2007) Protein secretion in gram-negative bacteria via the autotransporter pathway Annu Rev Microbiol 61:89–112
190. Basset C, Holton J, Bazeos A, Vaira D, Bloom S (2004) Are Helicobacter species and enterotoxigenic Bacteroides fragilis involved in inflammatory bowel disease? Dig Dis Sci 49:91425–91432
191. Prindiville TP, Sheikh RA, Cohen SH, Tang YJ, Cantrell MC, Silva J Jr (2000) Bacteroides fragilis enterotoxin gene sequences in patients with inflammatory bowel disease. Emerg Infect Dis 6:2171–2174
192. Wells CL, van de Westerlo EM, Jechorek RP, Feltis BA, Wilkins TD, Erlandsen SL (1996) Bacteroides fragilis enterotoxin modulates epithelial permeability and bacterial internalization by HT-29 enterocytes. Gastroenterology 110:51429–51437
193. Riegler M, Lotz M, Sears C, Pothoulakis C, Castagliuolo I, Wang CC, Sedivy R, Sogukoglu T, Cosentini E, Bischof G, Feil W, Teleky B, Hamilton G, LaMont JT, Wenzl E (1999) Bacteroides fragilis toxin 2 damages human colonic mucosa in vitro. Gut 44:4504–4510

194. Rhee KJ, Wu S, Wu X, Huso DL, Karim B, Franco AA, Rabizadeh S, Golub JE, Mathews LE, Shin J, Sartor RB, Golenbock D, Hamad AR, Gan CM, Housseau F, Sears CL (2009) Induction of persistent colitis by a human commensal, enterotoxigenic Bacteroides fragilis, in wild-type C57BL/6 mice. Infect Immun 77:41708–41718
195. Ferreira R, Alexandre MC, Antunes EN, Pinhao AT, Moraes SR, Ferreira MC, Domingues RM (1999) Expression of bacteroides fragilis virulence markers in vitro. J Med Microbiol 48:11999–1004
196. Medina C, Santana A, Llopis M, Paz-Cabrera MC, Antolin M, Mourelle M, Guarner F, Vilaseca J, Gonzalez C, Salas A, Quintero E, Malagelada JR (2005) Induction of colonic transmural inflammation by Bacteroides fragilis: implication of matrix metalloproteinases. Inflamm Bowel Dis 11:299–105
197. Pfeifer G, Schirmer J, Leemhuis J, Busch C, Meyer DK, Aktories K, Barth H (2003) Cellular uptake of Clostridium difficile toxin B. Translocation of the N-terminal catalytic domain into the cytosol of eukaryotic cells. J Biol Chem 278:4544535–4544541
198. Reineke J, Tenzer S, Rupnik M, Koschinski A, Hasselmayer O, Schrattenholz A, Schild H, Eichel-Streiber C (2007) Autocatalytic cleavage of *Clostridium difficile* toxin B. Nature 446:7134415–7134419
199. Egerer M, Giesemann T, Jank T, Satchell KJ, Aktories K (2007) Auto-catalytic cleavage of *Clostridium difficile* toxins A and B depends on cysteine protease activity. J Biol Chem 282:3525314–3525321
200. Triadafilopoulos G, Pothoulakis C, O'Brien MJ, LaMont JT (1987) Differential effects of *Clostridium difficile* toxins A and B on rabbit ileum. Gastroenterology 93:2273–2279
201. Riegler M, Sedivy R, Pothoulakis C, Hamilton G, Zacherl J, Bischof G, Cosentini E, Feil W, Schiessel R, LaMont JT (1995) Clostridium difficile toxin B is more potent than toxin A in damaging human colonic epithelium in vitro. J Clin Invest 95:52004–52011
202. Guslandi M, Mezzi G, Sorghi M, Testoni PA (2000) *Saccharomyces boulardii* in maintenance treatment of Crohn's disease. Dig Dis Sci 45:71462–71464
203. Garcia VE, De Lourdes De Abreu Ferrari, Oswaldo Da Gama TH, Guerra PA, Carolina Carneiro AA, Paiva MF, Marcos Andrade GE, Sales DC (2008) Influence of Saccharomyces boulardii on the intestinal permeability of patients with Crohn's disease in remission. Scand J Gastroenterol 43:7842–7848
204. Pothoulakis C (2009) Review article: anti-inflammatory mechanisms of action of *Saccharomyces boulardii*. Aliment Pharmacol Ther 30:8
205. Castagliuolo I, LaMont JT, Nikulasson ST, Pothoulakis C (1996) *Saccharomyces boulardii* protease inhibits *Clostridium difficile* toxin A effects in the rat ileum. Infect Immun 64: 125225–125232
206. Castagliuolo I, Riegler MF, Valenick L, LaMont JT, Pothoulakis C (1999) *Saccharomyces boulardii* protease inhibits the effects of *Clostridium difficile* toxins A and B in human colonic mucosa. Infect Immun 67:1302–1307
207. Schnoeller C, Rausch S, Pillai S, Avagyan A, Wittig BM, Loddenkemper C, Hamann A, Hamelmann E, Lucius R, Hartmann S (2008) A helminth immunomodulator reduces allergic and inflammatory responses by induction of IL-10-producing macrophages. J Immunol 180:64265–64272
208. Schierack P, Lucius R, Sonnenburg B, Schilling K, Hartmann S (2003) Parasite-specific immunomodulatory functions of filarial cystatin. Infect Immun 71:52422–52429
209. Cenac N, Coelho A, Nguyen C, Compton S, Andrade-Gordon P, MacNaughton W, Wallace J, Hollenberg M, Bunnett N, Garcia-Villar R, Bueno L, Vergnolle N (2002) Induction of intestinal inflammation in mouse by activation of proteinase-activated receptor-2. Am J Pathol 161:5
210. Vergnolle N, Cellars L, Mencarelli A, Rizzo G, Swaminathan S, Beck P, Steinhoff M, Andrade-Gordon P, Bunnett N, Hollenberg M, Wallace J, Cirino G, Fiorucci S (2004) A role for proteinase-activated receptor-1 in inflammatory bowel diseases. J Clin Invest 114:10

211. Nguyen C, Coelho AM, Grady E, Compton SJ, Wallace JL, Hollenberg MD, Cenac N, Garcia-Villar R, Bueno L, Steinhoff M, Bunnett NW, Vergnolle N (2003) Colitis induced by proteinase-activated receptor-2 agonists is mediated by a neurogenic mechanism. Can J Physiol Pharmacol 81:9920–9927
212. Shpacovitch V, Varga G, Strey A, Gunzer M, Mooren F, Buddenkotte J, Vergnolle N, Sommerhoff C, Grabbe S, Gerke V, Homey B, Hollenberg M, Luger T, Steinhoff M (2004) Agonists of proteinase-activated receptor-2 modulate human neutrophil cytokine secretion, expression of cell adhesion molecules, and migration within 3-D collagen lattices. J Leukoc Biol 76:2
213. Hyun E, Andrade-Gordon P, Steinhoff M, Vergnolle N (2008) Protease-activated receptor-2 activation: a major actor in intestinal inflammation. Gut 57:91222–91229
214. Vergnolle N (2005) Clinical relevance of proteinase activated receptors (pars) in the gut. Gut 54:6867–6874
215. Shpacovitch V, Feld M, Hollenberg M, Luger T, Steinhoff M (2008) Role of protease-activated receptors in inflammatory responses, innate and adaptive immunity. J Leukoc Biol 83:6
216. Chin A, Vergnolle N, MacNaughton W, Wallace J, Hollenberg M, Buret A (2003) Proteinase-activated receptor 1 activation induces epithelial apoptosis and increases intestinal permeability. Proc Natl Acad Sci U S A 100:19
217. Cenac N, Chin AC, Garcia-Villar R, Salvador-Cartier C, Ferrier L, Vergnolle N, Buret AG, Fioramonti J, Bueno L (2004) PAR2 activation alters colonic paracellular permeability in mice via IFN-gamma-dependent and -independent pathways. J Physiol 558(Pt):3913–3925
218. Coelho AM, Vergnolle N, Guiard B, Fioramonti J, Bueno L (2002) Proteinases and proteinase-activated receptor 2: a possible role to promote visceral hyperalgesia in rats. Gastroenterology 122:41035–41047
219. Auge C, Balz-Hara D, Steinhoff M, Vergnolle N, Cenac N (2009) Protease-activated receptor-4 (PAR 4): a role as inhibitor of visceral pain and hypersensitivity. Neurogastroenterol Motil 21:111189–e107
220. Walsh D, Greene C, Carroll T, Taggart C, Gallagher P, O'Neill S, McElvaney N (2001) Interleukin-8 up-regulation by neutrophil elastase is mediated by MyD88/IRAK/TRAF-6 in human bronchial epithelium. J Biol Chem 276:38
221. Tsujimoto H, Ono S, Majima T, Kawarabayashi N, Takayama E, Kinoshita M, Seki S, Hiraide H, Moldawer L, Mochizuki H (2005) Neutrophil elastase, MIP-2, and TLR-4 expression during human and experimental sepsis. Shock 23:1
222. Devaney J, Greene C, Taggart C, Carroll T, O'Neill S, McElvaney N (2003) Neutrophil elastase up-regulates interleukin-8 via toll-like receptor 4. FEBS Lett 544:1–3
223. Cario E, Golenbock D, Visintin A, Runzi M, Gerken G, Podolsky D (2006) Trypsin-sensitive modulation of intestinal epithelial MD-2 as mechanism of lipopolysaccharide tolerance. J Immunol 176:7
224. Rallabhandi P, Nhu QM, Toshchakov VY, Piao W, Medvedev AE, Hollenberg MD, Fasano A, Vogel SN (2008) Analysis of proteinase-activated receptor 2 and TLR4 signal transduction: a novel paradigm for receptor cooperativity. J Biol Chem 283:3624314–3624325
225. Nhu QM, Shirey KA, Pennini M, Stiltz J, and Vogel S (2011) Protease-activated receptor 2 activation promotes an anti-inflammatory and alternatively activated phenotype in LPS-stimulated murine macrophages. Innate Immun
226. Sun R, Iribarren P, Zhang N, Zhou Y, Gong W, Cho E, Lockett S, Chertov O, Bednar F, Rogers T, Oppenheim J, Wang J (2004) Identification of neutrophil granule protein cathepsin G as a novel chemotactic agonist for the G protein-coupled formyl peptide receptor. J Immunol 173:1
227. Anton P, Targan S, Shanahan F (1989) Increased neutrophil receptors for and response to the proinflammatory bacterial peptide formyl-methionyl-leucyl-phenylalanine in Crohn's disease. Gastroenterology 97:1

228. Anton P, O'Connell J, O'Connell D, Whitaker L, O'Sullivan G, Collins J, Shanahan F (1998) Mucosal subepithelial binding sites for the bacterial chemotactic peptide, formyl-methionyl-leucyl-phenylalanine (FMLP). Gut 42:3
229. Laforge M, Bidere N, Carmona S, Devocelle A, Charpentier B, Senik A (2006) Apoptotic death concurrent with CD3 stimulation in primary human CD8+ T lymphocytes: a role for endogenous granzyme B. J Immunol 176:7
230. Cullen S, Adrain C, Luthi A, Duriez P, Martin S (2007) Human and murine granzyme B exhibit divergent substrate preferences. J Cell Biol 176:4
231. Goping I, Barry M, Liston P, Sawchuk T, Constantinescu G, Michalak K, Shostak I, Roberts D, Hunter A, Korneluk R, Bleackley R (2003) Granzyme B-induced apoptosis requires both direct caspase activation and relief of caspase inhibition. Immunity 18:3
232. Martinvalet D, Dykxhoorn D, Ferrini R, Lieberman J (2008) Granzyme A cleaves a mitochondrial complex I protein to initiate caspase-independent cell death. Cell 133:4
233. Padrines M, Wolf M, Walz A, Baggiolini M (1994) Interleukin-8 processing by neutrophil elastase, cathepsin G and proteinase-3. FEBS Lett 352:2
234. Nufer O, Corbett M, Walz A (1999) Amino-terminal processing of chemokine ENA-78 regulates biological activity. Biochemistry 38:2
235. Sugawara S, Uehara A, Nochi T, Yamaguchi T, Ueda H, Sugiyama A, Hanzawa K, Kumagai K, Okamura H, Takada H (2001) Neutrophil proteinase 3-mediated induction of bioactive IL-18 secretion by human oral epithelial cells. J Immunol 167:116568–116575
236. Coeshott C, Ohnemus C, Pilyavskaya A, Ross S, Wieczorek M, Kroona H, Leimer A, Cheronis J (1999) Converting enzyme-independent release of tumor necrosis factor alpha and IL-1beta from a stimulated human monocytic cell line in the presence of activated neutrophils or purified proteinase 3. Proc Natl Acad Sci U S A 96:11
237. Robache-Gallea S, Morand V, Bruneau J, Schoot B, Tagat E, Realo E, Chouaib S, Roman-Roman S (1995) In vitro processing of human tumor necrosis factor-alpha. J Biol Chem 270:40
238. Scuderi P, Nez P, Duerr M, Wong B, Valdez C (1991) Cathepsin-G and leukocyte elastase inactivate human tumor necrosis factor and lymphotoxin. Cell Immunol 135:2
239. Bank U, Kupper B, Reinhold D, Hoffmann T, Ansorge S (1999) Evidence for a crucial role of neutrophil-derived serine proteases in the inactivation of interleukin-6 at sites of inflammation. FEBS Lett 461:3
240. Rawlings ND, Barrett AJ, and Bateman A (2010) MEROPS: the peptidase database. Nucleic Acids Res 38:Database issue:D227–D233
241. Finn RD, Mistry J, Tate J, Coggill P, Heger A, Pollington JE, Gavin OL, Gunasekaran P, Ceric G, Forslund K, Holm L, Sonnhammer EL, Eddy SR, and Bateman A (2010) The Pfam protein families database. Nucleic Acids Res 38:Database issue:D211–D222
242. He SH, Xie H (2004) Inhibition of tryptase release from human colon mast cells by protease inhibitors. World J Gastroenterol, 10:332–336

Proteinase-Activated Receptors and Arthritis

Fiona A. Russell and Jason J. McDougall

Abstract The novel family of proteinase-activated receptors (PARs) is activated through proteolytic cleavage by serine proteinases. This family of G protein-coupled receptors and their activating enzymes are found widely throughout the body. It has been known for some time that during arthritic conditions, high levels of serine proteinases are released from joint tissue and contribute to joint degradation. Expression of PARs is also enhanced in arthritic joints and act to mediate the intracellular signalling of the serine proteinases, leading to various inflammatory and painful effects. This chapter summarises what is known so far on all four of the currently known PARs with respect to their links with arthritis and discusses how these receptors may represent novel targets for the development of new arthritic treatments.

Keywords Cathepsin G • Inflammation • Joints • Mast cell tryptase • Osteoarthritis • Pain • Proteinase-activated receptors • Rheumatoid arthritis • Serine proteinases • Thrombin

1 Introduction

Arthritis is defined as "inflammation of the joint" and this term encompasses over 200 different joint diseases that are estimated to affect over 21% of adults in the USA [1]. Arthritic conditions can be broadly divided into two types, degenerative diseases such as osteoarthritis (OA) and inflammatory conditions, such as rheumatoid arthritis (RA). A major symptom of all arthritic conditions is persistent joint pain, which significantly reduces an individual's mobility and emotional

F.A. Russell • J.J. McDougall (✉)
Department of Physiology and Pharmacology, University of Calgary, 3330 Hospital Drive NW, Calgary, AB T2N 4N1, Canada
e-mail: mcdougaj@ucalgary.ca

N. Vergnolle and M. Chignard (eds.), *Proteases and Their Receptors in Inflammation*, Progress in Inflammation Research, DOI 10.1007/978-3-0348-0157-7_9,

well-being. In fact, musculoskeletal disorders are associated with a poorer quality of life score with respect to pain and bodily function than both cardiovascular and gastrointestinal disorders [2]. Arthritis progresses linearly over time and though primarily thought of as chronic disabling conditions, RA in particular is also associated with a plethora of comorbidities and reduced life expectancy [3]. Due to the common occurrence of joint pain, the prevalence of arthritis-related disability is extremely high and in fact is one of the major causes of long-term disability worldwide [4]. A review of productivity loss due to RA in North America and Europe found that the median number of work days missed in a year due to RA was 39 [5] and one UK study showed 33% of RA patients permanently stopped work within 5 years of disease onset compared to 3% of age-matched controls [6]. The impact of OA is even higher with it having a disability burden three times that of RA [7]. Thus, direct costs such as medication and rehabilitative care as well as the indirect costs due to disability mean that these diseases carry a huge economic burden, similar if not more than cancer [8]. Annual cost estimations for arthritis range from 1 to 2.5% of the gross national product of developed countries such as the USA, Canada and the UK [7] with OA alone thought to cost the USA more than $60 billion dollars per year [9]. A 2009 study estimated the annual health care costs of RA in the US to be $19.3 billion [10]. When costs associated with quality of life deterioration and premature death are factored in, the total annual societal cost of RA rises to $39.2 billion [10]. Worryingly, with the population ageing and the fact that the incidence of arthritis increases with age, this economic burden will continue to grow. Projections estimate that by 2030, 67 million people (25% of the adult population) in the USA will have some form of arthritis [11]. The scale of the problem is reflected in the endorsement by the United Nations and the World Health Organisation of the Bone and Joint Decade 2000–2010 to "raise awareness of the growing burden of musculo-skeletal disorders on society" [12] (http://www.boneandjointdecade.org).

Early treatment of arthritis seems to be critical for reducing disease severity and by corollary the economic burden [13–15]. Current treatment options are limited and therefore the need to develop new therapeutics for these diseases is critical. The recent discovery of proteinase-activated receptors (PARs) and their contribution to inflammation and pain has led to a surge in research in this area, and evidence is emerging to show the key roles these receptors play in arthritis.

1.1 Epidemiology of Rheumatoid Arthritis

Inflammatory joint diseases include gout, psoriatic arthritis, ankylosing spondylitis and RA. RA is one of the most common autoimmune diseases, estimated to affect approximately 1% of adults in the Western world, with three times as many females being affected compared to males [16]. The exact cause of RA is unknown but it is thought to be due to a complex interplay between genetic and environmental factors. Interestingly, several studies indicated a decline in the incidence of RA during the second half of the twentieth century [17]; however, a recent study in the USA showed that the incidence of RA in women increased from 1995 to 2007 [18]. It appears, therefore, that RA continues to be a major clinical problem. The reasons for an increased prevalence of RA are unclear but several environmental factors such as smoking, alcohol use, obesity and vitamin D deficiency have all been implicated [18].

RA normally tends to develop between the ages of 20 and 50 but can occur at any age. Typically RA is observed in the joints of the fingers, wrists, feet and ankles. Initially there is proliferation of synovial macrophages and fibroblasts, which instigate the process of inflammation leading to plasma extravasation and cellular infiltration. The consequent high influx of immune cells into the joint results in the release of large amounts of pro-inflammatory cytokines and chemokines, thus aggravating the inflammation further. Prolific angiogenesis occurs and the inflamed synovial tissue begins to grow and invades all regions of the joint. This pannus tissue gradually causes the erosion of joint cartilage and subchondral bone leading to joint destruction. Patients experience symptoms of prolonged joint stiffness, pain, swelling and heat which can begin to occur over weeks to months following disease initiation [19]. Treatment of RA is hindered by the fact that there is no simple test to confirm the presence of RA thereby delaying diagnosis.

1.2 Current Treatment Options for Rheumatoid Arthritis

There are several therapeutic approaches to treat RA but all are limited. Drugs currently used include non-steroidal anti-inflammatory drugs (NSAIDs), steroids and disease modifying anti-rheumatic drugs (DMARDs). NSAIDs alleviate the pain and swelling in RA by inhibiting prostaglandin synthesis through their ability to inhibit cyclo-oxygenase enzyme activity. NSAIDs do not affect disease progression and are normally used in the initial treatment of RA to reduce symptoms. Unwanted side effects of NSAIDs include gastrointestinal bleeding due to the inhibition of gastric cytoprotective prostaglandins. Mild gastric problems are fairly common with NSAID use but serious complications can also occur. In fact, it has been estimated that 16,500 NSAID-related deaths occur in arthritis patients every year in the USA making it one of the top 20 causes of death [20].

Corticosteroids such as prednisolone are often used to treat RA. These drugs are closely related to the hormone cortisol and attenuate inflammation by down-regulating pro-inflammatory cytokines such as TNFα, IL-1β and IL-6 [21]. Steroids are beneficial in that they relieve symptoms and slow joint damage but on the downside they are also responsible for the development of many unwanted side effects such as osteoporosis, diabetes and cataracts [22]. Therefore, the use of steroids tends to be restricted to low doses for limited periods of time.

DMARDs, such as methotrexate, hydroxychloroquine and gold containing compounds, can inhibit the progression of RA but do not alleviate pain. Methotrexate is the most commonly used DMARD and, although its precise mechanism of action is unknown, it is thought to target T cells [16]. The development of the new class of DMARD drugs termed "biologic" has arguably been the most significant advance in recent history with regards to RA treatment. Anti-TNFα drugs, such as etanercept and infliximab, have proven to be highly beneficial in RA patients, reducing disease severity and pain leading to a significant improvement in quality of life [19, 23–25]. In 2001, the drug anakinra, a recombinant form of IL-1 receptor antagonist which blocks the inflammatory effects of IL-1 was approved for clinical use for RA. However, anakinra has been seen to be less effective than anti-TNFα drugs, possibly due to its short half-life of 4–6 h [16]. Advantages of these biologics include their rapid onset of action compared to traditional DMARDs, their sustained efficacy and the fact that they are generally well tolerated. Disadvantages include a heightened risk of infection, the potential for these drugs to activate latent tuberculosis and probably most significant, the fact that these drugs are extremely expensive [24]. In addition, between 20 and 40% of patients do not respond to anti-TNFα drugs, though the reason for this is, as yet, unknown [26].

It is now believed that the optimal therapy for early onset RA is a combination of methotrexate with one of the biologics [27]. However, cost often prohibits this approach and unfortunately it is more common to treat with NSAIDs early on in the disease and only progress to the biologics once the disease has worsened considerably. Thus, the development of new cheaper therapeutic drugs that can be given during disease development is crucial to prevent disease progression.

1.3 Epidemiology of Osteoarthritis

OA is the most common type of arthritis accounting for almost half of arthritis sufferers, yet treatment options are even more limited than with RA. It is hard to estimate the overall incidence of OA but it is thought that approximately 10% of men and 18% of women worldwide over the age of 60 have symptomatic OA [12]. OA is classed as a degenerative disease characterised by the breakdown of cartilage in the joint. This so-called wear and tear disease was traditionally viewed as an inevitable consequence of ageing. However, this is now known to be an oversimplification of the disease process. Although symptoms typically begin to develop in people in their 40s and 50s, OA can occur at any age and progressively worsens as the patient gets older. Whilst ageing is undoubtedly a major contributing factor to the development of OA, many other factors are also involved including mechanical instability, joint injury, obesity and genetic factors [28]. OA commonly occurs in the knee, hands and hip joints and the major symptom is chronic pain in the joint and surrounding tissues. Other outcomes include reduced function, loss of mobility, stiffness as well as joint weakening and deformity [29]. Despite it not being classed as an inflammatory disease per se, inflammation can occur in some OA patients [28].

Once thought of as primarily a disorder of the articular cartilage, OA is far more complex than originally believed and affects other tissues such as the peri-articular and subchondral bone, ligaments, tendons, menisci and the synovium [30, 31]. Degradation of cartilage occurs due to an imbalance in the anabolic and catabolic metabolism of chondrocytes [30]. Anabolic mediators such as tumour growth factor (TGF)-β and fibroblast growth factors (FGFs) enhance the synthesis and repair of cartilage. However, during OA, pro-inflammatory cytokines such as IL-1β and TNFα and matrix metalloproteinases (MMPs) such as MMP-13 are secreted into the joint. MMPs are a large family of zinc-dependent endopeptidases that cause degradation of the extracellular matrix. Normally, the balance between MMPs and their inhibitors are tightly controlled; however, during OA an imbalance is often apparent leading to excessive cartilage degradation [32, 33]. The catabolic state in the joint is then exacerbated as these pro-inflammatory mediators can act on the chondrocytes stimulating further secretion of catabolic mediators. The loss of cartilage together with other important modifications such as osteophyte formation and subchondral bone remodelling all enhance joint deformity and reinforce abnormal loading of the joint, thus further augmenting the disease [31].

As seen with RA, there is no straightforward test for OA and a combination of physical assessment, medical history and radiographic information are required for diagnosis. Unfortunately, the use of radiography or other such joint imaging techniques for diagnosis is restricted by the intriguing fact that there is very little correlation between the severity of joint destruction and the reported pain [34, 35]. The reason for this has yet to be determined.

1.4 Current Treatment Options for Osteoarthritis

Current therapeutic approaches for OA are limited to the relief of symptoms and unlike RA, very few disease modifying drugs are available and efficacious. OA therapy is normally a combination of both pharmacological and non-pharmacological techniques. Non-pharmacological interventions as recommended by the American College of Rheumatology (ACR) include patient education, weight loss (if obesity is an issue), aerobic exercise programs, muscle strengthening exercises, assistive devices for mobility, bracing and occupational and physical therapy [31]. Pharmacological medications currently used to manage OA include analgesics, NSAIDs, steroids and some slow acting DMOADs. The first line of treatment for OA pain is normally a mild analgesic such as acetaminophen, then progressing to topical NSAIDs, steroids and if very severe pain is present, opioids. The same disadvantages occur with NSAIDs and steroids as previously noted. Non-prescribed capsaicin is used as a topical cream to locally desensitise peripheral nerve endings leading to reduced pain; however, capsaicin creams have low patient compliance due to the burning pain experienced on initial application [36]. Slow acting DMOADS include hyaluronic acid, glucosamine sulphate and chondroitin sulphate and are normally injected directly into the joint or taken orally. Hyaluronic acid is a large glycosaminoglycan and is a component of the synovial fluid. Glucosamine and chondroitin sulphate are both constituents of cartilage. The mechanisms responsible for how these agents provide pain relief in OA are unclear but could involve reduced inflammatory cytokine release [37]. Recent evaluations suggest that these drugs actually have no benefit over placebo [38, 39].

There are also surgical interventions that are recommended for OA. Total joint replacement is recommended for patients with hip or knee OA that do not respond to any of the non-pharmacological or pharmacological treatments [38]. Joint lavage and debridement to remove damaged tissue and loose debris from the joint have been used extensively but have limited benefit [39]. Joint fusion is seen as a last resort and only attempted when joint replacement has failed [38]. Considerable research is ongoing in the area of tissue engineering and gene therapy to try and promote cartilage repair as an alternative to total joint replacement or fusion [28].

2 Proteinase Activated Receptors and Arthritis

Levels of certain serine proteinases are elevated in the synovial fluid of arthritis patients [40–42] and were originally thought to have a purely degradative role leading to joint degeneration. Since the realisation that these proteolytic enzymes can also act as signalling molecules by activating PARs, the spotlight on their role in arthritis has attracted greater scrutiny. Compared to controls, the proteolytic activity in the synovial fluid from RA and OA patients is enhanced, with bioactivity in RA exudates being higher than OA. The largest difference in enzyme activity has been observed with thrombin which has an almost eightfold higher activity in RA samples [41]. This implies that thrombin is more relevant in the pathogenesis of RA than OA. Table 1 lists some of the major serine proteinases seen in arthritic joints and the PAR family member they activate.

It is perhaps not surprising that thrombin would play a role in RA pathogenesis as this coagulation factor causes fibrin deposition through the cleavage of

Table 1 The main serine proteinases found in arthritic joints and the PARs they activate

	Present in			
Serine proteinase	RA joint	OA joint	PAR activated	References
Thrombin	↑↑	↑	$PAR_1 > PAR_3 = PAR_4$	[41, 43, 44]
Tryptase	↑	↑	PAR_2	[40, 45]
Proteinase 3	↑		PAR_2	[46, 47]
Matriptase 1		↑	PAR_2	[48]
Cathepsin G	↑		$PAR_4 > PAR_1$	[49, 50]
Urokinase (uPA)	↑		Activates plasmin which can act on $PAR_4 > PAR_1$	[50, 51]
Tissue plasminogen activator (tPA)		↑	Activates plasmin which can act on $PAR_4 > PAR_1$	[50, 51]

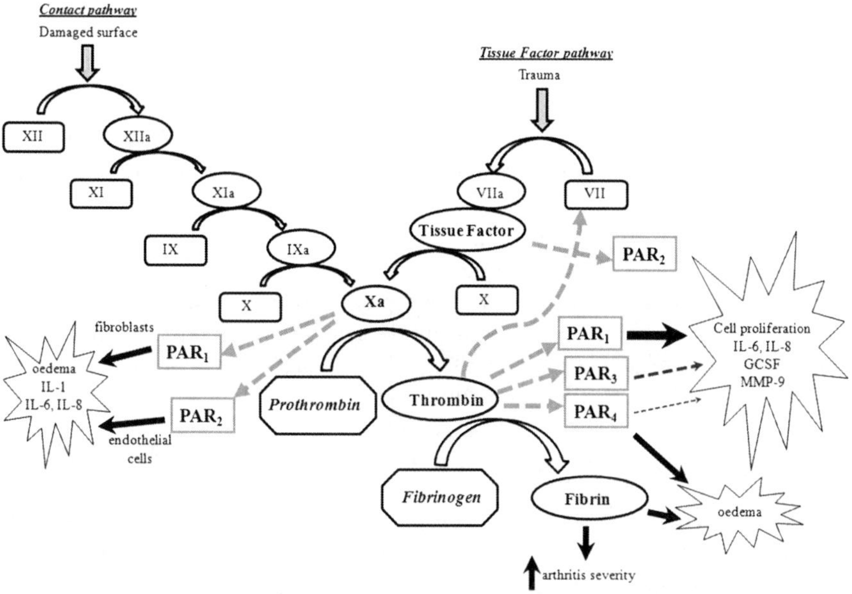

Fig. 1 A simplified diagram of the coagulation cascade and its links with PARs and inflammation. Two pathways are involved in coagulation, with the major one being the tissue factor pathway. Most of the coagulation factors are serine proteinases that are present as inactive zymogens (*rectangles*) that when activated (*ovals*; lowercase '**a**' indicates an active form) can catalyse the next reaction in the cascade. The two pathways converge to activate factor X, thrombin and fibrin. In an arthritic joint, high levels of many of these coagulation proteinases are seen [42] and fibrin deposition correlates with arthritis progression [52, 53]. Many of the proteinases in the cascade can also lead to pro-inflammatory signalling effects, thus further enhancing the pro-inflammatory state of the joint. These signalling effects are mediated via the proteolytic activation of PARs. Xa can activate PAR_1 and PAR_2 [54]. Thrombin mediates the majority of its effects via PAR_1 but can also activate PAR_3 and PAR_4. Whilst thrombin cannot directly activate PAR_2, it does have many feedback activation roles including the activation of factor VII which can then bind to tissue factor which can activate PAR_2. Tissue factor induced inflammation appears to require not only PAR_4 but also the formation of thrombin and fibrin [55]

fibrinogen (Fig. 1). Fibrin deposition in the synovial membrane is due to an imbalance between coagulation and fibrinolysis [42] and closely correlates with the extent of disease progression in both humans and rodent models of RA [52, 53]. The importance of thrombin in arthritis pathogenesis is confirmed by studies showing significantly reduced disease severity after thrombin inhibition in both collagen- and antigen-induced arthritis (AIA) models with reduced intra-articular fibrin seen after thrombin inhibition [43, 44]. Another interesting study in rats showed that articular chondrocytes express the prothrombin gene (the inactive precursor of thrombin) and this expression and thrombin levels increase after the joint is immobilised contributing to cartilage degeneration [56].

3 PAR_1

3.1 PAR_1, Thrombin and Arthritis

As well as playing a role in arthritis due to its proteolytic effects on fibrinogen, thrombin also initiates important cell signalling effects. Thrombin can cause synovial cell proliferation [57, 58], which is a key feature in the development of an invasive pannus in RA joints. In addition, thrombin induces the release of pro-inflammatory cytokines such as IL-6 and granulocyte colony stimulating factor from rheumatoid synovial fibroblasts [59], thereby contributing further to the inflammatory state within the joint. Thrombin mediates these effects through the cleavage and activation of the thrombin receptor, which was classified as proteinase activated receptor 1 (PAR_1), the first member of the PAR family to be discovered. Increased expression of PAR_1 is seen in the synovial tissue from RA patients with little or no receptor observed in normal and OA tissue [57, 60]. PAR_1 has also been detected in numerous other cell types involved in the pathogenesis of arthritis including endothelial cells, monocytes, fibroblasts, neurons and glial cells [61, 62] and can cause potent inflammatory effects such as oedema, increased vascular permeability, leukocyte recruitment and the release of potent pro-inflammatory cytokines such as TNFα and IL-1β [62–66].

Despite thrombin being able to activate PAR_3 and PAR_4, experimental evidence implies that the majority of the thrombin-induced effects during arthritis are mediated via its activation of PAR_1. The thrombin-induced proliferation of synovial fibroblasts from RA patients was shown to be mediated via PAR_1 [67]. The reduction in disease severity in the rodent arthritic models [43, 44] was achieved by the thrombin inhibitor hirudin, which is derived from the salivary glands of the leech *Hirudo medicinalis* [68]. Only PAR_3 contains a similar hirudin-binding domain to PAR_1 [69] and as will be discussed later, it is still unclear whether activation of PAR_3 initiates any cell signalling. PAR_4 expression was not observed in RA synovial fibroblasts, and when these cells were exposed to thrombin, there was an increase in expression of the chemokine, RANTES, known for its involvement in RA pathogenesis [70]. The authors used a reporter gene assay to show that this effect was mediated via PAR_1 and not PAR_3 [70]. Furthermore, a recent study using PAR_1 knock-out mice showed that these animals developed a milder form of AIA with lower levels of pro-inflammatory cytokines in the synovial fluid [71]. Thus, PAR_1 appears to be the dominant receptor mediating the effect of thrombin and therefore may be a promising target for the development of anti-arthritic drugs.

3.2 PAR_1 and Factor Xa

Factor Xa is another major proteinase associated with the coagulation cascade (Fig. 1). Similar to thrombin, it has recently been expounded that this enzyme has many pro-inflammatory effects [72]. Factor Xa can lead to the release of many pro-inflammatory cytokines such as IL-1, IL-6 and IL-8 from various different cell types including macrophages, fibroblasts and endothelial cells [73–75]. Furthermore, injection of Factor Xa in vivo leads to oedema in the rat hind paw [63]. Originally, these signalling effects of Xa were thought to be mediated via a receptor termed effector cell protease receptor 1 (EPR-1) [63]. However, it is unclear now whether this receptor exists [72] and it is more likely that Xa works via the PARs as it has been shown to cleave and activate both PAR_1 and PAR_2 [76]. It appears that in fibroblasts PAR_1 is the dominant receptor for Factor Xa, whereas in endothelial cells PAR_2 accounts for the majority of the signalling effects of Factor Xa [54]. There is little research looking specifically at the role of Factor Xa in arthritis; however, this enzyme is upstream of thrombin in the coagulation cascade (Fig. 1). Therefore, blocking Factor Xa would have the knock-on effect of inhibiting thrombin, thus potentially having a similar effect in arthritis models to thrombin inhibition as discussed previously. As Xa and thrombin both signal via PAR_1, antagonising this receptor could be a beneficial way of preventing the signalling effects of both these enzymes without affecting their coagulation properties.

3.3 PAR_1 Antagonism for the Treatment of Arthritis

PAR_1 antagonists have recently been developed as antithrombotic agents to help treat and prevent adverse cardiovascular events. Two compounds, SCH 530348 and E5555 are currently in phase II clinical trials for patients with acute coronary disease [77, 78]. The potential role of PAR_1 antagonists to treat arthritis is less clear. Yang et al. showed a reduction in inflammatory parameters in the antigen-induced model of arthritis in the $PAR_1^{-/-}$ mice, while Martin et al. observed no difference in oedema and leukocyte recruitment in these mice following collagen-induced arthritis [71, 79]. The discontinuity between the two results could be due to the differences in how the experimental arthritis was induced, with the collagen-induced model being a systemic immunisation model whereas the AIA model used an intra-articular injection of bovine serum albumin after systemic immunisation. A further explanation for the discrepancy could be that Yang et al. measured inflammatory parameters at a slightly later time point than Martin et al. Further work is needed, with perhaps the use of other arthritic models, to clarify the effect of PAR_1 activation during chronic inflammatory joint disease.

Activation of PAR_1 can have an anti-nociceptive effect, thus PAR_1 antagonism could potentially heighten arthritic joint pain. Intraplantar injections of the PAR_1-activating peptide into the rat hind paw cause mechanical and thermal analgesia and can attenuate inflammatory pain [80, 81] (Table 2). PAR_1 is expressed in nociceptive neurons, although an inhibitory role for PAR_1 in pain messages conveyed by primary afferents has never been observed [79]. The analgesic effect of PAR_1 is due, at least in part, to the release of endogenous opioids, as it has been shown in fibroblasts and keratinocytes [79]. In addition, $PAR_1^{-/-}$ mice exhibit an enhanced

Table 2 The activating peptides and antagonists of each of the four PARs and their effect in joints

PAR	Activating peptides	Effect in joint	Antagonists	Effect in joint
PAR_1	TFLLR-NH_2	Reduces referred pain in arthritic mice. Analgesia is opioid-dependent [79]	Hirudin SCH 530348 E5555	Reduces arthritis severity in rodent models [43, 44] Not tested
PAR_2	SLIGRL-NH_2	Causes hindlimb incapacitance and secondary allodynia/ hyperalgesia. Effect is TRPV1-dependent [82]	ENMD-1068 Monoclonal antibody SAM-11 p520	Reduces joint swelling in acute inflammatory model [83] Reduces osteoarthritic progression in rodent model [84]
PAR_3	None	N/A	N/A	N/A
PAR_4	AYPGKF-NH_2	Increases blood flow, promotes joint oedema, sensitises joint primary afferents, causes thermal and mechanical hyperalgesia [85, 86]	Pepducin P4pal-10	Reduces cellular infiltration and synovial hyperplasia in acute inflammatory model [85]

hyperalgesia compared to wild types after injections of trypsins which can activate PAR_1 [87], suggesting that in wild-type mice trypsins produce a mild analgesic effect via PAR_1 activation with this effect lacking in PAR_1 knock-out mice. Yang et al. did not study whether the $PAR_1^{-/-}$ mice with the milder form of arthritis had altered pain responses but another group saw that these mice exhibited heightened responses to both thermal and mechanical stimuli during collagen-induced arthritis [79]. The role of PAR_1 in controlling joint pain requires further elucidation before it can be considered a viable analgesic target.

4 PAR_2

4.1 *Preclinical Evidence of PAR_2 Involvement in Arthritis*

Ferrell et al. found that PAR_2 is crucial for the development of inflammatory arthritis. In chronic adjuvant-induced arthritis, $PAR_2^{-/-}$ mice exhibit almost zero joint damage and develop significantly less joint swelling than wild-type mice [88]. Elsewhere, Busso et al. found no role for PAR_2 in the same adjuvant-induced arthritis model [89], this discrepancy may be due to differences between the two strains of $PAR_2^{-/-}$ mice used. Ferrell et al. incorporated a β-galactosidase reporter gene into the disrupted PAR_2 gene, which could have been responsible for the reduced inflammatory response in the $PAR_2^{-/-}$ mice [88]. In AIA, Busso et al. observed reduced knee joint inflammation in $PAR_2^{-/-}$ mice [71, 89]. Two other arthritis models, viz., zymosan-induced arthritis (ZIA) and K/BxN serum-induced arthritis were also examined in the same study but no difference in severity was noted between $PAR_2^{-/-}$ and wild-type mice [89]. Thus, in their hands the only model to implicate PAR_2 in arthritis was the AIA model which unlike other models requires prior immunisation for induction. They then demonstrated a reduced antibody and T cell response in the $PAR_2^{-/-}$ animals leading them to the suggestion that PAR_2 is more important in antigen-specific immune responses rather than innate immune pathways of arthritis [89].

A very recent study has utilised $PAR_2^{-/-}$ mice in a model of OA induced by sectioning of the medial meniscotibial ligament, in order to study the effect of a lack of the PAR_2 receptor on OA [84]. $PAR_2^{-/-}$ mice were significantly protected from the development of OA pathology, though the mechanisms for this protection have yet to be investigated [84].

4.2 *Clinical Evidence of PAR_2 in Arthritis*

Studies using both RT-PCR and immunohistological methods show significantly higher expression of PAR_2 mRNA and protein in the synovia from RA patients than OA patients [89, 90]. This implies that PAR_2 may play more of an important role in RA than OA but crucially, neither study compared this expression to PAR_2 expression in healthy controls. In contrast, Xiang et al. proposed a role for PAR_2 in the pathogenesis of OA as both PAR_2 mRNA and protein levels were increased in articular chondrocytes from OA patients compared to controls [91]. In synovial fibroblasts and synoviocytes taken from RA patients, PAR_2 expression could be up-regulated by pro-inflammatory cytokines and growth factors and down-regulated by anti-inflammatory cytokines [91–93]. The effect on PAR_2 expression seems to depend on the current inflammatory state as the anabolic mediator TGF-β up-regulated PAR_2 in normal chondrocytes but down-regulated the receptor in OA chondrocytes [91]. TGF-β was also seen to suppress the enhancement of PAR_2 expression in human OA primary synovial cells [93]. This tight regulation of PAR_2 expression during inflammation suggests a key role for PAR_2 in arthritic conditions. Thus, further investigation is merited to clarify the relative expression of PAR_2 in RA, OA and healthy tissue and to follow its expression and function as disease becomes more advanced.

Recently, several studies have attempted to determine the functional consequences of PAR_2 up-regulation in arthritic tissues. It is known that PAR_2 activation can promote the release of inflammatory cytokines and activate the pro-inflammatory transcription factor NF-κB in human endothelial cells [94]. PAR_2 is highly expressed in endothelial cells in RA synovium as well as in macrophages and mast cells present in the tissue [95]. Furthermore, inhibition of PAR_2 in cultures of RA synovial tissue was shown to significantly inhibit the release of the pro-inflammatory cytokines, IL-1β and TNFα [95].

Boileau et al. confirmed that the expression of PAR_2 is enhanced in OA chondrocytes compared to normal and this could be further increased by stimulation with IL-1β, TNFα and most significantly by PAR_2 activating peptides [96], suggesting a positive feedback mechanism upon receptor activation. PAR_2 activation in these cells led to an increase in MMP-1, MMP-13 and COX-2, all key enzymes in OA pathology [96]. The signalling pathways induced following PAR_2 activation involved extracellular signal-related kinase1/2 (Erk 1/2) and p38 [96].

Since subchondral bone remodelling plays a key role in OA pathology, it is noteworthy that PAR_2 expression is also enhanced in human OA subchondral bone osteoblasts compared to healthy controls [97]. As with the chondrocytes, PAR_2 expression in the osteoblasts could be increased by the pro-inflammatory cytokines, IL-1β and TNFα, as well as the prostaglandin, PGE_2 [97]. However, no effect was seen after treatment with TGF-β, suggesting differential regulation between cell types. Bone remodelling is normally a tightly regulated balance between bone formation and resorption but in OA this balance is altered with increases seen initially in bone resorption [98, 99]. Thus, it is extremely interesting that PAR_2 activation in OA osteoblasts induced bone resorptive activity in these cells [97]. This was proposed to be due to the PAR_2-induced up-regulation of MMP-1, MMP-9, IL-6 and receptor activator of nuclear factor kappa B ligand (RANKL), all factors involved in bone resorption [97]. Amiable et al. also studied the intracellular signalling pathways involved upon PAR_2 activation and once again found activation of Erk 1/2 to be crucial, together with Jun N-terminal kinas (JNK). Unlike chondrocytes, the p38 pathway was not shown to be modulated by PAR_2 [97]. Thus, despite the differential PAR_2 signalling in chondrocytes and osteoblasts it appears that up-regulation of PAR_2 leads to detrimental effects in both cell types. This is important because if antagonists of PAR_2 are to be developed to modulate OA, they would inhibit all PAR_2 signalling regardless of the cell type expressing the receptor. Cross-over signalling between cartilage and bone tissue has been seen with the ephrin family of proteins. Ephrin B2 can inhibit the bone resorptive activity of osteoblasts but can also reduce the expression of PAR_2 in OA chondrocytes and thus inhibit the release of pro-inflammatory cytokines and MMPS, reducing the catabolic state of the cartilage [100, 101].

4.3 The Role of Endogenous PAR_2 Activating Proteinases in Arthritis

All the evidence discussed so far points towards an important role for PAR_2 in arthritis. Thus, which endogenous proteinases are present during arthritis to cause activation of PAR_2? The main activators of PAR_2 are trypsin and tryptase [61] which induce knee swelling in wild-type joints that is absent in $PAR_2^{-/-}$ mice [83, 102]. Tryptase can be released from connective tissue mast cells which reside in the joint synovium [103]. It is not surprising, therefore, that the number of mast cells in the synovium of patients with arthritis is increased, as is the level of tryptase in the synovial fluid [40, 104–106]. Two groups have observed that PAR_2 is expressed on synovial fibroblasts from RA patients that are in close proximity to tryptase positive mast cells in the synovium [45, 102]. Furthermore, tryptase inhibited the apoptosis of these fibroblasts via PAR_2 activation and it has been proposed that this process contributes to the proliferation of RA fibroblasts and the hyperplasia and eventual pannus formation that occurs during RA [45].

Another serine proteinase that can activate PAR_2 and is implicated in arthritis pathogenesis is the neutrophil derived enzyme, proteinase 3 which is secreted from activated neutrophils and can cleave and activate PAR_2 [46, 107]. Recent evidence has shown that proteinase 3 expression is increased in neutrophils from RA patients [47]. Interestingly, after treatment with an anti-TNFα antibody, proteinase 3 expression decreased and this correlated with an improvement in the overall condition of the patient. Although further work is required, this finding implies an involvement of proteinase 3 in neutrophil-dependent inflammatory pathways activated during RA. It still needs to be determined whether proteinase 3 is mediating its effects via PAR_2 but in human oral epithelial cells, it was seen that proteinase 3 caused the induction of pro-inflammatory cytokines via the activation of epithelial PAR_2 [46].

Matriptase 1 (also called membrane-type serine protease-1) is a member of the transmembrane family of serine proteinases and a recent study found significantly elevated levels of this enzyme in human OA chondrocytes [48]. Matriptase 1 can cleave and activate various MMPs from their inactive primordial forms, allowing these MMPs to cause the digestion of collagen thereby contributing to cartilage breakdown in OA. Interestingly, matriptase 1 was also seen to enhance the action of these MMPS in a less direct manner by inducing gene expression of these degradative enzymes. Of relevance here was the fact that matriptase 1 was shown to activate PAR_2 and inhibition of PAR_2 significantly reduced matriptase-induced collagenolysis [48]. Thus, it was suggested that the functional effects of matriptase 1 is mediated by PAR_2 activation and the consequent relevant intracellular signalling pathways [48].

4.4 PAR_2 Antagonism and Joint Disease

While PAR_2 has been shown to be involved in arthritis pathogenesis, the clinical translation of these findings has been hindered by problems with the selectivity of current PAR_2 antagonists. Most studies trying to examine the effect of PAR_2 inhibition have used large blocking antibodies, small interfering RNA technology to down-regulate the receptor or non-selective small molecule antagonists [83]. In 2006, a novel small molecule PAR_2 antagonist was developed called N3-methylbutyryl-N-6-aminohexanoyl-piperazine (ENMD-1068) [83]. ENMD-1068 is a disubstituted piperazine that was designed as a result of peptide antagonist screening [83]. Systemic administration of this antagonist dose-dependently attenuated knee joint swelling in an acute inflammatory model of arthritis (Table 2) [83] and inhibited release of TNFα and IL-1β in synovial tissue from RA patients

[95]. Current biological therapies targeting the inactivation of inflammatory cytokines have been extremely successful but are limited by cost [24]. Therefore, any small molecule agents that also lead to the reduction in levels of these cytokines are of great interest. A recent publication examining $PAR_2^{-/-}$ mice in an OA model makes use of a selective PAR_2 antagonist termed p520 [84]; however, very little information is provided for this antagonist (Table 2). It will be interesting to see if any PAR_2 antagonists become commercially available in the next few years.

4.5 PAR_2 and Pain

As discussed previously, chronic joint pain is the major symptom of arthritis, and unlike PAR_1, PAR_2 activation appears to lead to hyperalgesia, rather than analgesia [108–110]. Thus, in addition to any potential disease modifying effects, PAR_2 antagonism may also have the added benefit in reducing joint pain. It has already been proposed that PAR_2 antagonism has the potential to alleviate abdominal pain in patients with irritable bowel syndrome as trypsin and tryptase levels are elevated in colonic biopsies of these patients and these proteinases cause PAR_2-dependent visceral hyperalgesia [111]. In joints, it has recently been shown that intra-articular administration of a PAR_2 peptide caused hyperalgesia, secondary hyperalgesia/allodynia and impaired weight bearing in mice and rats [112]. These joint pain responses were found to be mediated by the transient receptor potential vanilloid-1 (TRPV1) ion channel which is a prominent integrator of noxious stimuli (Table 2). Furthermore, joint PAR_2 activation caused enhanced release of IL-1β into the synovial fluid of treated knees indicating that inflammatory cytokines are also involved in PAR_2 responses [112].

5 PAR_3

PAR_3 is the least well characterised of all the PARs. This is partly due to the failure in developing selective synthetic peptides that can activate PAR_3 [113, 114] and the lack of a G-protein coupling domain in the C-terminal region [61], suggesting that activation of PAR_3 does not lead to intracellular signalling. Instead, the role of PAR_3 may depend on its co-localisation with other PARs. For example, it appears that PAR_3 can act as a co-factor for PAR_4, by acting as a tethering protein for thrombin since PAR_3 contains a thrombin hirudin binding site which is absent in PAR_4 [69]. Thus, PAR_3 can facilitate PAR_4 activation by tethering thrombin close to the activation site of PAR_4 [69, 115]. Conversely, evidence in platelets shows that PAR_3 can cause a reduction in PAR_4 activation in response to plasmin [116], thus the regulatory role of PAR_3 on PAR_4 may differ depending on the proteinase involved, which in turn is determined by both the cellular location and physiological condition of the tissue. PAR_3 has also been seen to form heterodimers with

PAR_1 in human endothelial cells leading to co-dependent signalling [117]. Nevertheless, there is still speculation that PAR_3 has its own signalling role independent of the other PARs, since PAR_3 is abundantly expressed in the body and seen in cells that do not express any other PAR [118].

5.1 *PAR_3 and Arthritis*

Due to the problem of selectively activating PAR_3 with a peptide agonist and the absence of an obvious signalling process, it is perhaps not surprising that, as yet, there have been no direct studies examining a role for PAR_3 in arthritis. However, its ability to affect PAR_1 and PAR_4 signalling indicates that it is possible that PAR_3 could play some part in modulating arthritis progression. There are also a few lines of evidence that point towards PAR_3 having its own independent role. Expression of PAR_3 has been determined in various cell types that are involved in arthritis including nociceptive neurons, monocytes, mast cells and synovial fibroblasts [67, 118–120]. As discussed earlier, thrombin-induced effects such as synovial cell proliferation and cytokine release are largely thought to be mediated through PAR_1. However, a recent article by Ostrowska et al. implicated PAR_3 in causing the release of the pro-inflammatory cytokine IL-8 in various cell types after stimulation by thrombin [121]. In addition, thrombin-induced up-regulation of the degradative enzyme MMP-9 in monocytes was only fully inhibited by blockade of both PAR_1 and PAR_3 suggesting a role for both of these receptors [122]. MMP-9 is markedly elevated in the sera and synovial fluid of RA patients and appears to be released from synovial fibroblasts and macrophages [123]. This is interesting with respect to PAR_3 as both these cell types have been shown to express PAR_3 [67, 119]. Furthermore, the expression of PAR_3 is enhanced when monocytes differentiate into macrophages and thrombin itself causes increased PAR_3 expression [119, 124]. This raises the possibility that during normal conditions, PAR_3 is present at low levels and is relatively inactive, whereas during pathological conditions, such as arthritis, this receptor is up-regulated, and contributes to cytokine release and the MMP imbalance, thus augmenting the chronic inflammatory state.

6 PAR_4

PAR_4 was the fourth, and thus far, the last PAR to be identified [125]. As with the other PARs, PAR_4 has a broad tissue distribution and is expressed by many cells involved in the inflammatory process such as endothelial cells, neutrophils and sensory neurons [126, 127]. PAR_4 activation is known to lead to pro-inflammatory effects such as increased blood flow, leukocyte rolling and adhesion and acute oedema [85, 127–129]. Expression of PAR_4 is seen extensively throughout rodent knee joints including areas such as the subchondral bone, menisci, synovium and articular cartilage [85, 86]. Injection of the PAR_4 activating peptide AYPGKF-NH_2 into the mouse knee joint causes increased blood flow and oedema [85], implying that articular PAR_4 activation contributes to joint inflammation (Table 2). Significantly, pre-treatment with the selective PAR_4 antagonist, pepducin P4pal-10, reduced cellular infiltration and synovial hyperplasia in a model of acute joint inflammation [85].

A study using PAR_4 knockout mice examined the role of PAR_4 in tissue factor-induced inflammation [55]. Tissue factor is a glycoprotein that can bind and activate serine proteinase activated factor VII, an important initiator of coagulation (Fig. 1). Using knock-out mouse strains for all four PAR receptors, Busso et al. showed that tissue factor-induced inflammation is dependent only on the activation of PAR_4 and not the other three PARs [55]. Furthermore, it appears that PAR_4 expressed on platelets is responsible for tissue factor-induced oedema. Intra-plantar injection of tissue factor in PAR_4 knockout mice resulted in less fibrin deposition in the foot pad than controls. Since fibrin deposition correlates with RA disease progression, this observation suggests a strong link between PAR_4 activity and RA. A caveat is that mouse platelets only express PAR_4 and PAR_3 but not PAR_1 [125]. This is in contrast with human platelets that contain both PAR_1 and PAR_4 [130], indicating there may be species differences in the signalling pathways induced upon activation of PARs in platelets.

6.1 PAR_4 and Joint Pain

An intriguing aspect of PAR_4 in arthritis is that it has been shown to be involved in the generation of joint pain. Administration of a PAR_4 activating peptide to rat knee joints has been shown to cause peripheral sensitisation of articular primary afferents and secondary thermal and mechanical hyperalgesia in mice [85, 86]. Selective inhibition of PAR_4 prevents this effect suggesting antagonism of PAR_4 could be relevant for the alleviation of joint pain (Table 2). This observation is in contrast to what has been alluded to in other tissues where PAR_4 activation seems to have an analgesic effect. In the rat hind paw, the PAR_4 activating peptide AYPGKF-NH_2 increases thermal and mechanical nociceptive thresholds and alleviates inflammatory pain [126]. In addition, in the gut PAR_4 activation was seen to inhibit visceral pain and intestinal afferent hypersensitivity [131]. This is important with regards to arthritis as approximately 30% of patients with inflammatory bowel disease are also seen to have some form of joint inflammation and peripheral arthritis [132–134]. Development of a PAR_4 antagonist as an analgesic for joint pain would not be beneficial for these patients if it had the opposite effect in the gut. Thus, it would be imperative to carefully choose the route of administration of any such antagonist and perhaps favour a local over a systemic application.

It is unclear why PAR_4 is anti-nociceptive in some tissues and pro-nociceptive in others, but it may depend on the PAR_4 expressing cell type that is being activated. PAR_4 is expressed on nociceptive nerve fibres both in the joint and in the gut

[86, 131]. Auge et al. suggest that the inhibitory effect of PAR_4 activation is through direct activation of PAR_4 located on the peripheral nerve endings of gut afferents; however, this has not been confirmed experimentally [131]. It was observed that PAR_4 is expressed on 60% of afferent fibres innervating the knee joint [86], yet PAR_4 activation has a sensitising effect on 100% of the joint fibres suggesting that other cell types may be involved. The intracellular signalling pathways activated by PAR_4 have yet to be fully elucidated and the type of G protein linked to PAR_4 could well differ between cell types. In vascular endothelial cells, PAR_4 preferentially activates $G\alpha_{i/o}$ and induces nitric oxide production independently of Ca^{2+} signalling but dependent on the PI3K/Akt pathway [135]. However, in platelets PAR_4 seems to be coupled with $G\alpha q$ and $G\alpha_{12/13}$ and not $G\alpha_{i/o}$ [136, 137]. Thus, the anti- or pro-nociceptive effect of PAR_4 activation may depend upon which cell type is being activated and which G protein is linked with the receptor. The role of PAR_4 may also depend on its co-localisation with other PARs since PAR_3, for example, can either facilitate or reduce PAR_4 signalling [69, 116].

6.2 Serine Proteinases and PAR_4 in Arthritis

PAR_4 can be activated by the neutrophil derived serine proteinase, Cathepsin G [138]. Cathepsin G is released by activated neutrophils and acts as a monocyte chemoattractant [49]. This enzyme has been found in high levels in tissues and synovial fluid from RA patients [49, 139, 140]. Importantly, Cathepsin G knock-out mice develop less severe collagen-induced arthritis with reduced inflammatory cell accumulation and cytokine release compared to wild-type counterparts [141]. Whether Cathepsin G mediates these effects via PAR_4 activation is still unclear. It was originally believed that PAR_4 was the only PAR receptor to be activated by Cathepsin G although PAR_1 and PAR_2 both contain Cathepsin G recognition sites downstream of their tethered ligand sequence [50]. Cleavage at these remote sites disarms PAR_1 and PAR_2, since the tethered ligand is removed and therefore unavailable to bind to the active site on the receptor. While it would appear that Cathepsin G can inhibit PAR_1 and PAR_2 signalling, the enzyme has recently been found to cause PAR_1 activation since the chemotaxis of human monocytes induced by Cathepsin G was shown to be PAR_1 dependent [142]. Interestingly, blocking PAR_1 with an antibody did not completely abrogate the chemotaxis [142].

PAR_4 has also been shown to be activated by the fibrinolytic serine proteinase plasmin [143]. Plasmin is responsible for the lysis of fibrin clots and can be activated from its inactive precursor, plasminogen, by tissue plasminogen activator (tPA) or urokinase plasminogen activator (uPA) [144] (Fig. 2). The plasminogen activation system is known to be altered in arthritic joints with typically an increase in tPA seen in OA tissues and an increase in uPA in RA tissues [51, 145, 146]. Plasmin has a broad specificity and whilst it causes the degradation of fibrin in an arthritic joint, it can also have deleterious effects such as cartilage and bone matrix degradation and cytokine release [149]. Studies with knock-out mice lacking

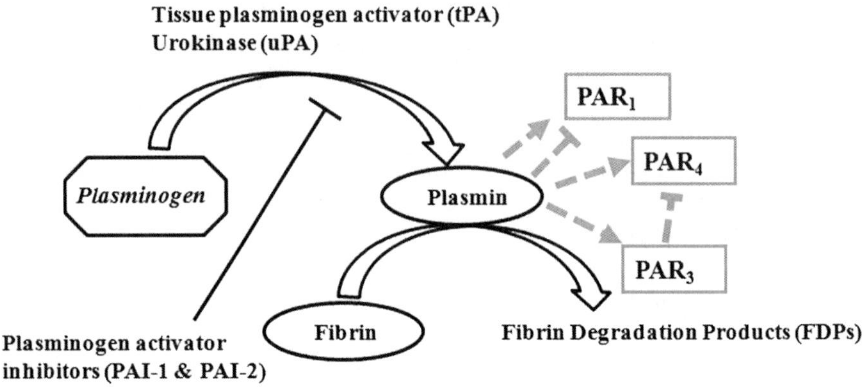

Fig. 2 A simplified diagram of fibrinolysis and its links with PARs. Plasminogen is activated by the enzymes tPA and uPA which are up-regulated in arthritic tissues [51, 145, 146]. PAs are inhibited by PAI-1 and PAI-2 that are also seen in high levels in arthritic patients [51, 145, 146] suggesting that dysregulation of this system contributes to the pathogenesis of arthritis. Plasmin causes fibrin degradation but can also both activate and disable PAR_1 [50, 147, 148] and activate PAR_4 [143]. In the presence of PAR_3, plasmin causes a reduction in PAR_4 activation [116]

various enzymes involved in the plasminogen activation system have shown differing results. Some mouse strains show resistance to some models of arthritis, but not to other models, whilst other mouse strains exhibit more severe arthritis (discussed in [150]). Therefore, the exact role of this system is still unclear but an imbalance in fibrinolysis is definitely a contributing factor to the pathophysiology of an arthritic joint. It still remains to be seen, however, if plasmin effects are mediated by PAR_4.

7 Conclusion

In conclusion, it is clear that PAR_1, PAR_2 and PAR_4, all play a major part in the pathogenesis of arthritis and joint pain. Further research is required to elucidate fully the relative contribution of the various PARs and determine the interactions between elevated serine proteinases and PAR activity in joints.

References

1. Helmick CG, Felson DT, Lawrence RC, Gabriel S, Hirsch R, Kwoh CK, Liang MH, Kremers HM, Mayes MD, Merkel PA et al (2008) Estimates of the prevalence of arthritis and other rheumatic conditions in the United States Part I. Arthritis Rheum 58:15–25
2. Sprangers MA, de Regt EB, Andries F, van Agt HM, Bijl RV, de Boer JB, Foets M, Hoeymans N, Jacobs AE, Kempen GI et al (2000) Which chronic conditions are associated with better or poorer quality of life? J Clin Epidemiol 53:895–907

3. Naz SM, Symmons DP (2007) Mortality in established rheumatoid arthritis. Best Pract Res Clin Rheumatol 21:871–883
4. Brooks PM (2006) The burden of musculoskeletal disease–a global perspective. Clin Rheumatol 25:778–781
5. Burton W, Morrison A, Maclean R, Ruderman E (2006) Systematic review of studies of productivity loss due to rheumatoid arthritis. Occup Med (Lond) 56:18–27
6. Barrett EM, Scott DG, Wiles NJ, Symmons DP (2000) The impact of rheumatoid arthritis on employment status in the early years of disease: a UK community-based study. Rheumatology (Oxford) 39:1403–1409
7. Reginster JY (2002) The prevalence and burden of arthritis. Rheumatology (Oxford) 41(Supp 1):3–6
8. Badley EM (1995) The economic burden of musculoskeletal disorders in Canada is similar to that for cancer, and may be higher. J Rheumatol 22:204–206
9. Buckwalter JA, Saltzman C, Brown T (2004) The impact of osteoarthritis: implications for research. Clin Orthop Relat Res:S6–S15
10. Birnbaum H, Pike C, Kaufman R, Maynchenko M, Kidolezi Y, Cifaldi M (2010) Societal cost of rheumatoid arthritis patients in the US. Curr Med Res Opin 26:77–90
11. Hootman JM, Helmick CG (2006) Projections of US prevalence of arthritis and associated activity limitations. Arthritis Rheum 54:226–229
12. Woolf AD, Pfleger B (2003) Burden of major musculoskeletal conditions. Bull World Health Organ 81:646–656
13. Schoels M, Wong J, Scott DL, Zink A, Richards P, Landewe R, Smolen JS, Aletaha D (2010) Economic aspects of treatment options in rheumatoid arthritis: a systematic literature review informing the EULAR recommendations for the management of rheumatoid arthritis. Ann Rheum Dis 69:995–1003
14. Pugner KM, Scott DI, Holmes JW, Hieke K (2000) The costs of rheumatoid arthritis: An international long-term view. Semin Arthritis Rheum 29:305–320
15. March L, Lapsley H (2001) What are the costs to society and the potential benefits from the effective management of early rheumatoid arthritis? Best Pract Res Clin Rheumatol 15:171–185
16. Sacre SM, Andreakos E, Taylor P, Feldmann M, Foxwell BM (2005) Molecular therapeutic targets in rheumatoid arthritis. Expert Rev Mol Med 7:1–20
17. Silman AJ (2002) The changing face of rheumatoid arthritis: why the decline in incidence? Arthritis Rheum 46:579–581
18. Myasoedova E, Crowson CS, Kremers HM, Therneau TM, Gabriel SE (2010) Is the incidence of rheumatoid arthritis rising? Results from Olmsted county, Minnesota, 1955–2007. Arthritis Rheum 62:1576–1582
19. Ignatavicius DD (2001) Rheumatoid arthritis and the older adult. Geriatr Nurs 22: 139–142
20. Wolfe MM, Lichtenstein DR, Singh G (1999) Gastrointestinal toxicity of nonsteroidal antiinflammatory drugs. N Engl J Med 340:1888–1899
21. Chikanza IC, Kozaci DL (2004) Corticosteroid resistance in rheumatoid arthritis: molecular and cellular perspectives. Rheumatology (Oxford) 43:1337–1345
22. Rindfleisch JA, Muller D (2005) Diagnosis and management of rheumatoid arthritis. Am Fam Physician 72:1037–1047
23. Williams RO, Feldmann M, Maini RN (1992) Anti-tumor necrosis factor ameliorates joint disease in murine collagen-induced arthritis. Proc Natl Acad Sci USA 89:9784–9788
24. Breedveld FC, Kalden JR (2004) Appropriate and effective management of rheumatoid arthritis. Ann Rheum Dis 63:627–633
25. Rankin EC, Choy EH, Kassimos D, Kingsley GH, Sopwith AM, Isenberg DA, Panayi GS (1995) The therapeutic effects of an engineered human anti-tumour necrosis factor alpha antibody (CDP571) in rheumatoid arthritis. Br J Rheumatol 34:334–342

26. Padyukov L, Lampa J, Heimburger M, Ernestam S, Cederholm T, Lundkvist I, Andersson P, Hermansson Y, Harju A, Klareskog L et al (2003) Genetic markers for the efficacy of tumour necrosis factor blocking therapy in rheumatoid arthritis. Ann Rheum Dis 62:526–529
27. van der Bijl AE, Goekoop-Ruiterman YP, de Vries-Bouwstra JK, Ten WS, Han KH, van Krugten MV, Allaart CF, Breedveld FC, Dijkmans BA (2007) Infliximab and methotrexate as induction therapy in patients with early rheumatoid arthritis. Arthritis Rheum 56:2129–2134
28. Goldring MB, Goldring SR (2007) Osteoarthritis. J Cell Physiol 213:626–634
29. Hunter DJ, McDougall JJ, Keefe FJ (2009) The symptoms of osteoarthritis and the genesis of pain. Med Clin North Am 93:83–100, xi
30. Goldring MB, Goldring SR (2010) Articular cartilage and subchondral bone in the pathogenesis of osteoarthritis. Ann N Y Acad Sci 1192:230–237
31. Altman RD (2010) Early management of osteoarthritis. Am J Manag Care 16 Suppl Management:S41–S47
32. Ram M, Sherer Y, Shoenfeld Y (2006) Matrix metalloproteinase-9 and autoimmune diseases. J Clin Immunol 26:299–307
33. Eck SM, Blackburn JS, Schmucker AC, Burrage PS, Brinckerhoff CE (2009) Matrix metalloproteinase and G protein coupled receptors: co-conspirators in the pathogenesis of autoimmune disease and cancer. J Autoimmun 33:214–221
34. Hannan MT, Felson DT, Pincus T (2000) Analysis of the discordance between radiographic changes and knee pain in osteoarthritis of the knee. J Rheumatol 27:1513–1517
35. McDougall JJ, Andruski B, Schuelert N, Hallgrimsson B, Matyas JR (2009) Unravelling the relationship between age, nociception and joint destruction in naturally occurring osteoarthritis of Dunkin Hartley guinea pigs. Pain 141:222–232
36. Reinharth D (2005) Capsaicin cream unpopular with patients. Arch Intern Med 165:702
37. Chou MM, Vergnolle N, McDougall JJ, Wallace JL, Marty S, Teskey V, Buret AG (2005) Effects of chondroitin and glucosamine sulfate in a dietary bar formulation on inflammation, interleukin-1beta, matrix metalloprotease-9, and cartilage damage in arthritis. Exp Biol Med 230:255–262
38. Zhang W, Moskowitz RW, Nuki G, Abramson S, Altman RD, Arden N, Bierma-Zeinstra S, Brandt KD, Croft P, Doherty M et al (2008) OARSI recommendations for the management of hip and knee osteoarthritis, Part II: OARSI evidence-based, expert consensus guidelines. Osteoarthr Cartil 16:137–162
39. Zhang W, Nuki G, Moskowitz RW, Abramson S, Altman RD, Arden NK, Bierma-Zeinstra S, Brandt KD, Croft P, Doherty M et al (2009) OARSI recommendations for the management of hip and knee osteoarthritis: part III: Changes in evidence following systematic cumulative update of research published through. Osteoarthr Cartil 18:476–499
40. Lavery JP, Lisse JR (1994) Preliminary study of the tryptase levels in the synovial fluid of patients with inflammatory arthritis. Ann Allergy 72:425–427
41. Nakano S, Ikata T, Kinoshita I, Kanematsu J, Yasuoka S (1999) Characteristics of the protease activity in synovial fluid from patients with rheumatoid arthritis and osteoarthritis. Clin Exp Rheumatol 17:161–170
42. So AK, Varisco PA, Kemkes-Matthes B, Herkenne-Morard C, Chobaz-Peclat V, Gerster JC, Busso N (2003) Arthritis is linked to local and systemic activation of coagulation and fibrinolysis pathways. J Thromb Haemost 1:2510–2515
43. Marty I, Peclat V, Kirdaite G, Salvi R, So A, Busso N (2001) Amelioration of collagen-induced arthritis by thrombin inhibition. J Clin Invest 107:631–640
44. Varisco PA, Peclat V, Bischof-Delaloye A, So A, Busso N (2000) Effect of thrombin inhibition on synovial inflammation in antigen induced arthritis. Ann Rheum Dis 59:781–787
45. Sawamukai N, Yukawa S, Saito K, Nakayamada S, Kambayashi T, Tanaka Y (2010) Mast cell-derived tryptase inhibits apoptosis of human rheumatoid synovial fibroblasts via rho-mediated signaling. Arthritis Rheum 62:952–959

46. Uehara A, Sugawara S, Muramoto K, Takada H (2002) Activation of human oral epithelial cells by neutrophil proteinase 3 through protease-activated receptor-2. J Immunol 169:4594–4603
47. Matsumoto T, Kaneko T, Seto M, Wada H, Kobayashi T, Nakatani K, Tonomura H, Tono Y, Ohyabu M, Nobori T et al (2008) The membrane proteinase 3 expression on neutrophils was downregulated after treatment with infliximab in patients with rheumatoid arthritis. Clin Appl Thromb Hemost 14:186–192
48. Milner JM, Patel A, Davidson RK, Swingler TE, Desilets A, Young DA, Kelso EB, Donell ST, Cawston TE, Clark IM et al (2010) Matriptase is a novel initiator of cartilage matrix degradation in osteoarthritis. Arthritis Rheum 62:1955–1966
49. Miyata J, Tani K, Sato K, Otsuka S, Urata T, Lkhagvaa B, Furukawa C, Sano N, Sone S (2007) Cathepsin G: the significance in rheumatoid arthritis as a monocyte chemoattractant. Rheumatol Int 27:375–382
50. Ramachandran R, Hollenberg MD (2008) Proteinases and signalling: pathophysiological and therapeutic implications via PARs and more. Br J Pharmacol 153(Suppl 1):S263–S282
51. Busso N, Peclat V, So A, Sappino AP (1997) Plasminogen activation in synovial tissues: differences between normal, osteoarthritis, and rheumatoid arthritis joints. Ann Rheum Dis 56:550–557
52. Carmassi F, de Negri F, Morale M, Song KY, Chung SI (1996) Fibrin degradation in the synovial fluid of rheumatoid arthritis patients: a model for extravascular fibrinolysis. Semin Thromb Hemost 22:489–496
53. Busso N, Peclat V, Van Ness K, Kolodziesczyk E, Degen J, Bugge T, So A (1998) Exacerbation of antigen-induced arthritis in urokinase-deficient mice. J Clin Invest 102:41–50
54. Camerer E, Kataoka H, Kahn M, Lease K, Coughlin SR (2002) Genetic evidence that protease-activated receptors mediate factor Xa signaling in endothelial cells. J Biol Chem 277:16081–16087
55. Busso N, Chobaz-Peclat V, Hamilton J, Spee P, Wagtmann N, So A (2008) Essential role of platelet activation via PAR-4 in tissue factor-initiated inflammation. Arthritis Res Ther 10: R42
56. Trudel G, Uhthoff HK, Laneuville O (2005) Prothrombin gene expression in articular cartilage with a putative role in cartilage degeneration secondary to joint immobility. J Rheumatol 32:1547–1555
57. Shin H, Nakajima T, Kitajima I, Shigeta K, Abeyama K, Imamura T, Okano T, Kawahara K, Nakamura T, Maruyama I (1995) Thrombin receptor-mediated synovial proliferation in patients with rheumatoid arthritis. Clin Immunol Immunopathol 76:225–233
58. Ohba T, Takase Y, Ohhara M, Kasukawa R (1996) Thrombin in the synovial fluid of patients with rheumatoid arthritis mediates proliferation of synovial fibroblast-like cells by induction of platelet derived growth factor. J Rheumatol 23:1505–1511
59. Shin H, Kitajima I, Nakajima T, Shao Q, Tokioka T, Takasaki I, Hanyu N, Kubo T, Maruyama I (1999) Thrombin receptor mediated signals induce expressions of interleukin 6 and granulocyte colony stimulating factor via NF-kappa B activation in synovial fibroblasts. Ann Rheum Dis 58:55–60
60. Morris R, Winyard PG, Brass LF, Blake DR, Morris CJ (1996) Thrombin receptor expression in rheumatoid and osteoarthritic synovial tissue. Ann Rheum Dis 55:841–843
61. Macfarlane SR, Seatter MJ, Kanke T, Hunter GD, Plevin R (2001) Proteinase-activated receptors. Pharmacol Rev 53:245–282
62. Naldini A, Sower L, Bocci V, Meyers B, Carney DH (1998) Thrombin receptor expression and responsiveness of human monocytic cells to thrombin is linked to interferon-induced cellular differentiation. J Cell Physiol 177:76–84
63. Cirino G, Cicala C, Bucci MR, Sorrentino L, Maraganore JM, Stone SR (1996) Thrombin functions as an inflammatory mediator through activation of its receptor. J Exp Med 183:821–827

64. D'Andrea MR, Rogahn CJ, ndrade-Gordon P (2000) Localization of protease-activated receptors-1 and -2 in human mast cells: indications for an amplified mast cell degranulation cascade. BiotechHistochem 75:85–90
65. de Garavilla L, Vergnolle N, Young SH, Ennes H, Steinhoff M, Ossovskaya VS, D'Andrea MR, Mayer EA, Wallace JL, Hollenberg MD et al (2001) Agonists of proteinase-activated receptor 1 induce plasma extravasation by a neurogenic mechanism. Br J Pharmacol 133:975–987
66. Mari B, Imbert V, Belhacene N, Far DF, Peyron JF, Pouyssegur J, Van Obberghen-Schilling E, Rossi B, Auberger P (1994) Thrombin and thrombin receptor agonist peptide induce early events of T cell activation and synergize with TCR cross-linking for CD69 expression and interleukin 2 production. J Biol Chem 269:8517–8523
67. Furuhashi I, Abe K, Sato T, Inoue H (2008) Thrombin-stimulated proliferation of cultured human synovial fibroblasts through proteolytic activation of proteinase-activated receptor-1. J Pharmacol Sci 108:104–111
68. Markwardt F (2002) Historical perspective of the development of thrombin inhibitors. Pathophysiol Haemost Thromb 32(Suppl 3):15–22
69. Nakanishi-Matsui M, Zheng YW, Sulciner DJ, Weiss EJ, Ludeman MJ, Coughlin SR (2000) PAR3 is a cofactor for PAR4 activation by thrombin. Nature 404:609–613
70. Hirano F, Kobayashi A, Hirano Y, Nomura Y, Fukawa E, Makino I (2002) Thrombin-induced expression of RANTES mRNA through protease activated receptor-1 in human synovial fibroblasts. Ann Rheum Dis 61:834–837
71. Yang YH, Hall P, Little CB, Fosang AJ, Milenkovski G, Santos L, Xue J, Tipping P, Morand EF (2005) Reduction of arthritis severity in protease-activated receptor-deficient mice. Arthritis Rheum 52:1325–1332
72. Krupiczojc MA, Scotton CJ, Chambers RC (2008) Coagulation signalling following tissue injury: focus on the role of factor Xa. Int J Biochem Cell Biol 40:1228–1237
73. Jones A, Geczy CL (1990) Thrombin and factor Xa enhance the production of interleukin-1. Immunology 71:236–241
74. Senden NH, Jeunhomme TM, Heemskerk JW, Wagenvoord R, van't Veer C, Hemker HC, Buurman WA (1998) Factor Xa induces cytokine production and expression of adhesion molecules by human umbilical vein endothelial cells. J Immunol 161:4318–4324
75. Bachli EB, Pech CM, Johnson KM, Johnson DJ, Tuddenham EG, McVey JH (2003) Factor Xa and thrombin, but not factor VIIa, elicit specific cellular responses in dermal fibroblasts. J Thromb Haemost 1:1935–1944
76. Camerer E, Huang W, Coughlin SR (2000) Tissue factor- and factor X-dependent activation of protease-activated receptor 2 by factor VIIa. Proc Natl Acad Sci USA 97:5255–5260
77. Smyth SS, Woulfe DS, Weitz JI, Gachet C, Conley PB, Goodman SG, Roe MT, Kuliopulos A, Moliterno DJ, French PA et al (2009) G-protein-coupled receptors as signaling targets for antiplatelet therapy. Arterioscler Thromb Vasc Biol 29:449–457
78. Macaulay TE, Allen C, Ziada KM (2010) Thrombin receptor antagonism -the potential of antiplatelet medication SCH 530348. Expert Opin Pharmacother 11:1015–1022
79. Martin L, Auge C, Boue J, Buresi MC, Chapman K, Asfaha S, Andrade-Gordon P, Steinhoff M, Cenac N, Dietrich G et al (2009) Thrombin receptor: An endogenous inhibitor of inflammatory pain, activating opioid pathways. Pain 146:121–129
80. Asfaha S, Brussee V, Chapman K, Zochodne DW, Vergnolle N (2002) Proteinase-activated receptor-1 agonists attenuate nociception in response to noxious stimuli. Br J Pharmacol 135:1101–1106
81. Kawabata A, Kawao N, Kuroda R, Tanaka A, Shimada C (2002) The PAR-1-activating peptide attenuates carrageenan-induced hyperalgesia in rats. Peptides 23:1181–1183
82. Helyes Z, Sandor K, Borbely E, Tekus V, Pinter E, Elekes K, Toth DM, Szolcsanyi J, McDougall JJ (2009) Involvement of transient receptor potential vanilloid 1 receptors in protease-activated receptor-2-induced joint inflammation and nociception. Eur J Pain 14:351-358

83. Kelso EB, Lockhart JC, Hembrough T, Dunning L, Plevin R, Hollenberg MD, Sommerhoff CP, McLean JS, Ferrell WR (2006) Therapeutic promise of proteinase-activated receptor-2 antagonism in joint inflammation. J Pharmacol Exp Ther 316:1017–1024
84. Ferrell WR, Kelso EB, Lockhart JC, Plevin R, McInnes IB (2010) Protease-activated receptor 2: a novel pathogenic pathway in a murine model of osteoarthritis. Ann Rheum Dis 69:2051–2054
85. McDougall JJ, Zhang C, Cellars L, Joubert E, Dixon CM, Vergnolle N (2009) Triggering of proteinase-activated receptor 4 leads to joint pain and inflammation in mice. Arthritis Rheum 60:728–737
86. Russell FA, Veldhoen VE, Tchitchkan D, McDougall JJ (2010) Proteinase-activated receptor-4 (PAR4) activation leads to sensitization of rat joint primary afferents via a bradykinin B2 receptor-dependent mechanism. J Neurophysiol 103:155–163
87. Knecht W, Cottrell GS, Amadesi S, Mohlin J, Skaregarde A, Gedda K, Peterson A, Chapman K, Hollenberg MD, Vergnolle N et al (2007) Trypsin IV or mesotrypsin and p23 cleave protease-activated receptors 1 and 2 to induce inflammation and hyperalgesia. J Biol Chem 282:26089–26100
88. Ferrell WR, Lockhart JC, Kelso EB, Dunning L, Plevin R, Meek SE, Smith AJ, Hunter GD, McLean JS, McGarry F et al (2003) Essential role for proteinase-activated receptor-2 in arthritis. J Clin Invest 111:35–41
89. Busso N, Frasnelli M, Feifel R, Cenni B, Steinhoff M, Hamilton J, So A (2007) Evaluation of protease-activated receptor 2 in murine models of arthritis. Arthritis Rheum 56:101–107
90. Nakano S, Mishiro T, Takahara S, Yokoi H, Hamada D, Yukata K, Takata Y, Goto T, Egawa H, Yasuoka S et al (2007) Distinct expression of mast cell tryptase and protease activated receptor-2 in synovia of rheumatoid arthritis and osteoarthritis. Clin Rheumatol 26:1284–1292
91. Xiang Y, Masuko-Hongo K, Sekine T, Nakamura H, Yudoh K, Nishioka K, Kato T (2006) Expression of proteinase-activated receptors (PAR)-2 in articular chondrocytes is modulated by IL-1beta, TNF-alpha and TGF-beta. Osteoarthr Cartil 14:1163–1173
92. Abe K, Aslam A, Walls AF, Sato T, Inoue H (2006) Up-regulation of protease-activated receptor-2 by bFGF in cultured human synovial fibroblasts. Life Sci 79:898–904
93. Tsai SH, Sheu MT, Liang YC, Cheng HT, Fang SS, Chen CH (2009) TGF-beta inhibits IL-1beta-activated PAR-2 expression through multiple pathways in human primary synovial cells. J Biomed Sci 16:97
94. Shpacovitch VM, Brzoska T, Buddenkotte J, Stroh C, Sommerhoff CP, Ansel JC, Schulze-Osthoff K, Bunnett NW, Luger TA, Steinhoff M (2002) Agonists of proteinase-activated receptor 2 induce cytokine release and activation of nuclear transcription factor kappaB in human dermal microvascular endothelial cells. J Invest Dermatol 118:380–385
95. Kelso EB, Ferrell WR, Lockhart JC, Elias-Jones I, Hembrough T, Dunning L, Gracie JA, McInnes IB (2007) Expression and proinflammatory role of proteinase-activated receptor 2 in rheumatoid synovium: ex vivo studies using a novel proteinase-activated receptor 2 antagonist. Arthritis Rheum 56:765–771
96. Boileau C, Amiable N, Martel-Pelletier J, Fahmi H, Duval N, Pelletier JP (2007) Activation of proteinase-activated receptor 2 in human osteoarthritic cartilage upregulates catabolic and proinflammatory pathways capable of inducing cartilage degradation: a basic science study. Arthritis Res Ther 9:R121
97. Amiable N, Tat SK, Lajeunesse D, Duval N, Pelletier JP, Martel-Pelletier J, Boileau C (2009) Proteinase-activated receptor (PAR)-2 activation impacts bone resorptive properties of human osteoarthritic subchondral bone osteoblasts. Bone 44:1143–1150
98. Hayami T, Pickarski M, Wesolowski GA, McLane J, Bone A, Destefano J, Rodan GA, le Duong T (2004) The role of subchondral bone remodeling in osteoarthritis: reduction of cartilage degeneration and prevention of osteophyte formation by alendronate in the rat anterior cruciate ligament transection model. Arthritis Rheum 50:1193–1206

99. Kwan Tat S, Lajeunesse D, Pelletier JP, Martel-Pelletier J (2010) Targeting subchondral bone for treating osteoarthritis: what is the evidence? Best Pract Res Clin Rheumatol 24:51–70
100. Kwan Tat S, Pelletier JP, Amiable N, Boileau C, Lajeunesse D, Duval N, Martel-Pelletier J (2008) Activation of the receptor EphB4 by its specific ligand ephrin B2 in human osteoarthritic subchondral bone osteoblasts. Arthritis Rheum 58:3820–3830
101. Kwan Tat S, Pelletier JP, Amiable N, Boileau C, Lavigne M, Martel-Pelletier J (2009) Treatment with ephrin B2 positively impacts the abnormal metabolism of human osteoarthritic chondrocytes. Arthritis Res Ther 11:R119
102. Palmer HS, Kelso EB, Lockhart JC, Sommerhoff CP, Plevin R, Goh FG, Ferrell WR (2007) Protease-activated receptor 2 mediates the proinflammatory effects of synovial mast cells. Arthritis Rheum 56:3532–3540
103. Krishnaswamy G, Kelley J, Johnson D, Youngberg G, Stone W, Huang SK, Bieber J, Chi DS (2001) The human mast cell: functions in physiology and disease. Front Biosci 6: D1109–D1127
104. Crisp AJ, Chapman CM, Kirkham SE, Schiller AL, Krane SM (1984) Articular mastocytosis in rheumatoid arthritis. Arthritis Rheum 27:845–851
105. Godfrey HP, Ilardi C, Engber W, Graziano FM (1984) Quantitation of human synovial mast cells in rheumatoid arthritis and other rheumatic diseases. Arthritis Rheum 27:852–856
106. Nigrovic PA, Lee DM (2007) Synovial mast cells: role in acute and chronic arthritis. Immunol Rev 217:19–37
107. Muller A, Voswinkel J, Gottschlich S, Csernok E (2007) Human proteinase 3 (PR3) and its binding molecules: implications for inflammatory and PR3-related autoimmune responses. Ann N Y Acad Sci 1109:84–92
108. Vergnolle N, Bunnett NW, Sharkey KA, Brussee V, Compton SJ, Grady EF, Cirino G, Gerard N, Basbaum AI, Andrade-Gordon P et al (2001) Proteinase-activated receptor-2 and hyperalgesia: A novel pain pathway. Nat Med 7:821–826
109. Kawabata A, Kawao N, Kuroda R, Tanaka A, Itoh H, Nishikawa H (2001) Peripheral PAR-2 triggers thermal hyperalgesia and nociceptive responses in rats. Neuroreport 12:715–719
110. Kawabata A, Kawao N, Itoh H, Shimada C, Takebe K, Kuroda R, Masuko T, Kataoka K, Ogawa S (2002) Role of N-methyl-D-aspartate receptors and the nitric oxide pathway in nociception/hyperalgesia elicited by protease-activated receptor-2 activation in mice and rats. Neurosci Lett 329:349–353
111. Cenac N, Andrews CN, Holzhausen M, Chapman K, Cottrell G, Andrade-Gordon P, Steinhoff M, Barbara G, Beck P, Bunnett NW et al (2007) Role for protease activity in visceral pain in irritable bowel syndrome. J Clin Invest 117:636–647
112. Helyes Z, Sandor K, Borbely E, Tekus V, Pinter E, Elekes K, Toth DM, Szolcsanyi J, McDougall JJ (2010) Involvement of transient receptor potential vanilloid 1 receptors in protease-activated receptor-2-induced joint inflammation and nociception. Eur J Pain 14:351–358
113. Ishihara H, Connolly AJ, Zeng D, Kahn ML, Zheng YW, Timmons C, Tram T, Coughlin SR (1997) Protease-activated receptor 3 is a second thrombin receptor in humans. Nature 386:502–506
114. Hansen KK, Saifeddine M, Hollenberg MD (2004) Tethered ligand-derived peptides of proteinase-activated receptor 3 (PAR3) activate PAR1 and PAR2 in Jurkat T cells. Immunology 112:183–190
115. Owen WG (2003) PAR-3 is a low-affinity substrate, high affinity effector of thrombin. Biochem Biophys Res Commun 305:166–168
116. Mao Y, Jin J, Daniel JL, Kunapuli SP (2009) Regulation of plasmin-induced protease-activated receptor 4 activation in platelets. Platelets 20:191–198
117. McLaughlin JN, Patterson MM, Malik AB (2007) Protease-activated receptor-3 (PAR3) regulates PAR1 signaling by receptor dimerization. Proc Natl Acad Sci USA 104:5662–5667

118. Zhu WJ, Yamanaka H, Obata K, Dai Y, Kobayashi K, Kozai T, Tokunaga A, Noguchi K (2005) Expression of mRNA for four subtypes of the proteinase-activated receptor in rat dorsal root ganglia. Brain Res 1041:205–211
119. Colognato R, Slupsky JR, Jendrach M, Burysek L, Syrovets T, Simmet T (2003) Differential expression and regulation of protease-activated receptors in human peripheral monocytes and monocyte-derived antigen-presenting cells. Blood 102:2645–2652
120. Moormann C, Artuc M, Pohl E, Varga G, Buddenkotte J, Vergnolle N, Brehler R, Henz BM, Schneider SW, Luger TA et al (2006) Functional characterization and expression analysis of the proteinase-activated receptor-2 in human cutaneous mast cells. J Invest Dermatol 126:746–755
121. Ostrowska E, Reiser G (2008) The protease-activated receptor-3 (PAR-3) can signal autonomously to induce interleukin-8 release. Cell Mol Life Sci 65:970–981
122. Chang CJ, Hsu LA, Ko YH, Chen PL, Chuang YT, Lin CY, Liao CH, Pang JH (2009) Thrombin regulates matrix metalloproteinase-9 expression in human monocytes. Biochem Biophys Res Commun 385:241–246
123. Gruber BL, Sorbi D, French DL, Marchese MJ, Nuovo GJ, Kew RR, Arbeit LA (1996) Markedly elevated serum MMP-9 (gelatinase B) levels in rheumatoid arthritis: a potentially useful laboratory marker. Clin Immunol Immunopathol 78:161–171
124. Bretschneider E, Spanbroek R, Lotzer K, Habenicht AJ, Schror K (2003) Evidence for functionally active protease-activated receptor-3 (PAR-3) in human vascular smooth muscle cells. Thromb Haemost 90:704–709
125. Kahn ML, Zheng YW, Huang W, Bigornia V, Zeng D, Moff S, Farese RV Jr, Tam C, Coughlin SR (1998) A dual thrombin receptor system for platelet activation. Nature 394:690–694
126. Asfaha S, Cenac N, Houle S, Altier C, Papez MD, Nguyen C, Steinhoff M, Chapman K, Zamponi GW, Vergnolle N (2007) Protease-activated receptor-4: a novel mechanism of inflammatory pain modulation. Br J Pharmacol 150:176–185
127. Vergnolle N, Derian CK, D'Andrea MR, Steinhoff M, Andrade-Gordon P (2002) Characterization of thrombin-induced leukocyte rolling and adherence: a potential proinflammatory role for proteinase-activated receptor-4. J Immunol 169:1467–1473
128. Hollenberg MD, Saifeddine M, Sandhu S, Houle S, Vergnolle N (2004) Proteinase-activated receptor-4: evaluation of tethered ligand-derived peptides as probes for receptor function and as inflammatory agonists in vivo. Br J Pharmacol 143:443–454
129. Houle S, Papez MD, Ferazzini M, Hollenberg MD, Vergnolle N (2005) Neutrophils and the kallikrein-kinin system in proteinase-activated receptor 4-mediated inflammation in rodents. Br J Pharmacol 146:670–678
130. Shapiro MJ, Weiss EJ, Faruqi TR, Coughlin SR (2000) Protease-activated receptors 1 and 4 are shut off with distinct kinetics after activation by thrombin. J Biol Chem 275:25216–25221
131. Auge C, Balz-Hara D, Steinhoff M, Vergnolle N, Cenac N (2009) Protease-activated receptor-4 (PAR(4)): a role as inhibitor of visceral pain and hypersensitivity. Neurogastroenterol Motil 21:1189
132. Rodriguez-Reyna TS, Martinez-Reyes C, Yamamoto-Furusho JK (2009) Rheumatic manifestations of inflammatory bowel disease. World J Gastroenterol 15:5517–5524
133. De Vos M Joint involvement in inflammatory bowel disease: managing inflammation outside the digestive system. *Expert Rev Gastroenterol Hepatol* 4: 81–89
134. Larsen S, Bendtzen K, Nielsen OH (2010) Extraintestinal manifestations of inflammatory bowel disease: epidemiology, diagnosis, and management. Ann Med 42:97–114
135. Hirano K, Nomoto N, Hirano M, Momota F, Hanada A, Kanaide H (2007) Distinct Ca2+ requirement for NO production between proteinase-activated receptor 1 and 4 (PAR1 and PAR4) in vascular endothelial cells. J Pharmacol Exp Ther 322:668–677
136. Holinstat M, Voss B, Bilodeau ML, McLaughlin JN, Cleator J, Hamm HE (2006) PAR4, but not PAR1, signals human platelet aggregation via Ca2+ mobilization and synergistic P2Y12 receptor activation. J Biol Chem 281:26665–26674

137. Faruqi TR, Weiss EJ, Shapiro MJ, Huang W, Coughlin SR (2000) Structure-function analysis of protease-activated receptor 4 tethered ligand peptides. Determinants of specificity and utility in assays of receptor function. J Biol Chem 275:19728–19734
138. Sambrano GR, Huang W, Faruqi T, Mahrus S, Craik C, Coughlin SR (2000) Cathepsin G activates protease-activated receptor-4 in human platelets. J Biol Chem 275:6819–6823
139. Velvart M, Fehr K (1987) Degradation in vivo of articular cartilage in rheumatoid arthritis and juvenile chronic arthritis by cathepsin G and elastase from polymorphonuclear leukocytes. Rheumatol Int 7:195–202
140. Nordstrom D, Lindy O, Konttinen YT, Lauhio A, Sorsa T, Friman C, Pettersson T, Santavirta S (1996) Cathepsin G and elastase in synovial fluid and peripheral blood in reactive and rheumatoid arthritis. Clin Rheumatol 15:35–41
141. Adkison AM, Raptis SZ, Kelley DG, Pham CT (2002) Dipeptidyl peptidase I activates neutrophil-derived serine proteases and regulates the development of acute experimental arthritis. J Clin Invest 109:363–371
142. Wilson TJ, Nannuru KC, Singh RK (2009) Cathepsin G recruits osteoclast precursors via proteolytic activation of protease-activated receptor-1. Cancer Res 69:3188–3195
143. Quinton TM, Kim S, Derian CK, Jin J, Kunapuli SP (2004) Plasmin-mediated activation of platelets occurs by cleavage of protease-activated receptor 4. J Biol Chem 279:18434–18439
144. Syrovets T, Simmet T (2004) Novel aspects and new roles for the serine protease plasmin. Cell Mol Life Sci 61:873–885
145. Ronday HK, Smits HH, Van Muijen GN, Pruszczynski MS, Dolhain RJ, Van Langelaan EJ, Breedveld FC, Verheijen JH (1996) Difference in expression of the plasminogen activation system in synovial tissue of patients with rheumatoid arthritis and osteoarthritis. Br J Rheumatol 35:416–423
146. Brommer EJ, Dooijewaard G, Dijkmans BA, Breedveld FC (1992) Plasminogen activators in synovial fluid and plasma from patients with arthritis. Ann Rheum Dis 51:965–968
147. Kimura M, Andersen TT, Fenton JW 2nd, Bahou WF, Aviv A (1996) Plasmin-platelet interaction involves cleavage of functional thrombin receptor. Am J Physiol 271:C54–C60
148. Kuliopulos A, Covic L, Seeley SK, Sheridan PJ, Helin J, Costello CE (1999) Plasmin desensitization of the PAR1 thrombin receptor: kinetics, sites of truncation, and implications for thrombolytic therapy. Biochemistry 38:4572–4585
149. Busso N, Hamilton JA (2002) Extravascular coagulation and the plasminogen activator/plasmin system in rheumatoid arthritis. Arthritis Rheum 46:2268–2279
150. Hamilton JA (2008) Plasminogen activator/plasmin system in arthritis and inflammation: friend or foe? Arthritis Rheum 58:645–648

Proteases, Coagulation, and Inflammation

Giuseppe Cirino and Mariarosaria Bucci

Abstract It is now widely accepted that inflammation and coagulation are two intimately linked processes. Pre-clinical evidence on this cross-talk have accumulated since 1961 when it was demonstrated that coagulation factors could cause an inflammatory response in in vivo pre-clinical studies. The discovery of thrombin receptors has been instrumental in clarifying several molecular aspects at the basis of the cross-talk between the coagulation and inflammation pathways. At the present stage we know that the coagulation–inflammation axis plays an important pathological role in many cardiovascular inflammatory-based disease. This chapter will address the role played by the thrombin receptors in the inflammation–coagulation axis and summarize the more recent pre-clinical and clinical findings.

Keywords Coagulation • Inflammation • Proteinase-activated receptor • Sch 530348 • Thrombin receptors

Undoubtedly, the concept that inflammation and coagulation are two aspects of the same, unique defensive host response, during the last 20 years has been well established. The first pre-clinical study that has demonstrated in vivo a link between inflammation and coagulation is dated 1961. It was demonstrated that the local administration in the rat hind paw of different purified pro-coagulant agents was able to induce an inflammatory response [1]. Using the same experimental animal model, in 1968, was demonstrated fibrin involvement, too [2]. However, the real definition of the molecular basis of this cross-talk has been possible only in recent years following the discovery and cloning of the thrombin receptors [3]

G. Cirino (✉) • M. Bucci
Department of Experimental Pharmacology, University of Naples "Federico II", via Domenico Montesano 49, 80131 Naples, Italy
e-mail: cirino@unina.it

N. Vergnolle and M. Chignard (eds.), *Proteases and Their Receptors in Inflammation*, Progress in Inflammation Research, DOI 10.1007/978-3-0348-0157-7_10,

and through the definition that thrombin can act through its receptor as an inflammatory mediator [4].

The endothelium can be considered the knot between inflammation and coagulation since damaged endothelium, consequent to the inflammatory response, represents the surface where the interaction between the proteins involved in both coagulation and development of inflammation take place. Activation of coagulation cascade, culminating in fibrin deposition, can be considered a key event in the inflammatory response to the site of injury or following infection. At the same time, platelets become activated releasing several mediators that modify tissue integrity. Thus, inflammation-induced coagulation participates to disease development, as shown by the coagulopathy associated to endotoxemia [5] and by the atherosclerotic plaque ruptured containing inflammatory cells [6]. During the 1990, several studies have been demonstrated that cytokines, such as tumor necrosis factor alpha (TNF_{α}), interleukin (IL)-6 and IL-1, modulate the coagulation system modifying the balance between procoagulant and anticoagulant activities [7–10] and that systemic activation of coagulation and inflammation can drive organ/tissue specific failure in critical pathological conditions such as endotoxemia [11]. Almost in the same time-frame proteinase-activated receptors (PARs) were discovered and cloned [12–15]. Serine proteases are a large family of enzymes involved in several biological functions ranging from hemostasis/fibrinolysis balancing to digestion of dietary proteins [16, 17]. Proteases like thrombin, trypsin, and tissue kallikreins regulate cell signaling by cleaving and activating a family of these G-protein-coupled PARs (PARs 1–4) via exposure of a tethered receptor-triggering ligand [18]. Thrombin cleaves the N-terminal domain of PAR-1, PAR-3, and PAR-4 being 10–100 times more potent at PAR-1 than at PAR-4 [19]. Trypsin, tryptase [20], and Factor Xa [21], instead, are able to activate specifically PAR-2. Since activation of PARs by proteolytic cleavage is an irreversible event, the PARs-induced signal transduction is switched-off by internalizing the cleaved receptors [22]. This peculiar mechanism has allowed the possibility to use small synthetic peptides (PARs-AP) that have the same sequence of the neo-amino terminus exposed in order to investigate the involvement of PARs (with the only exception of PAR-3) in several biological systems [23]. The thrombin-specific PARs can also be classified according to the presence of a hirudin-like domain: indeed, this binding site is present in PAR-1 and PAR-3 conferring to these receptors high affinity to thrombin [24]. Conversely, the hirudin-like domain is absent in PAR-4 with consequent low-affinity of this receptor with thrombin [24, 25]. PARs are widely distributed in several tissues: they are expressed in the vasculature on endothelial cells, mononuclear cells, platelets, fibroblasts, and smooth muscle cells [26].

In the vessels wall PAR-1 and PAR-2 induce responses involved in contractility, inflammation, and repair [27]. In healthy arteries, activation of PAR-1 causes nitric oxide release with consequent smooth muscle cell relaxation [28], however in presence of atherosclerotic lesions PAR-1 activation induces vasoconstriction [28]. The atherosclerotic process now is considered an inflammatory state of arteries where the cross-talk between coagulation and inflammation is manifest. The pro-coagulant factors (FVIIa, FXa, thrombin, FXIIa), fibrinogen and fibrin activate intracellular signaling, inducing the release of several pro-inflammatory mediators such as cytokines, chemokines, and growth factors triggering an hyper-inflammatory state associated to platelet and complement activations. The finding that PAR-1 is expressed in both normal and atherosclerotic vessels strongly implies that thrombin participates to the atherosclerotic lesion development through activation of this receptor [16].

In summary there are several experimental data that have confirmed earlier pre-clinical results that have hypothesized a cross-talk between inflammation and coagulation. Therefore, the concept that a link exists between the coagulation and inflammation pathways is now well established. It applies to several inflammatory-based diseases where the involvement of protease-activated receptors have been established. The present chapter will mainly deal with the role of thrombin receptors, namely, protease-activated receptor 1, 3, and 4.

1 Thrombin Receptors at Interface Between Coagulation and Inflammation

The findings that PAR can modulate the vascular inflammatory process has given an enormous boost to the research on the thrombin receptors and to the development of selective antagonists. Here, we will focus on the pre-clinical key discoveries on the thrombin receptors that have led to development of some drug candidates.

PAR-1 is a high affinity receptor for platelet activation at low thrombin concentrations, whereas PAR-4 appears to play a supportive "backup" function for thrombin-induced platelet activation mediating signaling only at high thrombin concentrations. Thrombin-induced PAR-1 activation in endothelial cells induces von Willebrand Factor release and promotes rolling and adhesion of platelets and leukocytes as well as access of plasma protein to extravascular space [29]. In addition, endothelial PAR1 activation stimulates chemokines production [30] and COX2 activation [31]. PAR-1 also serves as a receptor for tissue factor-Factor VIIa complex and Factor Xa in human and non-human primate [32], whereas such role is taken by PAR-3 in rodents [15]. A dominant role for platelets PAR-1 activation is well established in different thrombin-dependent arterial thrombosis models in primates where, blockade of the high affinity PAR-1 receptor results in a therapeutic effect [33, 34]. Of particular interest is the finding that blockade of PAR-1-induced platelet activation following thrombin stimulation, reduces platelet-mediated thrombosis without increasing bleeding risk. This finding is corroborated by the fact that several pre-clinical studies have shown that PAR-1-induced platelet activation seems not necessary for hemostasis processes [33, 35, 36]. These data have opened the possibility to develop safer anticoagulant drugs that could fulfill the need of reducing the bleeding risk associated to currently used therapies. Indeed, anti-thrombin drugs based on this concept should inhibit platelet activation/aggregation without increased risk of bleeding. There are several approaches that have been attempted to develop such drugs and at the present stage there are some promising results. Peptide mimetic compounds namely RWJ-56110 and RWJ-58259 have shown potent and selective PAR-1 antagonist properties [37–39]. They do not display any PAR-1 agonist activity and do not interact directly with thrombin. RWJ-56110 and RWJ-58259 following their binding to PAR-1 interfere with calcium mobilization and cellular functions associated to PAR-1 only. Interestingly, RWJ-58259 has demonstrated antirestenotic activity in a rat balloon angioplasty model and antithrombotic activity in a cynomologus monkey arterial injury model [40]. The pharmacological modulation operated by these selective thrombin antagonists, is not only a clear evidence that the coagulation–inflammation axis exists, but also that it represents a feasible and promising therapeutic target. More recently, new PAR-1 selective antagonists have been identified starting from the observation that N-palmitoylated peptides, based on the third intracellular loop of certain G-protein coupled receptors (GPCRs), can cause activation and/or inhibition of G protein signaling only in the presence of the parent GPCR [41]. Attachment of a palmitate lipid to peptides based on the N-terminal portion of the third loop of PAR-1 yielded P1pal-12 pepducin, whose sequence is pal-RCLSSSAVANRS-NH_2. This peptide lacked agonist activity, but was a full antagonist of PAR-1-dependent inositol phosphate production and Ca^{2+} signaling in platelets and recombinant systems [42]. In order to obtain non-peptide inhibitors to be exploited also in clinical studies several chemical approaches have been attempted. Among the non-peptide thrombin receptor antagonists the most interesting is the pyrroloquinazoline analogues SCH-79797 developed in 1999 [43]. SCH-79797 inhibited PAR-1-AP-induced platelet aggregation with an IC_{50} of 56 nM but only transiently inhibited platelet aggregation induced by thrombin. In addition, it has been described a cardioprotective effect of SCH-79797 related to its ability in inhibiting PAR-1. SCH-79797 was also shown to attenuate myocardial injury and dysfunction related to myocardial ischemia/reperfusion injury in an in vivo experimental pre-clinical study. Interestingly, the compound was active both when given before (preventive protocol) or during ischemia (therapeutic protocol) [44]. Following the synthesis and pharmacological characterization of these pilot compounds it has been developed a class of non-peptide-based PAR-1 antagonists based on the core structure of the natural product himbacine [45–47]. These derivatives have a selective high affinity for PAR-1. Further optimization of himbacine-derived PAR-1 antagonists have led to the discovery of SCH-530348, that showed $K_i = 8.1 \pm 1.1$ nM in the in vitro binding assays [48].

SCH-530348 is characterized by an excellent oral bioavailability in multiple species (33% in rats and 86% in monkeys) and has a potent oral activity in an ex vivo cynomologus monkey model of platelet aggregation. SCH-530348 is currently in clinical development and it has been evaluated in patients undergoing non-urgent percutaneous coronary intervention. SCH-530348 was generally well tolerated and did not cause increased thrombolysis in myocardial infarction (TIMI) bleeding, even when administered in association with aspirin and clopidogrel [49]. These results are a direct proof of the importance of the link between the inflammation and coagulation pathway and of its critical role in human pathologies. These results have allowed the design of a large phase III program, TRA•CER, a multicenter, randomized, double-blind, placebo-controlled study [50]. A part of this study, named TRA2°P-TIMI 50 trial, has been designed in order to evaluate the efficacy and safety of SCH-530348 during long-term treatment of patients with established atherosclerotic disease receiving standard therapy (up to 27,000). The outcome of this study will give information on whether the PAR-1 receptor is a valuable target for reducing major cardiovascular events with a favorable safety profile in patients with established atherosclerosis [51]. Recently Hezi-Yamit et al. [52] have described methods and medical devices used to deliver SCH-530348 locally to vasculature in treating and/or preventing cardiovascular conditions including, but not limited to, restenosis, in-stent restenosis, thrombosis, and in-stent thrombosis [52]. However, it is important to notice that PAR-1 activation is not always detrimental. It is known, in fact, that the anti-inflammatory and cytoprotective effects of activated protein C (APC) are elicited by APC-mediated PAR-1 activation [53, 54].

The presence of PAR-4, other than PAR-1, on human platelets has raised the possibility to act on PAR-4-induced signaling for the development of antithrombotic drugs. PAR-4 requires higher concentrations of thrombin for its activation compared to PAR-1 and this finding may suggest that proteases, other thrombin, could induce platelet activation via PAR-4 [29]. In fact, it has been shown that platelet activation by cathepsin G, a granzyme derived from activated neutrophils, is mediated by PAR-4 activation [55, 56] and that YD-3, a non-peptide selective PAR-4 antagonist, inhibits platelet aggregation and thromboxane formation aggregation induced by both purified cathepsin G and activated neutrophils [57, 58]. However, YD-3 inhibits also thrombin-induced signal via Ras/extracellular-signal-regulated kinase (ERK), which critically influences cell proliferation in vascular smooth muscle cells in vitro, attenuating, also, the restenosis after balloon angioplasty in vivo [59]. PAR-4 has been also shown to be involved in inflammation. Its role in inflammatory processes has been recently revisited testing the effects of PAR-4 inhibition in a model of systemic inflammation and disseminated intravascular coagulation [60]. This study has provided circumstantial evidence that the primary cellular target of PAR-4 antagonism may be neutrophils, rather than platelets or endothelial cells. Thus, a current hypothesis is that the beneficial effect of PAR-4 inhibition could be based on a reduced neutrophil infiltration into the site of the inflammation. This latter hypothesis suggests a possible development for PAR-4 antagonists in the treatment of systemic inflammation.

PAR-3 is usually considered as a cofactor for PAR-4 activation in human platelets that might serve to enhance the specificity of thrombin's actions [61]. Data obtained from mouse platelets suggest that PAR-3 does not by itself activate a signal transduction but instead acts as a cofactor that localizes thrombin to the surface of the mouse platelets to promote cleavage and activation of PAR-4 [29]. The fact that PAR-3 has been so far considered as "accessory" receptor has in someway impaired the research and at present there are few papers that have attempted to investigate and possibly define a role for this receptor. However, recently it has been shown that in the presence of pro-inflammatory cytokines, such as TNFα or IL 1β, thrombin interacts with PAR-3. Indeed, in this experimental condition, resembling a pathological condition sustained by an inflammatory state, PAR-3 is highly expressed in lung fibroblasts. This enhanced expression results in the induction of granulocyte-macrophage colony stimulating-factor to a level that is found in asthma. This observation has made the basis for a possible use of PAR-3 antagonists in the treatment of asthma [62].

2 Conclusions

Coagulation and inflammation are two intimately related processes whose molecular basis have started to be clarified only in the past 10 years following the discovery of the thrombin receptors. The key players in this cross-talk are the endothelium and the platelets. The endothelium represents the active surface where proteins involved in both coagulation and development of inflammation interacts. Platelets are an important source of key mediators involved in the coagulation–inflammation axis [27]. The recent clinical data that have been obtained with selective PAR-1 antagonists confirm that the coagulation–inflammation axis is not anymore an interesting pre-clinical evidence but that it represents an important therapeutic target that could lead to safer and more efficacious drugs.

References

1. Jancso' N (1961) Inflammation and the inflammatory mechanisms. J Pharm Pharmacol 13:577–594
2. Wiseman EH, Chang YH (1968) The role of fibrin in the inflammatory response to carrageenin. J Pharmacol Exp Ther 159:206–210
3. Vu TK, Hung DT, Wheaton VI, Coughlin SR (1991) Molecular cloning of a functional thrombin receptor reveals a novel proteolytic mechanism of receptor activation. Cell 64:1057–1068
4. Cirino G, Cicala C, Bucci MR, Sorrentino L, Maraganore JM, Stone SR (1996) Thrombin functions as an inflammatory mediator through activation of its receptor. J Exp Med 183(3): 821–827
5. Opal SM, Esmon CT (2003) Bench to bedside review: functional relationship between coagulation and innate immune response and their respective roles in pathogenesis of sepsis. Crit Care 7:23–38
6. Libby P, Aikawa M (2002) Stabilization of atherosclerotic plaque: new mechanisms and clinical targets. Nat Med 8:1257–1262
7. van der Poll T, Levi M, Hack CE, ten Cate H, van Deventer SJ, Eerenberg AJ, de Groot ER, Jansen J, Gallati H, Büller HR et al (1994) Elimination of interleukine-6 attenuates coagulation activation in experimental endotoxemia in chimpanzees. J Exp Med 179:1253–1259
8. van Deventer SJ, Buller HR, ten Cate JW, Aarden LA, Hack CE, Sturk A (1990) Experimental endotoxemia in humans: analysis of cytokine release and coagulation, fibrinolytic, and complement pathways. Blood 76:2520–2526

9. Boermeester MA, Leeuwen PA, Coyle SM, Wolbink GJ, Hack CE, Lowry SF (1995) Interleukin-1 blockade attenuates mediator release and dysregulation of the hemostatic mechanism during human sepsis. Arch Surg 130:739–748
10. Cicala C, Cirino G (1998) Linkage between inflammation and coagulation: an update on the molecular basis of the crosstalk. Life Sci 62(20):1817–1824
11. Aird WC (2001) Vacular bed-specific hemostasis:role of endothelium in sepsis pathogenesis. Crit Care Med 29:S28–S34
12. Vu TK, Hung DT, Wheaton VI, Coughlin SR (1991) Molecular cloning of a functional thrombin receptor reveals a novel proteolytic mechanism of receptor activation. Cell 64:1057–1068
13. Nystedt S, Emilsson K, Wahlestedt C, Sundelin J (1994) Molecular cloning of a potential proteinase activated receptor. Proc Nat Acad Sci USA 91:9208–9212
14. Ishiara H, Connolly AJ, Zeng D, Kahn ML, Zheng YW, Timmons C, Tram T, Coughlin SR (1997) Protease-activated receptor 3 is a second thrombin receptor in humans. Nature 386:502–506
15. Kahn ML, Zheng YW, Huang W, Bigornia V, Zeng D, Moff S, Farese RV Jr, Tam C, Coughlin SR (1998) A dual thrombin receptor system for platelet activation. Nature 394:690–694
16. Cirino G, Napoli C, Bucci M, Cicala C (2000) Inflammation-coagulation network: are serine protease receptors the knot? Tips 21:170–172
17. Hollenberg MD, Oikonomopoulou K, Hansen KK, Saifeddine M, Ramachandran R, Diamandis EP (2008) Kallikreins and proteinase-mediated signaling: proteinase-activated receptors (PARs) and the pathophysiology of inflammatory diseases and cancer. Biol Chem 389:643–651
18. Macfarlane SR, Seatter MJ, Kanke T, Hunter GD, Plevin R (2001) Proteinase-activated receptors. Pharmacol Rev 53:245–282
19. Holinstat M, Voss B, Bilodeau ML, Hamm HE (2007) Protease-activated receptors differentially regulate human platelet activation through a phosphatidic acid-dependent pathway. Mol Pharmacol 71:686–694
20. Molino M, Barnathan ES, Numerof R, Clark J, Dreyer M, Cumashi A, Hoxie JA, Schechter N, Woolkalis M, Brass LF (1997) Interactions of mast cell tryptase with thrombin receptors and PAR-2. J Biol Chem 27:4043–4049
21. Fox MT, Harriott P, Walker B, Stone SR (1997) Identification of potential activators of proteinase-activated receptor-2. FEBS Lett 417:267–269
22. Cottrell GS, Amadesi S, Schmidlin F, Bunnett N (2003) Protease-activated receptor 2: activation, signalling and function. Biochem Soc Trans 31(Pt 6):1191–1197
23. Déry O, Corvera CU, Steinhoff M, Bunnett NW (1998) Proteinase-activated receptors: novel mechanisms of signaling by serine proteases. Am J Physiol 274:C1429–C1452
24. De Candia E, Hall SW, Rutella S, Landolfi R, Andrews RK, De Cristofaro R (2001) Binding of thrombin to glycoprotein Ib accelerates the hydrolysis of Par-1 on intact platelets. J Biol Chem 276:4692–4698
25. Nakanishi-Matsui M, Zheng YW, Sulciner DJ, Weiss EJ, Ludeman MJ, Coughlin SR (2000) PAR-3 is a cofactor for PAR-4 activation by thrombin. Nature 404:609–613
26. Coughlin SR (2000) Thrombin signalling and protease-activated receptors. Nature 407:258–264
27. Jennings LK (2009) Mechanisms of platelet activation: need for new strategies to protect against platelet-mediated atherothrombosis. Thromb Haemost 102:248–257
28. Leger AJ, Covic L, Kuliopulos A (2006) Protease-activated receptors in cardiovascular diseases. Circulation 14:1070–1077
29. Coughlin SR (2005) Protease-activated receptors in hemostasis, thrombosis and vascular biology. J Thromb Haemost 3:1800–1814
30. Johnson K, Choi Y, DeGroot E, Samuels I, Creasey A, Aarden L (1998) Potential mechanisms for proinflammatory vascular cytokine response to coagulation activation. J Immunol 160:5130–5135

31. Houliston RA, Keogh RJ, Sugden D, Dudhia J, Carter TD, Wheeler-Jones CP (2002) Protease-activated receptors upregulate cyclooxygenase-2 expression in human endothelial cells. Thromb Hemost 88:321–328
32. Levi M, van der Poll T (2010) Inflammation and coagulation. Crit Care Med 38:s26–s32
33. Derian CK, Damiano BP, Addo MF, Darrow AL, D'Andrea MR, Nedelman M, Zhang HC, Maryanoff BE, Andrade-Gordon P (2003) Blockade of the thrombin receptor protease-activated receptor-1 with a small molecule antagonist prevents thrombus formation and vascular occlusion in non human primates. J Pharmacol Exp Ther 304:855–861
34. Covic L, Misra M, Badar J, Singh C, Kuliopulos A (2002) Pepducin-based intervention of thrombin-receptor signaling and systemic platelet activation. Nat Med 8:1161–1165
35. Kato Y, Kita Y, Hirasawa-Taniyama Y, Nishio M, Mihara K, Ito K, Yamanaka T, Seki J, Miyata S, Mutoh S (2003) Inhibition of arterial thrombosis by protease-activated receptor antagonist, FR171113, in the guinea pig. Eur J Pharmacol 473:163–169
36. Vandendries ER, Hamilton JR, Coughlin SR, Furie B, Furie BC (2007) PAR-4 is required for platelet thrombus propagation but not fibrin generation in a mouse model of thrombosis. Proc Natl Acad Sci USA 104:288–292
37. Andrade-Gordon P, Maryanoff BE, Derian CK, Zhang HC, Addo MF, Darrow AL, Eckardt AJ, Hoekstra WJ, McComsey DF, Oksenberg D et al (1999) Design, synthesis, and biological characterization of a peptide-mimetic antagonist for a tethered-ligand receptor. Proc Natl Acad Sci USA 96:12257–12262
38. Zhang HC, McComsey DF, White KB, Addo MF, Andrade-Gordon P, Derian CK, Oksenberg D, Maryanoff BE (2001) Thrombin receptor (PAR-1) antagonists. Solid-phase synthesis of indole-based peptide mimetics by anchoring to a secondary amide. Bioorg Med Chem Lett 11:2105–2109
39. Zhang HC, Derian CK, Andrade-Gordon P, Hoekstra WJ, McComsey DF, White KB, Poulter BL, Addo MF, Cheung WM, Damiano BP et al (2001) Discovery and optimization of a novel series of thrombin receptor (PAR-1) antagonists: potent, selective peptide mimetics based on indole and indazole templates. J Med Chem 44:1021–1024
40. Maryanoff BE, Zhang H, Andrade-Gordon P, Derian CK (2003) Discovery of potent peptide-mimetic antagonists for the human thrombin receptor, protease-activated receptor-1 (PAR-1). Curr Med Chem 1:13–36
41. Kuliopulos A, Covic L (2005) G protein coupled receptor (GPCR) agonists and antagonists and methods of activating and inhibiting GPCR using the same. US 6,864,229
42. Covic L, Gresser AL, Talavera J, Swift S, Kuliopulos A (2002) Activation and inhibition of G protein-coupled receptors by cell-penetrating membrane tethered peptides. Proc Natl Acad Sci USA 99:643–648
43. Ahn HS, Arik L, Boykow G, Burnett DA, Caplen MA, Czarniecki M, Domalski MS, Foster C, Manna M, Stamford AW, Wu Y (1999) Structure-activity relationships of pyrroloquinazolines as thrombin receptor antagonists. Bioorg Med Chem Lett 9:2073–2078
44. Strande JL, Hsu A, Su J, Fu X, Gross GJ, Baker JE (2007) SCH 79797, a selective PAR-1 antagonist, limits myocardial ischemia/reperfusion injury in rat hearts. Basic Res Cardiol 102:350–358
45. Chackalamannil S, Asberom T, Xia Y (2000) Preparation of himbacine analogs as thrombin receptor antagonists. US 6,063,847
46. Chackalamannil S, Chelliah MV, Clasby MC, Xia Y (2003) Preparation of himbacine analogues as thrombin receptor antagonists. WO 2,003,033,501
47. Chelliah MV, Chackalamannil S, Xia Y (2005) Preparation of constrained himbacine analogs as thrombin receptor antagonists. US 2,005,267,155
48. Chackalamannil S, Xia Y, Greenlee WJ, Clasby M, Doller D, Tsai H, Asberom T, Czarniecki M, Ahn HS, Boykow G, Foster C, Agans-Fantuzzi J, Bryant M, Lau J, Chintala M (2008) Discovery of a novel, orally active himbacine-based thrombin receptor antagonist (SCH 530348) with potent antiplatelet activity. J Med Chem 51:3061–3064

49. Becker RC, Moliterno DJ, Jennings LK, Pieper KS, Pei J, Niederman A, Ziada KM, Berman G, Strony J, Joseph D, Mahaffey KW, Van de Werf F, Veltri E, Harrington RA (2009) Safety and tolerability of SCH 530348 in patients undergoing non-urgent percutaneous coronary intervention: a randomised, double-blind, placebo-controlled phase II study. Lancet 14:919–928
50. The TRA•CER Executive and Steerin (2009) The thrombin receptor antagonist for clinical event reduction in acute coronary syndrome (TRA*CER) trial: study design and rationale. Am Heart J 158:327–334
51. Morrow D, Scirica B, Fox K, Berman G, Strony J, Veltri E, Bonaca MP, Fish P, McCabe CH, Braunwald E (2009) Evaluation of a novel antiplatelet agent for secondary prevention in patients with a history of atherosclerotic disease: design and rationale for the Thrombin-Receptor Antagonist in Secondary Prevention of Atherothrombotic Ischemic Events (TRA 2 degrees P)-TIMI 50 trial. Am Heart J 158:335–341
52. Hezi-Yamit A, Wong J (2009) Local delivery of PAR-1 antagonists to treat vascular complications. US 2,009,297,576
53. Riewald M, Petrovan RJ, Donner A, Ruf W (2003) Activated protein C signals through the thrombin receptor PAR-1 in endothelial cells. J Endotoxin Res 9:317–321
54. Feistritzer C, Riewald M (2005) Endothelial barrier protection by activated protein C through PAR-1-dependent sphingosine 1-phosphate receptor-1 crossactivation. Blood 105:3178–3184
55. Selak MA, Chignard M, Smith JB (1988) Cathepsin G is a strong platelet agonist released by neutrophils. Biochem J 251:293–299
56. Sambrano GR, Huang W, Faruqi T, Mahrus S, Craik C, Coughlin SR (2000) Cathepsin G activates protease-activated receptor 4 in human platelets. J Biol Chem 275:6819–6823
57. Wu CC, Hwang TL, Liao CH, Kuo SC, Lee CY, Teng CM (2002) Selective inhibition of protease-activated receptor 4-dependent platelet activation by YD-3. Thromb Haemost 87:1026–1033
58. Wu CC, Hwang TL, Liao CH, Kuo SC, Lee FY, Lee CY, Teng CM (2003) The role of PAR-4 in thrombin-induced thromboxane production in human platelets. Thromb Haemost 90:299–308
59. Peng CY, Pan SL, Guh JH, Liu YN, Chang YL, Kuo SC, Lee FY, Teng CM (2004) The indazole derivative YD-3 inhibits thrombin-induced vascular smooth muscle cell proliferation and attenuates intimal thickening after balloon injury. Thromb Haemost 92:1232–1239
60. Slofstra SH, Bijlsma MF, Groot AP, Reitsma PH, Lindhout T, ten Cate H, Spek CA (2007) Protease-activated receptor-4 inhibition protects from multiorgan failure in a murine model of systemic inflammation. Blood 110:3176–3182
61. Cirino G, Severino B (2010) Thrombin receptors and their antagonists: an update on the patent literature. Expert Opin Ther Pat 20:875–884
62. Beri R (2006) The use and identification of antagonists of protease-activated receptor 3 (PAR-3) for the treatment of asthma. WO 2,006,054,931

Proteases and Inflammatory Pain

Nicolas Cenac

Abstract Inflammation mediators are known to signal neurons and provoke pain. In this context, proteases can signal sensory neurons at the site of inflammation and modulate pain transmission. In this chapter, we will perform an overview of the multiple mechanisms whereby proteases can signal nerve endings. Proteases can directly activate proteases activated receptor at the membrane of the nerve endings to modulate pain signaling. They can not only be implicated in the secretion of molecules implicated in pain modulation by the nerve endings such as substance P or CGRP but also by other cell type such as fibroblast able to secrete opioid under protease stimulation. Proteases act on pain signal by cleaving proteins implicated in pain sensation, e.g., kalikreins can convert kininogen in bradykinin which acts on pain modulation. Proteins of nerve can be cleaved by proteases and provoke an increase in pain sensation following inflammation. These different mechanisms are explained in this chapter in the context of inflammatory pain.

Keywords Aminopeptidase • Calcium channels • Inflammation • Kinins • Nerve degradation • Neuromediators • Pain • Protease • Protease-activated receptor • Transient receptor potential vanilloid

Abbreviations

APN	AminoPeptidase N
ASIC	Acid-sensing ion channels
B1R B2R	Bradykinin receptor 1, 2

N. Cenac (✉)
INSERM, U1043, 31024, Toulouse, France

CNRS, U5282, 31024, Toulouse, France

Université de Toulouse, UPS, 31300 Toulouse, France
e-mail: nicolas.cenac@inserm.fr

N. Vergnolle and M. Chignard (eds.), *Proteases and Their Receptors in Inflammation*, Progress in Inflammation Research, DOI 10.1007/978-3-0348-0157-7_11,

BK	BradyKinin
CCI	Chronic constriction injury
CD13	Cluster of differentiation
CGRP	Calcitonin gene-related peptide
CNS	Central nervous system
COX	CycloOXygenase
DAG	DiAcylGlycerol
DRG	Dorsal root ganglia
GABA	Gamma-aminobutyric acid
GCP	Glutamate carboxypeptidase
HEK	Human embryonic kidney
IBS	Irritable Bowel syndrome
IL	InterLeukin
KLP	Kallidin-like peptide
LPA	LysoPhosphAtidic acid
MAG	Myelin-associated glycoprotein
mGlu	Metabotropic glutamate
MPO	MyeloPerOxidase
NAAG	*N*-AcetylAspartylGlutamate
NFL	Neurofilament light chain
NK	NeuroKinin
NMDA	*N*-Methyl-D-aspartic acid
PAR	Protease-activated receptor
PIP2	Phosphatidyl Inositol 4,5-bisPhosphate
PKC	Protein kinase C
PLC	PhosphoLipase C
PMPA	Phosphonomethyl pentanedioic acid
SP	Substance P
TNF	Tumor necrosis factor
TRP	Transient receptor potential
TRPA	Transient receptor potential ankyrin
TRPV	Transient receptor potential vanilloid

1 Introduction

The International Association for the Study of Pain (IASP) has defined inflammatory pain as a reference to pain in response to an insult to tissues, at a cellular level [1]. Damage to the cells leads to release of molecules that activate pain receptors locally, at the site of injury and that participate to the inflammatory response. These inflammatory mediators can directly activate nociceptors to induce a pain response, or may sensitize them to mechanical or thermal stimuli. Similarly, some inflammatory mediators may sensitize pain receptors to cell response to other mediators of inflammation. Thereby, in addition to spontaneous pain sensation, increased sensitivity to noxious stimuli (thermal or mechanical) known as hyperalgesia and responses to innocuous stimuli known as allodynia, are associated with inflammatory states. In that setting, proteases occupy a place of choice, being released by all the cellular actors of inflammation (damaged cells, immune cells, resident cells such as neurons, fibroblasts, epithelial cell, endothelium, or even pathogens), they are able to either cause or modulate pain signals. Proteases can act on pain perception by activating families of receptors such as the Protease-Activated Receptors (PARs); they can also degrade neuronal components or cleave molecules active on pain pathways (Fig. 1). Taken together, the results discussed in this chapter highlight proteases as true signaling molecules of inflammatory pain pathways. Considering the amount of proteases released upon inflammation and their ubiquitous effects, we need to consider proteases as important potential targets for the development of analgesic drugs in the treatment of inflammatory pain.

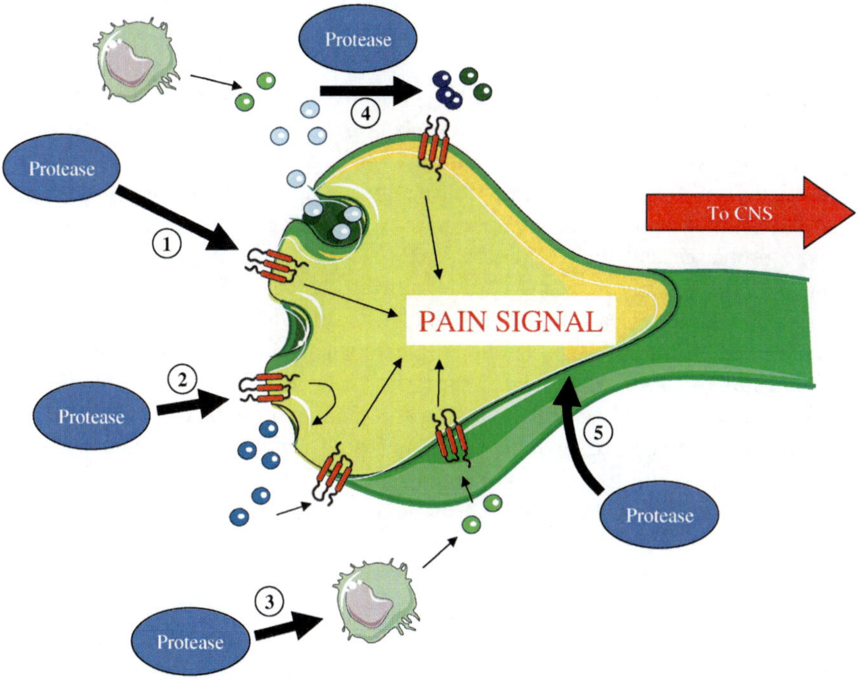

Fig. 1 Protease signaling to modulate pain signal. Proteases can modulate pain perception through several mechanisms: they can directly modulate pain signal by activation of receptor (1), by the release of neuromediators (2), by the nerve ending or by the release of molecules implicated in pain signaling by other cell types (3). Proteases can also be implicated in the modulation of pain signal by the conversion of neuromediators or other inflammatory molecules (4). Finally, proteases can directly be implicated in nerve degradation provoking an increase in pain response (5)

2 Protease and Protease-Activated Receptor

Serine proteases, which are released upon inflammation, can activate PARs. A number of studies have focused on the effect of PAR activation on activation or inhibition of pain pathways, using in vivo, sub-inflammatory doses of PAR agonist peptides [2]. As a whole, these different studies suggest that activation of PAR_1, PAR_2 and PAR_4 can interfere with the conduction of pain, independent of their pro-inflammatory effects [2]. The function of each PAR in the modulation of pain pathways seems to be very different and sometimes opposite. PAR_2 can be considered as a pro-nociceptive receptor causing pain and hypersensitivity, while PAR_1 and PAR_4 are able to decrease nociceptive signals, causing analgesia [2]. Only few studies have determined the implication of PARs in inflammatory pain. Depletion of connective tissue mast cells by repeated and chronic administration of compound 48/80 allowed ruling out the role of this cell type and its mediators in PAR_2-induced thermal hyperalgesia. The nociceptive response to thermal stimulation after the administration of an inflammatory dose of PAR_2-specific agonist was similar in naïve and connective tissue mast cell depleted animals [3]. This clearly indicates that mast cell activation is not a necessary step downstream from PAR_2 activation to induce activation of pain pathways. However, it is interesting to note that thermal and mechanical hyperalgesia caused by connective tissue mast cell degranulation (mice that have received an intraplantar injection of compound 48/80) were totally abolished in PAR_2-deficient mice, compared to wild-type controls [4]. These results confirmed the potential of mast cell proteases to activate PAR_2 and the major role of PAR_2 activation in mast cell induced hyperalgesia (Fig. 2). Release of mast cell derived proteases and their effect on PAR_2 activation might represent a crucial element in the bi-directional regulatory pathways between mast cells and sensory neurons. In human diseases, such as irritable bowel syndrome (IBS), a pathology characterized by chronic abdominal pain is associated sometimes with the presence of micro-inflammation in colonic tissues [5]; it has been shown that the distance between mast cells and enteric neurons was significantly reduced in patients with IBS, compared to healthy subjects [6]. This favored crosstalk between neurons and mast cells in the setting of a painful pathology could implicate the release of mast cell proteases and further activation of PAR_2 on sensory neurons of those patients. Such hypothesis has been addressed, at least in part, in a study published by Cenac et al. Biopsies from IBS patients were taken during colonoscopy and mediators that are released by those fresh tissues into culture media were analyzed [7]. Increased proteolytic activity was found in the culture media of biopsies harvested from IBS patients compared to control patients. Further, this proteolytic activity was able to signal sensory neurons and reproduce hypersensitivity symptoms when injected into the mouse colonic lumen [7]. Finally, it was shown that calcium signals in sensory neurons and hypersensitivity symptoms in mice (both hyperalgesia and allodynia), induced by supernatants obtained from IBS patient biopsies, were entirely PAR_2 dependent. Neurons from PAR_2-deficient mice failed to respond to those supernatants, and PAR_2-deficient mice did not develop hypersensitivity

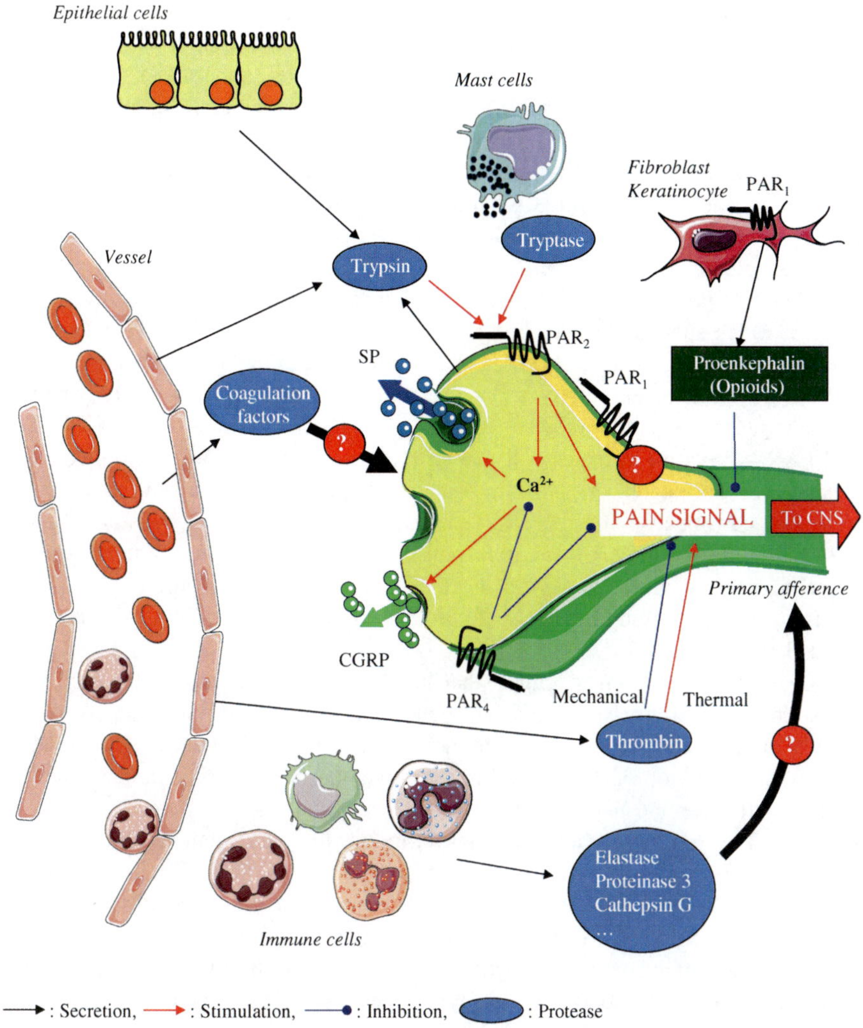

Fig. 2 Protease signaling through PARs: implication for nociceptive pathways. Proteases from the coagulation cascade, immune cells such as mast cells, endothelial cells and epithelial cells can cleave PARs on a variety of different cells types implicated in pain transmission. These proteases are often generated during inflammation, and PARs coordinate responses that modify pain signals from the periphery to the central nervous system

symptoms when administered with supernatants from IBS patient biopsies [7]. In addition, the pivotal role of PAR_2 in IBS patient biopsies-induced hypersensitivity was similarly observed whether experiments were performed with biopsies from patients with diarrhea-predominant, constipation-predominant, or alternating-type IBS [7]. This study identified PAR_2 as the first common possible mediator in all IBS patients and raised evidence that PAR_2-activating proteases are released in human diseases associated with visceral pain symptoms. The nature of those proteases still has to be fully investigated, but increased levels of tryptase and trypsin, two PAR_2-activating proteases, were observed in biopsies and biopsy supernatants [7] (Fig. 2). In addition, a recent study has confirmed that trypsinogen IV expression was enhanced in tissues from IBS patients [8].

Mast cells are in close proximity to nerve fibers containing substance P (SP) and calcitonin gene-related peptide (CGRP) in many tissues [9]. Activation of PAR_2 by trypsin or PAR_2-specific peptidic agonist on primary afferent neurons leads to the local release of SP and CGRP (Fig. 2). SP degranulates mast cells, leading to the release of proteases (such as tryptase) that in turn are not only able to degrade neuropeptides such as CGRP, but also can potentially further activate PAR_2 on neurons or other adjacent cells [10, 11]. Additionally, CGRP potentiates the release of SP from spinal afferent nerves [12] and inhibits the degradation of SP by neutral endopeptidase [13], adding further complexity to the interactions among CGRP, SP and PAR_2-activating proteases. It is also interesting to note that PAR_2 activation seems to be implicated not only in mast cell degranulation-induced inflammatory hyperalgesia, but also in other pain models. Formalin-induced hyperalgesia was reduced in PAR_2-deficient mice, in response to both mechanical and thermal stimuli [4]. Similarly, a study by Ferrell et al. has shown the implication for PAR_2 in arthritis, where all parameters of inflammation were reduced in PAR_2-deficient mice [14].

Together, these findings suggest that PAR_2 activation can signal the peripheral projections of primary spinal afferent neurons to trigger the release of neuropeptides from their central projections inducing hyperalgesia. In fact, PAR_2 activation induces sensitization of dorsal horn nociceptive neurons, which is mediated by the central activation of neurokinin-1 (NK-1) receptors and the release of prostaglandins. This proposed pathway of PAR_2-induced hyperalgesia is in good accord with previous studies that have documented that activation of sensory afferent nerves leads to the release of SP within laminae I and II of the spinal cord, and that binding of SP to NK-1 receptors can elicit the release of prostaglandins, which in turn can sensitize spinal nociceptive neurons [15, 16]. Intrathecal administration of PAR_2 agonist provokes thermal hyperalgesia after paw stimulation, and this reduction in hindpaw withdrawal latency (i.e. hyperalgesia) was dependent on spinal COX-1 and -2 activation [17]. These results further support the hypothesis that spinal release of prostaglandins is involved in PAR_2-induced activation of primary afferents. Moreover, intrathecal administration of PAR_2 agonists enhances peripherally induced inflammatory allodynia and hyperalgesia through activation of DRG sensory neurons [18].

In contrast to PAR_2, activation of PAR_1 or PAR_4 in models of paw inflammation seemed to be protective against the development of hyperalgesia. Activation of PAR_1 by selective agonists attenuated nociception, caused analgesia in non-inflammatory conditions and reduced inflammatory hyperalgesia independent of the inflammatory response [19]. Interestingly, in this study the authors demonstrate that the agonist peptide of PAR_1 was able to decrease inflammatory mechanical and thermal hyperalgesia induced by intraplantar injection of carrageenan but thrombin had an opposite effects on nociception in response to a thermal stimulus [19]. In fact, intraplantar injection of thrombin also inhibited carrageenan-induced mechanical hyperalgesia but exacerbated the thermal carrageenan-induced hyperalgesia [19] (Fig. 2). The hyperalgesic effects of thrombin cannot be explained by a pro-inflammatory effect of thrombin because doses of thrombin that caused hyperalgesia did not increase MPO activity or cause oedema [19]. Co-injection of thrombin with carrageenan did not amplify the inflammatory response caused by carrageenan [19]. Moreover, if inflammation was responsible for thrombin hyperalgesic effect in response to a thermal noxious stimulus, hyperalgesia instead of analgesia would also have been observed in response to a mechanical stimulus. The hyperalgesic effect of thrombin may be explained by the fact that thrombin is not a selective agonist for PAR_1. Thrombin's actions are mediated not only by activation of PAR_1, but also by activation of PAR_3 and PAR_4. Thrombin's actions can also be mediated by its non-catalytic site [20–22]. Thus, it is possible that thrombin causes thermal hyperalgesia by activating a receptor different from PAR_1 and located on nociceptive fibres. This study highlights one of the major efforts that need to be taken to understand the implication of PARs in inflammatory pain: define which endogenous proteases are present at inflammatory sites and can modulate pain pathways. Even though PAR_1 is expressed in primary culture sensory neurons extracted from dorsal root ganglia of mice, its activation has no effect on calcium mobilization [23]. Interestingly, PAR_1 agonist peptide triggers the production of proenkephalin and the activation of opioid receptors, suggesting that PAR_1 endogenously controls inflammatory pain, by activating opioid pathways. In non-inflammatory condition, activation of PAR_1 induces an increase in opioid mRNA in the rat paw [23]. Several studies have demonstrated that endogenous opioids are extensively produced by immune cells [24, 25] and also by cells from the skin [25, 26]. PAR_1 activation did not induce proenkephalin expression in isolated dendritic cells and macrophages in vitro, both in inflammatory and non-inflammatory conditions [23] (Fig. 2). However, proenkephalin expression was increased in keratinocytes and fibroblasts after PAR_1 activation in vitro, signifying that these cells could be the source of proenkephalin release into mouse paws following PAR_1 activation [23]. Nevertheless, secretion of proenkephalin by other types of immune cells present at the site of inflammation cannot be excluded.

Similar to PAR_1-specific agonists, administration of sub-inflammatory doses or PAR_4 peptidic agonists (but not control peptides) was able to significantly reduce carrageenan-induced thermal and mechanical hyperalgesia [27]. However, in contrast to PAR_1, PAR_4 was functional on DRG neurons, its activation was able to reduce capsaicin, PAR_2, TRPV4 or KCl-induced calcium mobilization [27, 28] (Fig. 2). Study of neuronal currents following PAR_4 agonist peptide treatment provides a direct support for the concept of a direct effect on neurons and unequivocal evidence that suppression of intrinsic excitability can contribute to the observed antinociceptive PAR_4 actions [29].

In conclusion, agonist peptides of the different PARs helped us to understand the implication of each receptor in inflammatory pain. Those studies have demonstrated opposite effects of the different PARs on inflammatory pain signaling pathways, PAR_2 being pro-nociceptive and PAR_1 and PAR_4 being antinociceptive. Further, evidence has been raised for a possible direct effect for PAR_2 and PAR_4 activation on sensory neurons (respectively, activating or inhibiting nociceptors), and indirect effect for PAR1, which seemed to act through the release of endogenous opioids by resident cells. One major and still open question is the nature of endogenous proteases that are released at inflammatory sites and active on pain pathways (Fig. 2). Characterization of these proteases and their effects on inflammatory pain need to be the future direction of research in this field. This would be essential for the classification of PARs as potential therapeutic target in chronic inflammatory pain.

3 Interaction of Protease-Activated Receptors with Other Receptors and Implication in Inflammatory Pain

The molecular mechanism of PAR_2-mediated hyperalgesia is largely unknown. However, it is likely to involve signaling events that regulate activity or expression of ion channels. The first serious candidate is the transient receptor potential vanilloid-1 (TRPV1). In several models, PAR_2 co-expresses with TRPV1; it is the case in nerve fibers [30, 31], in the bladder [32] or in the pancreas [33]. TRPV1 is a member of the transient receptor potential (TRP) family of channels. TRPV1 is a non-selective cation channel that is activated by protons, elevated temperature and certain lipids, as well as by exogenous vanilloid molecules such as capsaicin [34, 35]. On the nerve fibers arising from the pancreas, PAR_2 and TRPV1 are also functionally coupled; Response to capsaicin in PAR_2-sensitized DRG neurons is enhanced as measured by capsaicin evoked CGRP release [33]. The TRPV1 antagonist capsazepine blocks PAR_2-mediated thermal hyperalgesia, but not spontaneous nociceptive behavior or Fos expression, showing that sensitization/transactivation of TRPV1 by PAR_2 might be involved in thermal hyperalgesia but not in the primary pain message [36]. In a model of visceral pain induced by intracolonic administration of capsaicin, intracolonic injection of PAR_2 agonist, 6 or 18 h before capsaicin, produces a delayed sensitization of capsaicin receptors, resulting in facilitation of visceral pain and referred hyperalgesia in response to von Frey filaments stimuli [37]. In the same study, the authors have shown that PAR_1 activation appears to play an antinociceptive role in the processing of visceral pain in this model [37]. To study the relationship between PAR_2 and TRPV1, Amadesi and colleagues used four different approaches: calcium signaling in HEK cells co-transfected with PAR_2 and TRPV1, calcium signals in DRGs, electrophysiology approach and in vivo experiments. All these different approaches pointed to the same mechanism, demonstrating that PAR_2 activation potentiates the capsaicin response in a PKC-dependent manner [30, 38] (Fig. 3). In HEK cells and DRG neurons, PAR_2 and TRPV1 are functionally coupled; application of PAR_2 specific agonists, trypsin or tryptase potentiates capsaicin-induced

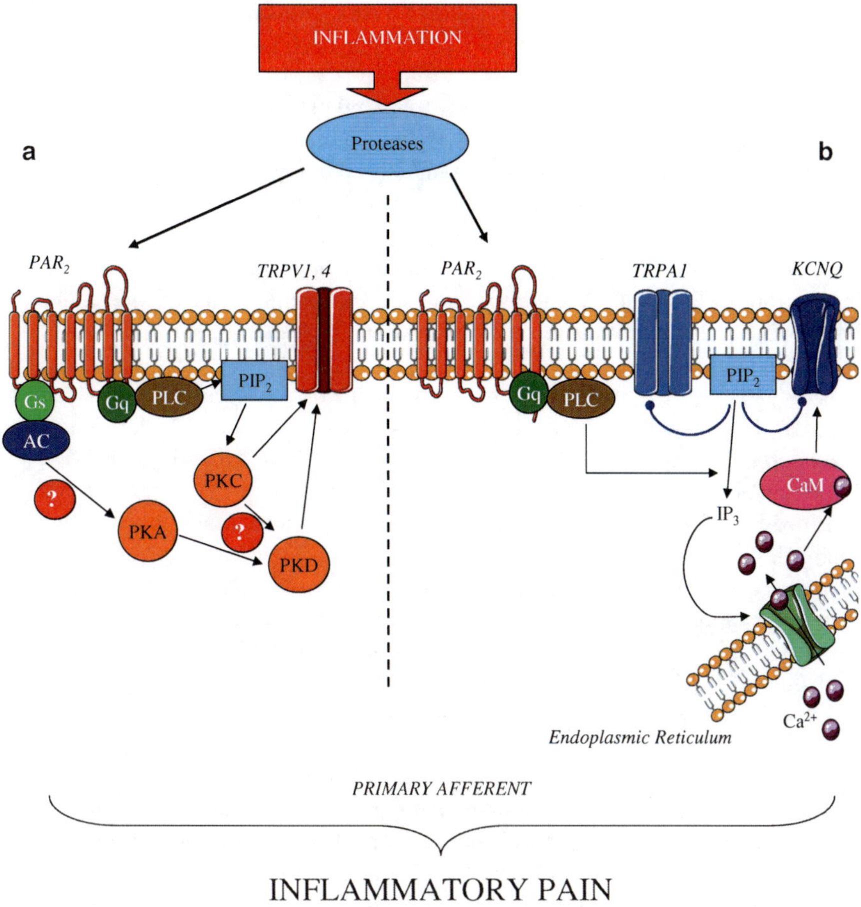

Fig. 3 Proposed model of PAR_2 interaction with ion channels. (**a**) Activation of PAR_2 by proteases activates PLCβ through its coupling with $G_{\alpha Q11}$. This pathway induces activation of PKC which can phosphorylate TRPV1 or TRPV4. PKC may also activate other downstream kinases, such as PKD, which also phosphorylates and sensitizes TRPV1 and TRPV4. In addition, PAR_2 agonists also activate PKA, possibly by coupling to $G_{\alpha s}$, which phosphorylates and sensitizes TRPV1 and 4 directly or via PKD. (**b**) PAR_2 activation induces the conversion of PIP2 to IP3 and DAG through $G_{\alpha Q}$ and PLC activation. Metabolization of PIP2 releases the inhibition of TRPA1 and KCNQ from plasma membrane PIP2. IP3 released by this cleavage induces calcium secretion from the endoplasmic reticulum to the cytosol. Calcium binds to and activates calmodulin which increases KCNQ channel sensitivity

currents [30, 39]. Dai and colleagues confirmed a functional interaction between TRPV1 and PAR_2 in HEK293 and in mouse DRG neurons by a patch-clamp technique [31]. Further, Amadesi et al. have shown that PAR_2 agonists in sensory neurons promoted translocation of the epsilon form of PKC and protein kinase A catalytic subunits from the cytosol, to the plasma membrane [39] (Fig. 3). In vivo experiments, showed that thermal hyperalgesia mediated by intraplantar

PAR_2-specific agonist injection is prevented by antagonism of TRPV1 and is absent in mice deficient for TRPV1 [30, 39]. Thus, antagonism of PAR_2, TRPV1 or PKC may abrogate protease-induced thermal hyperalgesia. However, Fos expression in mice lacking TRPV1 was not completely inhibited [31], suggesting the existence of other mechanisms downstream from PAR_2 activation [40]. As TRPV1 is implicated mainly in thermal hyperalgesia, the residual Fos expression in mice lacking TRPV1 could relate to mechanical hyperalgesia mediated by PAR_2 activation [30, 39]. All these results suggest that activated PAR_2 is coupled to PKC, which may phosphorylate and activate TRPV1, thereby inducing hyperalgesia. Studies on the interactions of PARs and channels such as the TRP family could be the challenges and opportunities of future research in molecular mechanism of PARs-regulated inflammatory pain.

PAR_2 also co-expresses with other members of the TRP family in sensory neurons: TRPV4 [41, 42] and TRPA1 [43, 44]. In sensory neuron, PAR_2 activation potentiates the response of those cells to agonists of TRPA1 [44] or TRPV4 [42]. Inhibitors of Protein kinase A, C and D were able to suppress PAR_2 agonist-induced sensitization of TRPV4-mediated calcium signals in sensory neurons [42] (Fig. 3). Interestingly, the mechanism of TRPA1 sensitization by PAR_2 activation is independent of PKC [44]. In sensory neurons, the increased TRPA1 sensitivity is due to the activation of PLC, which releases the inhibition of TRPA1 from plasma membrane PIP2 [44] (Fig. 3). Direct activation of TRPV4 or TRPA1 causes somatic and visceral hypersensitivity to mechanical stimuli [41–44]. Moreover, TRPV4 and TRPA1 mediate PAR_2-induced somatic and visceral hyperalgesia in vivo [41–44]. For TRPV4, the interaction between this receptor and PAR_2 is confirmed by the fact that direct responses of splanchnic colonic afferents to PAR_2 activation are totally lost in TRPV4-deficient mice [45]. However, the same group observed that the loss of TRPA1 did not affect the magnitude or proportion of direct responses to PAR_2-specific agonist [46]. The authors suggest that the strong interaction between TRPV4 and PAR_2 in sensory neurons projecting from the colon is due to tight colocalization, whereas TRPA1 and PAR_2 are localized discretely so that TRPA1 is not accessible by downstream products of PAR_2 activation [46]. Thus, while electrophysiological recordings from splanchnic afferents did not reveal that PAR_2 sensitizes TRPA1 [46], behavioral studies of awake animals showed sensitization [43]. As all of these studies have been performed with activation of PAR_2 by the agonist peptide, we cannot extrapolate these results to the pathophysiology of proteases in inflammatory pain. In fact, if trypsin is used as an example, trypsin can activate both PAR_2 and PAR_4. On sensory neurons, will trypsin sensitize TRPV4 through PAR_2 activation or inhibit TRPV4 signal through PAR_4 activation [47]? Studies of the effect of endogenous protease on this sensitization pathway are necessary to understand such mechanisms and to develop potential therapeutic drug against inflammatory pain.

Two others channels have been described as potential partners of PAR_2 in sensory neurons. A first study shows that PAR_2 activation significantly increased the excitability of pulmonary sensory neurons in response to acid stimulation through the potentiation of both acid-evoked ASIC- and TRPV1-like whole cell inward currents independent of PLC and PKC activation [48]. The potentiation of acid signaling by PAR_2 agonists in the airway sensory neurons should be important and relevant to pathophysiological conditions such as airway inflammatory diseases in which acidification and PAR_2 activation may occur simultaneously [48]. M current conducted by the Kv7 (KCNQ) potassium channels is a mechanism used throughout the nervous system to maintain neurons at the resting membrane potential and to resist depolarizing inputs [49]. Several studies have demonstrated that M current inhibition is responsible for neuronal excitability increase [50]. Linley and colleagues have demonstrated that PAR_2 activation by agonist peptide and also by trypsin was able to inhibit M current in the sensory neurons by a PLC- and Ca^{2+}/PIP2 pathway [51] (Fig. 3).

In the hippocampus, thrombin, via activation of PAR_1, potentiates NMDA receptor activity [52]. This potentiation was attenuated in mice lacking PAR_1 and mimicked by the agonist peptide of PAR_1 [52]. In opposition with the action of thrombin in the hippocampus, transient mechanical hyperalgesia induced by intrathecally injected NMDA in mice were inhibited by an intrathecal administration of PAR_1 agonist peptide as well as thrombin [53]. Incubation of astrocytes with thrombin provokes the liberation of endothelin-1 [54], which is antinociceptive in mice when injected intrathecally [55] or intracerebroventricularly [56]. Fang and colleagues showed that the antinociceptive effect of PAR_1 activation in the central nervous system implicates endothelin type A receptor [53]. Endothelin-1 could be implicated in the mediation of central PAR_1 analgesic effect.

4 Protease and Nerve Degradation

Based on the observation that repeated acute stimulation of C-fibers due to ongoing peripheral inflammation increases the excitability of nociceptive neurons in the spinal cord leading to pain [57], Kunz and colleagues have investigated time-dependent protein expression patterns in the lumbar spinal cord following the induction of an ongoing hind paw inflammation. For protein expression, they have used two-dimensional gel electrophoresis (2D PAGE) combined with matrix-assisted laser desorption ionisation time of flight mass spectrometry (MALDI-TOFMS) [58]. They observed a modification of ten protein spots on the 2D analysis, with a dramatic decrease in neurofilament light chain (NFL) 96 h after zymosan injection [58]. NFL is a scaffolding protein exclusively expressed in neurons implicated in the assembly of neurofilaments and the maintenance of myelinated fibers diameter [59]. Studies with transgenic mice suggest that neurofilaments disorganization can lead to neurodegenerative diseases [60]. Interestingly, NFL is a substrate for calpains, a family of ubiquitously expressed calcium-dependent cysteine proteases [61]. The calpain system initially comprised two Ca^{2+}-dependent proteases, μ-calpain and m-calpain; both of them are heterodimers containing an identical 28- and a 80-kDa subunit that shares 55–65% sequence homology between the two proteases [62]. Since 1989, cDNA cloning has identified 12 additional mRNAs in mammals that encode polypeptides homologous to domains IIa and IIb of the 80-kDa subunit of μ- and m-calpain [62]. It is a highly evolutionarily conserved protease with homologues present in invertebrates, plants, fungi and mammals, and consists of 15 ubiquitous and tissue-specific isoforms in humans [62]. The catalytic subunits found in the CNS include isoforms 1, 2, 3, 5 and 10 (P. S. Vosler. Mol Neurobiol. 2008). A single dose of calpain inhibitor III attenuates the development of zymosan-induced thermal hyperalgesia and paw inflammation in rats. Treatment with calpain inhibitor was associated with inhibition of neurofilament L breakdown in the spinal cord and DRGs [58]. A model of chronic constriction nerve injury (CCI) has been used to study the effects of calpain in inflammatory conditions caused by nerve injury [63]. CCI has been described to cause an increase of IL1β and TNFα expression [64, 65]. Treatment of mice with MDL-28170, a calpain inhibitor, prevents TNFα and IL1β increased by CCI [63]. Similarly, lysophosphatidic acid (LPA) receptor activation induced calpain activation in the dorsal root and down-regulation of myelin-associated glycoprotein (MAG) [66]. The pre-treatment with calpain inhibitors attenuated LPA-induced neuropathic pain behaviors such as hyperalgesia and allodynia [66].

Concerning inflammatory pain per se, there is no other evidences of the implication of proteases in nerve degradation or in processes of re- or de-myelination. Lysosomal protease cathepsin D, proteases of the mitochondria or caspase which have been extensively studied in the context of neurodegenerative disease [67] due to their action on cell death and cellular remodeling have not yet been considered in the context of inflammatory pain. However, considering their demonstrated role in neuronal generation, they might also contribute significantly to nociceptive dysfunctions associated with chronic inflammatory pain.

5 Conversion of Molecules Implicated in Pain Transmission

5.1 The Kalikrein/Kinin System

Bradykinin (BK) and C-terminal related kinins are powerful pro-inflammatory peptides known to cause vasodilation and increase vascular permeability and extravasation of proteins and fluids, thereby leading to local edema [68]. In addition, bradykinin (BK) and C-terminal related kinin peptides provoke the release by different cell types of other mediators of inflammation [69]. BK content in inflamed tissue has been associated with the edema produced by the injection of carrageenan in rat hind paws [70]. Kinins act through two receptors: B1R and B2R receptors [71]. The specific role of B1R and B2R in inflammatory pain perception is correlated to the amplitude and the kinetics of their expression [71]. B2R expression has been described on primary sensory neurons and BK activates these nociceptors to contribute to the acute pain response [72], through the release of diacylglycerol (DAG) [73], protein kinase C activation [74] and transcient receptor

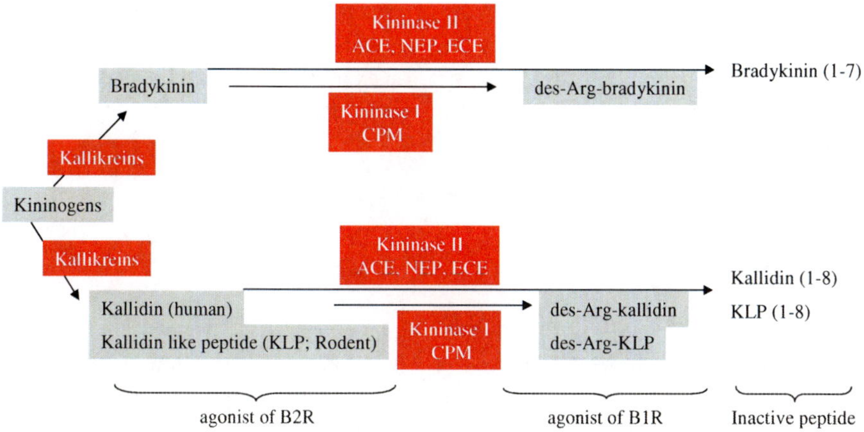

Fig. 4 Biosynthesis and metabolism of kinins. Kininogen is converted by proteolytic cleavage into bradykinin and kallidin (KLP for rodent) by kallikreins. Kallidins, KLP and bradykinin are converted in des-Arg-bradykinin, -kallidin, -KLP by Kininase I. All the kinins are inactivated by proteolytic cleavage of Kininase II. *ACE* angiotensin converting enzyme, *NEP* neprilysin (endopeptidase 24.11), *ECE* endothelin-converting enzyme, *CPM* carboxypeptidase-M. *Red*: Proteases

potential channels [46]. One of the most promising applications for the use of kinin receptor antagonists in clinics seem to be analgesia. A number of animal models of inflammatory pain have been used to show that B1R or B2R antagonist treatments have analgesic effects [75–77]. The tissular expression of the B1R mRNA in a zymosan model of inflammatory pain applied to the rat paw supports the hypothesis that B1R pro-nociceptive effects are generated in peripheral tissues, and therefore that B1R antagonists might not exert their analgesic effects within the central nervous system [76]. Interestingly, kinins generation and degradation are under the control of proteases (Fig. 4). The two classical pathways are the plasma kallikrein–kinin system that initiates the activation of the intrinsic coagulation pathway and tissue kallikrein pathway [78]. Kininogens are metabolized in Bradykinin and Kallidin (human) or Kallidin-like peptide (KLP, rodent) by the proteolytic actions of kallikreins [79] (Fig. 3). Kallidin can be converted into bradykinin by a plasma aminopeptidase [79]. All the kinins are strong agonists of B2R [80], and to a lesser extent, activators of the B1 receptor. Kininase I (carboxypeptidase-N) and carboxypeptidase-M remove arginine from the carboxyl terminus of the kinins and generate their des-Arg-derivatives [79], which are agonists mainly of B1R [80] (Fig. 4). The uses of kininogen-deficient rats [81] and recent progress with the help of experiments using gene-targeted transgenic or knockout mice make obvious that the kallikrein-kinin system is a major element of inflammatory pain reaction [82].

Alternative proteolytic pathways have been described for the metabolization of kininogen into kinin. Therefore, it has been demonstrated that the combined action of mast cell tryptase and human neutrophil elastase leads to the release of kinins from oxidized kininogens [83]. Similarly, kininogenase activity was assigned to the plasma kallikrein/neutrophil elastase couple [84]. Other proteases such as calpains may also act as kininogenases [85]. Finally, cathepsin family members may also actively contribute to kinin degradation. As an example, cathepsin L can metabolize kininogen into bradykinin and cathepsin K can degrade bradykinin into inactive peptides [86].

5.2 Peptidase

Aminopeptidase N/CD13 (APN/CD13) is attached to the plasma membrane by the C terminus facing extracellularly and is found as a dimer with a molecular mass of 160 kDa [87–89]. The protein is composed of an amino acid sequence His-Glu-Xaa-Xaa-His which is a Znþþ binding motif found in one class of metallo-peptidases [90]. APN/CD13 preferentially cleaves neutral amino acids (with the exception of proline) from the unsubstituted N terminus of oligopeptides [87, 91]. Biologically active peptide substrates cleaved by APN/CD13 include neuropeptides that are heavily involved in the transmission or inhibition of not only pain messages such as Met- and Leu-enkephalins, neurokinin A, Met-lys-bradykinin and endorphins such as spinorphin [92–95], but also vasoactive peptides [96] and chemotactic peptides (monocyte chemotactic protein/MCP-1 and *N*-formyl methionine leucine phenylalanine/f-MLP) [88, 97]. Interestingly, overexpression of APN/CD13 in T lymphocytes or neutrophils has been quantified in several inflammatory diseases associated with chronic inflammatory pain (various forms of joint effusions, rheumatoid arthritis, heart diseases) [98–101]. Considering the fact that lymphocytes are major producers of endogenous opioids [102], it seems reasonable to think that overexpressed APN/CD13 in those cells degrade natural opioids, thereby contributing to maintaining chronic pain symptoms. As for the transmission of pain to the central nervous system, centrally produced enkephalins are known to be inactivated by APN/CD13 and the membrane-bound protease neutral endopeptidase 24.11/CD10 [103]. Taken together, those results suggest that inhibition of aminopeptidase could be considered as a new therapeutic option for chronic inflammatory pain treatment. Indeed, actinonin as well as the dual inhibitors RB101 and RB120 exhibited analgesic properties against chronic pain [104, 105].

N-Acetylaspartylglutamate (NAAG) is the most released and distributed peptide transmitter in the mammalian nervous system [106]. NAAG activates the metabotropic glutamate Group II (mGlu) receptors, and particularly mGlu3 receptor [107]. NAAG acts on presynaptic receptors to inhibit the release of neurotransmitters, such as GABA and glutamate [108, 109]. NAAG is principally inactivated by two peptidases, glutamate carboxypeptidase II (GCPII) and GCPIII, which have been cloned from rat and human cDNA [110–113]. Interestingly, systemic injection of Group II mGlu receptor agonist, LY379268, reduced inflammatory hyperalgesia in a model of carrageenan-induced paw inflammation, and reduced the pain associated with the subcutaneous injection of capsaicin (agonist of TRPV1) [114]. Another Group II mGlu receptor agonist (APDC) was also able to decrease the perception of inflammatory pain induced by prostaglandin E_2 [115]. The hypothesis that increased endogenous NAAG at the synaptic level produces analgesia has been tested in rats by intrathecal administration of 2-PMPA (Phosphonomethyl pentane-dioic acid), a potent and selective inhibitor of glutamate carboxypeptidase II. This administration was effective at reducing the perception of inflammatory pain in both the carrageenan and the formalin models [116, 117]. New NAAG peptidase inhibitors (ZJ11, ZJ17 and ZJ43) were effective at reducing inflammation in both carrageenan and formalin models when administered intrathecally or systemically [118, 119].

Inhibition of peptidases such as GCPII and GCPIII or APN/CD13 present a strong therapeutic potential for the treatment of pain associated with disorders of the human nervous system. The synthesis of additional peptidase inhibitors with structures that facilitate movement across the blood–brain barrier will be important in the future development of this therapeutic approach. Moreover, concerning inflammatory pain more bench-to-bedside studies are needed to determine the importance of these peptidases in human pain pathways.

6 Conclusions

The ability of proteases to act at different levels of inflammatory pain pathways demonstrates their value as potential therapeutic targets in this phenomenon. Based on the studies performed mostly on neurodegenerative diseases, several proteolytic targets should also be considered for the treatment of inflammatory pain. Most studies have concentrated so far at understanding the role of proteases as signaling molecules to sensory neurons; however, the role of proteases on other cell types involved in neuronal support and survival, such as astrocytes or glial cells has to be considered now in the context of inflammatory pain. Moreover, even if basic science results point to a crucial role for proteases in inflammatory pain, there is a severe lack of knowledge concerning the nature of these proteases released upon human pain associated diseases. Future directions in this field would have to include the determination of the nature and specific functions of the protease released in human pathologies associated with inflammatory pain.

References

1. Pain terms: a list with definitions and notes on usage. Recommended by the IASP Subcommittee on Taxonomy. Pain (1979) 6:249
2. Vergnolle N (2009) Protease-activated receptors as drug targets in inflammation and pain. Pharmacol Ther 123:292–309
3. Kawabata A, Kawao N, Kuroda R, Tanaka A, Itoh H, Nishikawa H (2001) Peripheral PAR-2 triggers thermal hyperalgesia and nociceptive responses in rats. Neuroreport 12:715–719
4. Vergnolle N, Bunnett NW, Sharkey KA, Brussee V, Compton S, Grady EF, Cirino G, Gerard N, Basbaum A, Andrade-Gordon P, Hollenberg MD, Wallace JL (2001) Proteinase-activated receptor-2 and hyperalgesia: a novel pain pathway. Nat Med 7:821–826

5. Salzmann JL, Peltier-Koch F, Bloch F, Petite JP, Camilleri JP (1989) Morphometric study of colonic biopsies: a new method of estimating inflammatory diseases. Lab Invest 60:847–851
6. Barbara G, Stanghellini V, De GR, Cremon C, Cottrell GS, Santini D, Pasquinelli G, Morselli-Labate AM, Grady EF, Bunnett NW, Collins SM, Corinaldesi R (2004) Activated mast cells in proximity to colonic nerves correlate with abdominal pain in irritable bowel syndrome. Gastroenterology 126:693–702
7. Cenac N, Andrews CN, Holzhausen M, Chapman K, Cottrell G, Andrade-Gordon P, Steinhoff M, Barbara G, Beck P, Bunnett NW, Sharkey KA, Ferraz JG, Shaffer E, Vergnolle N (2007) Role for protease activity in visceral pain in irritable bowel syndrome. J Clin Invest 117:636–647
8. Kerckhoffs AP, Ter Linde JJ, Akkermans LM, Samsom M (2008) Trypsinogen IV, serotonin transporter transcript levels and serotonin content are increased in small intestine of irritable bowel syndrome patients. Neurogastroenterol Motil 20:900–907
9. Stead RH, Tomioka M, Quinonez G, Simon GT, Felten SY, Bienenstock J (1987) Intestinal mucosal mast cells in normal and nematode-infected rat intestines are in intimate contact with peptidergic nerves. Proc Natl Acad Sci USA 84:2975–2979
10. Tam EK, Caughey GH (1990) Degradation of airway neuropeptides by human lung tryptase. Am J Respir Cell Mol Biol 3:27–32
11. Shanahan F, Denburg JA, Fox J, Bienenstock J, Befus D (1985) Mast cell heterogeneity: effects of neuroenteric peptides on histamine release. J Immunol 135:1331–1337
12. Oku R, Satoh M, Fujii N, Otaka A, Yajima H, Takagi H (1987) Calcitonin gene-related peptide promotes mechanical nociception by potentiating release of substance P from the spinal dorsal horn in rats. Brain Res 403:350–354
13. Le Greves P, Nyberg F, Terenius L, Hokfelt T (1985) Calcitonin gene-related peptide is a potent inhibitor of substance P degradation. Eur J Pharmacol 115:309–311
14. Ferrell WR, Lockhart JC, Kelso EB, Dunning L, Plevin R, Meek SE, Smith AJ, Hunter GD, McLean JS, McGarry F, Ramage R, Jiang L, Kanke T, Kawagoe J (2003) Essential role for proteinase-activated receptor-2 in arthritis. J Clin Invest 111:35–41
15. McCarson KE, Goldstein BD (1991) Release of substance P into the superficial dorsal horn following nociceptive activation of the hindpaw of the rat. Brain Res 568:109–115
16. Hua XY, Chen P, Polgar E, Nagy I, Marsala M, Phillips E, Wollaston L, Urban L, Yaksh TL, Webb M (1998) Spinal neurokinin NK1 receptor down-regulation and antinociception: effects of spinal NK1 receptor antisense oligonucleotides and NK1 receptor occupancy. J Neurochem 70:688–698
17. Koetzner L, Gregory JA, and Yaksh TL (2004) Intrathecal protease-activated receptor stimulation produces thermal hyperalgesia through spinal cyclooxygenase activity. J Pharmacol Exp Ther
18. Alier KA, Endicott JA, Stemkowski PL, Cenac N, Cellars L, Chapman K, Andrade-Gordon P, Vergnolle N, Smith PA (2008) Intrathecal administration of proteinase-activated receptor-2 agonists produces hyperalgesia by exciting the cell bodies of primary sensory neurons. J Pharmacol Exp Ther 324:224–233
19. Asfaha S, Brussee V, Chapman K, Zochodne DW, Vergnolle N (2002) Proteinase-activated receptor-1 agonists attenuate nociception in response to noxious stimuli. Br J Pharmacol 135:1101–1106
20. Bar-Shavit R, Kahn A, Wilner G, Fenton J (1983) Monocyte chemotaxis: stimulation by specific exosite region in thrombin. Science 220:728–731
21. Bar-Shavit R, Kahn A, Mudd M, Wilner G, Mann K, Fenton J (1984) Localization of a chemotactic domain in human thrombin. Biochemistry 23:397–400
22. Herbert J, Dupuy E, Laplace M, Zini J, Bar-Shavit R, Tobelem G (1994) Thrombin induces endothelial cell growth via both a proteolytic and non-proteolytic pathway. Biochem J 303:227–231
23. Martin L, Auge C, Boue J, Buresi MC, Chapman K, Asfaha S, Andrade-Gordon P, Steinhoff M, Cenac N, Dietrich G, Vergnolle N (2009) Thrombin receptor: An endogenous inhibitor of inflammatory pain, activating opioid pathways. Pain 146:121–129

24. Jaume M, Laffont S, Chapey E, Blanpied C, Dietrich G (2007) Opioid receptor blockade increases the number of lymphocytes without altering T cell response in draining lymph nodes in vivo. J Neuroimmunol 188:95–102
25. Nissen JB, Kragballe K (1997) Enkephalins modulate differentiation of normal human keratinocytes in vitro. Exp Dermatol 6:222–229
26. Bottger A, Spruce BA (1995) Proenkephalin is a nuclear protein responsive to growth arrest and differentiation signals. J Cell Biol 130:1251–1262
27. Asfaha S, Cenac N, Houle S, Altier C, Papez MD, Nguyen C, Steinhoff M, Chapman K, Zamponi GW, Vergnolle N (2007) Protease-activated receptor-4: a novel mechanism of inflammatory pain modulation. Br J Pharmacol 150:176–185
28. Auge C, Balz-Hara D, Steinhoff M, Vergnolle N, and Cenac N (2009) Protease-activated receptor-4 (PAR(4)): a role as inhibitor of visceral pain and hypersensitivity, Neurogastroenterol. Motil
29. Karanjia R, Spreadbury I, Bautista-Cruz F, Tsang ME, Vanner S (2009) Activation of protease-activated receptor-4 inhibits the intrinsic excitability of colonic dorsal root ganglia neurons. Neurogastroenterol Motil 21:1218–1221
30. Amadesi S, Nie J, Vergnolle N, Cottrell GS, Grady EF, Trevisani M, Manni C, Geppetti P, McRoberts JA, Ennes H, Davis JB, Mayer EA, Bunnett NW (2004) Protease-activated receptor 2 sensitizes the capsaicin receptor transient receptor potential vanilloid receptor 1 to induce hyperalgesia. J Neurosci 24:4300–4312
31. Dai Y, Moriyama T, Higashi T, Togashi K, Kobayashi K, Yamanaka H, Tominaga M, Noguchi K (2004) Proteinase-activated receptor 2-mediated potentiation of transient receptor potential vanilloid subfamily 1 activity reveals a mechanism for proteinase-induced inflammatory pain. J Neurosci 24:4293–4299
32. Dattilio A, Vizzard MA (2005) Up-regulation of protease activated receptors in bladder after cyclophosphamide induced cystitis and colocalization with capsaicin receptor (VR1) in bladder nerve fibers. J Urol 173:635–639
33. Hoogerwerf WA, Zou L, Shenoy M, Sun D, Micci MA, Lee-Hellmich H, Xiao SY, Winston JH, Pasricha PJ (2001) The proteinase-activated receptor 2 is involved in nociception. J Neurosci 21:9036–9042
34. Caterina MJ, Schumacher MA, Tominaga M, Rosen TA, Levine JD, Julius D (1997) The capsaicin receptor: a heat-activated ion channel in the pain pathway. Nature 389: 816–824
35. Julius D, Basbaum AI (2001) Molecular mechanisms of nociception. Nature 413:203–210
36. Kawao N, Shimada C, Itoh H, Kuroda R, Kawabata A (2002) Capsazepine inhibits thermal hyperalgesia but not nociception triggered by protease-activated receptor-2 in rats. Jpn J Pharmacol 89:184–187
37. Kawao N, Ikeda H, Kitano T, Kuroda R, Sekiguchi F, Kataoka K, Kamanaka Y, Kawabata A (2004) Modulation of capsaicin-evoked visceral pain and referred hyperalgesia by protease-activated receptors 1 and 2. J Pharmacol Sci 94:277–285
38. Amadesi S, Cottrell GS, Divino L, Chapman K, Grady EF, Bautista F, Karanjia R, Barajas-Lopez C, Vanner S, Vergnolle N, and Bunnett NW (2006) Protease-activated receptor 2 sensitizes TRPV1 by protein kinase C- (epsilon) and A-dependent mechanisms in rats and mice. J Physiol
39. S. Amadesi, G. S. Cottrell, L. Divino, K. Chapman, E. F. Grady, F. Bautista, R. Karanjia, C. Barajas-Lopez, S. Vanner, N. Vergnolle, and N. W. Bunnett, Protease-activated receptor 2 sensitizes TRPV1 by protein kinase C- (epsilon) and A-dependent mechanisms in rats and mice., J. Physiol, (2006).
40. Seeliger S, Derian CK, Vergnolle N, Bunnett NW, Nawroth R, Schmelz M, Der Weid PY, Buddenkotte J, Sunderkotter C, Metze D, Andrade-Gordon P, Harms E, Vestweber D, Luger TA, Steinhoff M (2003) Proinflammatory role of proteinase-activated receptor-2 in humans and mice during cutaneous inflammation in vivo. FASEB J 17:1871–1885

41. Cenac N, Altier C, Chapman K, Liedtke W, Zampon G, Vergnolle N (2008) Transient receptor potential vanilloid-4 has a major role in visceral hypersensitivity symptoms. Gastroenterology 135:937–946
42. Grant AD, Cottrell GS, Amadesi S, Trevisani M, Nicoletti P, Materazzi S, Altier C, Cenac N, Zamponi GW, Bautista-Cruz F, Lopez CB, Joseph EK, Levine JD, Liedtke W, Vanner S, Vergnolle N, Geppetti P, Bunnett NW (2007) Protease-activated receptor 2 sensitizes the transient receptor potential vanilloid 4 ion channel to cause mechanical hyperalgesia in mice. J Physiol 578:715–733
43. F. Cattaruzza, I. Spreadbury, M. Miranda-Morales, E. F. Grady, S. J. Vanner, and N. W. Bunnett, Transient Receptor Potential Ankyrin-1 has a Major Role in Mediating Visceral Pain in Mice, Am. J. Physiol Gastrointest. Liver Physiol, (2009)
44. Dai Y, Wang S, Tominaga M, Yamamoto S, Fukuoka T, Higashi T, Kobayashi K, Obata K, Yamanaka H, Noguchi K (2007) Sensitization of TRPA1 by PAR2 contributes to the sensation of inflammatory pain. J Clin Invest 117:1979–1987
45. Brierley SM, Page AJ, Hughes PA, Adam B, Liebregts T, Cooper NJ, Holtmann G, Liedtke W, Blackshaw LA (2008) Selective role for TRPV4 ion channels in visceral sensory pathways. Gastroenterology 134:2059–2069
46. Brierley SM, Hughes PA, Page AJ, Kwan KY, Martin CM, O'Donnell TA, Cooper NJ, Harrington AM, Adam B, Liebregts T, Holtmann G, Corey DP, Rychkov GY, Blackshaw LA (2009) The ion channel TRPA1 is required for normal mechanosensation and is modulated by algesic stimuli. Gastroenterology 137:2084–2095
47. C. Auge, D. Balz-Hara, M. Steinhoff, N. Vergnolle, and N. Cenac, Protease-activated receptor-4 (PAR(4)): a role as inhibitor of visceral pain and hypersensitivity, Neurogastroenterol. Motil., (2009).
48. Gu Q, Lee LY (2010) Regulation of acid signaling in rat pulmonary sensory neurons by protease-activated receptor-2. Am J Physiol Lung Cell Mol Physiol 298:L454–L461
49. Delmas P, Wanaverbecq N, Abogadie FC, Mistry M, Brown DA (2002) Signaling microdomains define the specificity of receptor-mediated InsP(3) pathways in neurons. Neuron 34:209–220
50. Passmore GM, Selyanko AA, Mistry M, Al Qatari M, Marsh SJ, Matthews EA, Dickenson AH, Brown TA, Burbidge SA, Main M, Brown DA (2003) KCNQ/M currents in sensory neurons: significance for pain therapy. J Neurosci 23:7227–7236
51. Linley JE, Rose K, Patil M, Robertson B, Akopian AN, Gamper N (2008) Inhibition of M current in sensory neurons by exogenous proteases: a signaling pathway mediating inflammatory nociception. J Neurosci 28:11240–11249
52. Gingrich MB, Junge CE, Lyuboslavsky P, Traynelis SF (2000) Potentiation of NMDA receptor function by the serine protease thrombin. J Neurosci 20:4582–4595
53. Fang M, Kovacs KJ, Fisher LL, Larson AA (2003) Thrombin inhibits NMDA-mediated nociceptive activity in the mouse: possible mediation by endothelin. J Physiol 549:903–917
54. Ehrenreich H, Costa T, Clouse KA, Pluta RM, Ogino Y, Coligan JE, Burd PR (1993) Thrombin is a regulator of astrocytic endothelin-1. Brain Res 600:201–207
55. Kamei J, Hitosugi H, Kawashima N, Misawa M, Kasuya Y (1993) Antinociceptive effects of intrathecally administered endothelin-1 in mice. Neurosci Lett 153:69–72
56. Nikolov R, Semkova I, Maslarova J, Moyanova S (1993) Antinociceptive effect of centrally administered endothelin-1 and endothelin-3 in the mouse. Methods Find Exp Clin Pharmacol 15:447–453
57. Yaksh TL, Hua XY, Kalcheva I, Nozaki-Taguchi N, Marsala M (1999) The spinal biology in humans and animals of pain states generated by persistent small afferent input. Proc Natl Acad Sci USA 96:7680–7686
58. Kunz S, Niederberger E, Ehnert C, Coste O, Pfenninger A, Kruip J, Wendrich TM, Schmidtko A, Tegeder I, Geisslinger G (2004) The calpain inhibitor MDL 28170 prevents inflammation-induced neurofilament light chain breakdown in the spinal cord and reduces thermal hyperalgesia. Pain 110:409–418

59. Sakaguchi T, Okada M, Kitamura T, Kawasaki K (1993) Reduced diameter and conduction velocity of myelinated fibers in the sciatic nerve of a neurofilament-deficient mutant quail. Neurosci Lett 153:65–68
60. Lariviere RC, Julien JP (2004) Functions of intermediate filaments in neuronal development and disease. J Neurobiol 58:131–148
61. Schlaepfer WW, Lee C, Lee VM, Zimmerman UJ (1985) An immunoblot study of neurofilament degradation in situ and during calcium-activated proteolysis. J Neurochem 44:502–509
62. Goll DE, Thompson VF, Li H, Wei W, Cong J (2003) The calpain system. Physiol Rev 83:731–801
63. Uceyler N, Tscharke A, Sommer C (2007) Early cytokine expression in mouse sciatic nerve after chronic constriction nerve injury depends on calpain. Brain Behav Immun 21:553–560
64. Kleinschnitz C, Brinkhoff J, Zelenka M, Sommer C, Stoll G (2004) The extent of cytokine induction in peripheral nerve lesions depends on the mode of injury and NMDA receptor signaling. J Neuroimmunol 149:77–83
65. Kleinschnitz C, Brinkhoff J, Sommer C, Stoll G (2005) Contralateral cytokine gene induction after peripheral nerve lesions: dependence on the mode of injury and NMDA receptor signaling. Brain Res Mol Brain Res 136:23–28
66. Xie W, Uchida H, Nagai J, Ueda M, Chun J, Ueda H (2010) Calpain-mediated down-regulation of myelin-associated glycoprotein in lysophosphatidic acid-induced neuropathic pain. J Neurochem 113:1002–1011
67. Jellinger KA (2009) Recent advances in our understanding of neurodegeneration. J Neural Transm 116:1111–1162
68. Dray A, Bevan S (1993) Inflammation and hyperalgesia: highlighting the team effort. Trends Pharmacol Sci 14:287–290
69. Geppetti P (1993) Sensory neuropeptide release by bradykinin: mechanisms and pathophysiological implications. Regul Pept 47:1–23
70. Decarie A, Drapeau G, Closset J, Couture R, Adam A (1994) Development of digoxigenin-labeled peptide: application to chemiluminoenzyme immunoassay of bradykinin in inflamed tissues. Peptides 15:511–518
71. Moreau ME, Garbacki N, Molinaro G, Brown NJ, Marceau F, Adam A (2005) The kallikrein-kinin system: current and future pharmacological targets. J Pharmacol Sci 99:6–38
72. Dray A, Patel IA, Perkins MN, Rueff A (1992) Bradykinin-induced activation of nociceptors: receptor and mechanistic studies on the neonatal rat spinal cord-tail preparation in vitro. Br J Pharmacol 107:1129–1134
73. Yusuf S, Sleight P, Pogue J, Bosch J, Davies R, Dagenais G (2000) Effects of an angiotensin-converting-enzyme inhibitor, ramipril, on cardiovascular events in high-risk patients. The Heart Outcomes Prevention Evaluation Study Investigators. N Engl J Med 342:145–153
74. Dray A (1997) Kinins and their receptors in hyperalgesia. Can J Physiol Pharmacol 75:704–712
75. Couture R, Harrisson M, Vianna RM, Cloutier F (2001) Kinin receptors in pain and inflammation. Eur J Pharmacol 429:161–176
76. Belichard P, Landry M, Faye P, Bachvarov DR, Bouthillier J, Pruneau D, Marceau F (2000) Inflammatory hyperalgesia induced by zymosan in the plantar tissue of the rat: effect of kinin receptor antagonists. Immunopharmacology 46:139–147
77. Burgess GM, Perkins MN, Rang HP, Campbell EA, Brown MC, McIntyre P, Urban L, Dziadulewicz EK, Ritchie TJ, Hallett A, Snell CR, Wrigglesworth R, Lee W, Davis C, Phagoo SB, Davis AJ, Phillips E, Drake GS, Hughes GA, Dunstan A, Bloomfield GC (2000) Bradyzide, a potent non-peptide B(2) bradykinin receptor antagonist with long-lasting oral activity in animal models of inflammatory hyperalgesia. Br J Pharmacol 129:77–86
78. Kaplan AP, Joseph K, Silverberg M (2002) Pathways for bradykinin formation and inflammatory disease. J Allergy Clin Immunol 109:195–209

79. Bhoola KD, Figueroa CD, Worthy K (1992) Bioregulation of kinins: kallikreins, kininogens, and kininases. Pharmacol Rev 44:1–80
80. Kakoki M, Smithies O (2009) The kallikrein-kinin system in health and in diseases of the kidney. Kidney Int 75:1019–1030
81. Damas J (1996) The brown Norway rats and the kinin system. Peptides 17:859–872
82. Ueno A, Oh-ishi S (2003) Roles for the kallikrein-kinin system in inflammatory exudation and pain: lessons from studies on kininogen-deficient rats. J Pharmacol Sci 93:1–20
83. Kozik A, Moore RB, Potempa J, Imamura T, Rapala-Kozik M, Travis J (1998) A novel mechanism for bradykinin production at inflammatory sites. Diverse effects of a mixture of neutrophil elastase and mast cell tryptase versus tissue and plasma kallikreins on native and oxidized kininogens. J Biol Chem 273:33224–33229
84. Sato F, Nagasawa S (1989) Kinin release from human LMW-kininogen by the cooperative action of human plasma kallikrein and leukocyte elastase. Adv Exp Med Biol 247A:493–498
85. Higashiyama S, Ishiguro H, Ohkubo I, Fujimoto S, Matsuda T, Sasaki M (1986) Kinin release from kininogens by calpains. Life Sci 39:1639–1644
86. Veillard F, Lecaille F, Lalmanach G (2008) Lung cysteine cathepsins: intruders or unorthodox contributors to the kallikrein-kinin system? Int J Biochem Cell Biol 40:1079–1094
87. C. Antczak, M. De, I, and B. Bauvois, Ectopeptidases in pathophysiology, Bioessays, 23 (2001) 251-260
88. Lendeckel U, Kahne T, Riemann D, Neubert K, Arndt M, Reinhold D (2000) Review: the role of membrane peptidases in immune functions. Adv Exp Med Biol 477:1–24
89. Sjostrom H, Noren O, Olsen J (2000) Structure and function of aminopeptidase N. Adv Exp Med Biol 477:25–34
90. Hooper NM (1994) Families of zinc metalloproteases. FEBS Lett 354:1–6
91. C. Antczak, M. De, I, and B. Bauvois, Transmembrane proteases as disease markers and targets for therapy, J. Biol. Regul. Homeost. Agents, 15 (2001) 130-139
92. Lendeckel U, Arndt M, Frank K, Wex T, Ansorge S (1999) Role of alanyl aminopeptidase in growth and function of human T cells (review). Int J Mol Med 4:17–27
93. Giros B, Gros C, Solhonne B, Schwartz JC (1986) Characterization of aminopeptidases responsible for inactivating endogenous (Met5)enkephalin in brain slices using peptidase inhibitors and anti-aminopeptidase M antibodies. Mol Pharmacol 29:281–287
94. Yamamoto Y, Kanazawa H, Shimamura M, Ueki M, Hazato T (1998) Inhibitory action of spinorphin, an endogenous regulator of enkephalin-degrading enzymes, on carrageenan-induced polymorphonuclear neutrophil accumulation in mouse air-pouches. Life Sci 62:1767–1773
95. Wang LH, Ahmad S, Benter IF, Chow A, Mizutani S, Ward PE (1991) Differential processing of substance P and neurokinin A by plasma dipeptidyl(amino)peptidase IV, aminopeptidase M and angiotensin converting enzyme. Peptides 12:1357–1364
96. Robertson MJ, Cunoosamy MP, Clark KL (1992) Effects of peptidase inhibition on angiotensin receptor agonist and antagonist potency in rabbit isolated thoracic aorta. Br J Pharmacol 106:166–172
97. Ahmad S, Wang L, Ward PE (1992) Dipeptidyl(amino)peptidase IV and aminopeptidase M metabolize circulating substance P in vivo. J Pharmacol Exp Ther 260:1257–1261
98. Dan H, Tani K, Hase K, Shimizu T, Tamiya H, Biraa Y, Huang L, Yanagawa H, Sone S (2003) CD13/aminopeptidase N in collagen vascular diseases. Rheumatol Int 23: 271–276
99. Hafler DA, Hemler ME, Christenson L, Williams JM, Shapiro HM, Strom TB, Strominger JL, Weiner HL (1985) Investigation of in vivo activated T cells in multiple sclerosis and inflammatory central nervous system diseases. Clin Immunol Immunopathol 37:163–171
100. Riemann D, Schwachula A, Hentschel M, Langner J (1993) Demonstration of CD13/ aminopeptidase N on synovial fluid T cells from patients with different forms of joint effusions. Immunobiology 187:24–35

101. Shimizu T, Tani K, Hase K, Ogawa H, Huang L, Shinomiya F, Sone S (2002) CD13/ aminopeptidase N-induced lymphocyte involvement in inflamed joints of patients with rheumatoid arthritis. Arthritis Rheum 46:2330–2338
102. Boué Jérôme, Blanpied Catherine, Brousset Pierre, Vergnolle Nathalie, and Dietrich Gilles, Endogenous opioid-mediated analgesia is dependent on adaptive T cell response in mice, J Immunol, (2011)
103. Bauvois B, Dauzonne D (2006) Aminopeptidase-N/CD13 (EC 3.4.11.2) inhibitors: chemistry, biological evaluations, and therapeutic prospects. Med Res Rev 26:88–130
104. Fournie-Zaluski MC, Chaillet P, Bouboutou R, Coulaud A, Cherot P, Waksman G, Costentin J, Roques BP (1984) Analgesic effects of kelatorphan, a new highly potent inhibitor of multiple enkephalin degrading enzymes. Eur J Pharmacol 102:525–528
105. Fournie-Zaluski MC, Coric P, Turcaud S, Lucas E, Noble F, Maldonado R, Roques BP (1992) "Mixed inhibitor-prodrug" as a new approach toward systemically active inhibitors of enkephalin-degrading enzymes. J Med Chem 35:2473–2481
106. Neale JH, Olszewski RT, Gehl LM, Wroblewska B, Bzdega T (2005) The neurotransmitter *N*-acetylaspartylglutamate in models of pain, ALS, diabetic neuropathy, CNS injury and schizophrenia. Trends Pharmacol Sci 26:477–484
107. Wroblewska B, Wroblewski JT, Pshenichkin S, Surin A, Sullivan SE, Neale JH (1997) *N*-acetylaspartylglutamate selectively activates mGluR3 receptors in transfected cells. J Neurochem 69:174–181
108. Zhao J, Ramadan E, Cappiello M, Wroblewska B, Bzdega T, Neale JH (2001) NAAG inhibits KCl-induced [(3)H]-GABA release via mGluR3, cAMP, PKA and L-type calcium conductance. Eur J Neurosci 13:340–346
109. Sanabria ER, Wozniak KM, Slusher BS, Keller A (2004) GCP II (NAALADase) inhibition suppresses mossy fiber-CA3 synaptic neurotransmission by a presynaptic mechanism. J Neurophysiol 91:182–193
110. Bzdega T, Turi T, Wroblewska B, She D, Chung HS, Kim H, Neale JH (1997) Molecular cloning of a peptidase against *N*-acetylaspartylglutamate from a rat hippocampal cDNA library. J Neurochem 69:2270–2277
111. Bzdega T, Crowe SL, Ramadan ER, Sciarretta KH, Olszewski RT, Ojeifo OA, Rafalski VA, Wroblewska B, Neale JH (2004) The cloning and characterization of a second brain enzyme with NAAG peptidase activity. J Neurochem 89:627–635
112. Carter RE, Feldman AR, Coyle JT (1996) Prostate-specific membrane antigen is a hydrolase with substrate and pharmacologic characteristics of a neuropeptidase. Proc Natl Acad Sci USA 93:749–753
113. Luthi-Carter R, Berger UV, Barczak AK, Enna M, Coyle JT (1998) Isolation and expression of a rat brain cDNA encoding glutamate carboxypeptidase II. Proc Natl Acad Sci USA 95:3215–3220
114. Sharpe EF, Kingston AE, Lodge D, Monn JA, Headley PM (2002) Systemic pre-treatment with a group II mGlu agonist, LY379268, reduces hyperalgesia in vivo. Br J Pharmacol 135:1255–1262
115. Yang D, Gereau RW (2002) Peripheral group II metabotropic glutamate receptors (mGluR2/ 3) regulate prostaglandin E2-mediated sensitization of capsaicin responses and thermal nociception. J Neurosci 22:6388–6393
116. Yamamoto T, Nozaki-Taguchi N, Sakashita Y (2001) Spinal *N*-acetyl-alpha-linked acidic dipeptidase (NAALADase) inhibition attenuates mechanical allodynia induced by paw carrageenan injection in the rat. Brain Res 909:138–144
117. Yamamoto T, Nozaki-Taguchi N, Sakashita Y, Inagaki T (2001) Inhibition of spinal *N*-acetylated-alpha-linked acidic dipeptidase produces an antinociceptive effect in the rat formalin test. Neuroscience 102:473–479
118. Yamamoto T, Hirasawa S, Wroblewska B, Grajkowska E, Zhou J, Kozikowski A, Wroblewski J, Neale JH (2004) Antinociceptive effects of *N*-acetylaspartylglutamate

(NAAG) peptidase inhibitors ZJ-11, ZJ-17 and ZJ-43 in the rat formalin test and in the rat neuropathic pain model. Eur J Neurosci 20:483–494

119. Kozikowski AP, Zhang J, Nan F, Petukhov PA, Grajkowska E, Wroblewski JT, Yamamoto T, Bzdega T, Wroblewska B, Neale JH (2004) Synthesis of urea-based inhibitors as active site probes of glutamate carboxypeptidase II: efficacy as analgesic agents. J Med Chem 47:1729–1738

Microbial Proteases: Relevance to the Inflammatory Response

Takahisa Imamura and Jan Potempa

Abstract As virulence factors, microbial proteases often exert dual activities, enhancing the inflammatory response and affecting leukocyte functions. By activating the kallikrein-kinin system, they induce the release of kinins, which cause vascular leakage, and by inducing C5a production from the fifth complement component, they trigger the release of histamine. Leukocytes also accumulate in response to the production of the chemoattractant C5a and the secretion of chemokines from various cells, leading to exaggerated inflammatory reactions. Conversely, microbial proteases degrade chemoattractants, including chemokines, and the leukocyte receptors essential for leukocyte infiltration and the immune responses, thereby contributing to the bacterial evasion of the host defense system and the survival of the microorganism. In this way, they contribute to modulation of the inflammation caused by microbial infections, and their activity may exacerbate infectious diseases. Therefore, the specific inhibition of microbial proteases constitutes a valid therapeutic strategy for infectious diseases.

Keywords Bacteria • C5a • Chemokine • Chemotaxis • Cytokine • Gingipain • Histamine • Immune evasion • Inflammation • Kinin • Leukocyte • PAR • Protease • Vascular leakage

---

T. Imamura
Department of Molecular Pathology, Kumamoto University School of Medicine, Kumamoto 860-8556, Japan

J. Potempa (✉)
Department of Microbiology, Faculty of Biochemistry, Biophysics and Biotechnology, Jagiellonian University, Kraków 30-387, Poland

Oral Health and Systemic Diseases Research Group, University of Louisville School of Dentistry, Louisville, KY 40202, USA
e-mail: potempa@uga.edu

N. Vergnolle and M. Chignard (eds.), *Proteases and Their Receptors in Inflammation*, Progress in Inflammation Research, DOI 10.1007/978-3-0348-0157-7_12,

Microbial infections cause a variety of diseases in which the pathological outcome is related to the exaggerated host inflammatory response to certain microbial components, including secreted substances. Inflammation is an essential component of innate immunity, a nonspecific host defense reaction against invading microorganisms. An initial event in the inflammatory process is vascular leakage, which supplies blood-plasma-derived antimicrobial factors to the infection site, such as antibodies and the constituents of the complement system. This is followed by the infiltration of leukocytes, which phagocytose and kill the pathogens, thereby eliminating the harmful invaders. However, the inflammatory process is inherently associated with pathological events, such as edema, tissue destruction, and shock. Therefore, excessive inflammatory reactions exacerbate infectious diseases and seriously damage the host in concert with the actions of microbial toxins.

Many microbial species release various proteases that are important virulence factors [1, 2]. The virulence activities of microbial proteases are exerted through the cleavage of host tissue proteins at the infection site, ultimately resulting in tissue destruction, which further enhances the inflammatory response. These proteases can also act on proteinaceous inflammatory mediators leading to their inactivation (e.g., chemokines, anaphylatoxins), and their precursors (e.g., kininogens, complement factors) leading to the uncontrolled generation of bioactive mediators. Microbial proteases can also cleave leukocyte membrane proteins (e.g., receptors and binding proteins), thereby activating or suppressing leukocyte functions (e.g., migration and phagocytosis). Thus, bacterial proteases can modulate the inflammatory process.

In this chapter, we discuss the proinflammatory activities of microbial proteases and their involvement in pathways that alter the host inflammatory response.

1 Vascular Leakage

Vascular leakage (VL) occurs in microvessels, mostly venules, and results in the outflow of plasma only, without blood cells, through openings in the adhesive junctions of endothelial cells. Kinins and histamine induce VL by creating interendothelial cell openings, when these inflammatory mediators bind to their corresponding receptors on endothelial cells. The activation of the kallikrein-kinin system liberates kinins from kininogens [3], whereas anaphylatoxins, products of the activation of the complement system, cause histamine to be released from mast cells [4]. In both systems, the release of active mediators relies on the limited cascade-like proteolysis of proteins. Therefore, microbial proteases can potentially participate in the activation processes at any step, and eventually lead to the uncontrolled release of kinins or histamine. Consequently, these proteases can induce excessive VL, exaggerating the infectious disease with complications, such as severe edema at the infection site, and shock in the circulation. We describe below the microbial proteases that cause VL via the release of kinins or histamine, and the mechanisms for their release.

1.1 Kinins

The kallikrein-kinin system in plasma consists of Hageman factor (HF, coagulation factor XII), prekallikrein (PK), and kininogens [3]. The activation of the system is initiated by the activation of HF on a negatively charged surface, followed by the production of kallikrein from PK by active HF. Plasma kallikrein releases a nanopeptide, bradykinin (BK; RPPGFSPFR), from high-molecular-weight kininogen (HMWK), whereas tissue kallikrein liberates kallidin (lysyl-BK) from low-molecular-weight kininogen (LMWK), which is ultimately converted to BK by aminopeptidase-mediated cleavage [5]. Various biological activities of the kinins, such as the induction of hypotension, VL, smooth muscle contraction, and the sensation of pain [5] are mediated by kinin receptors. BK and kallidin bind to the BK B_2 receptor constitutively expressed on various cells, whereas desArg-BK (RPPGFSPF) specifically binds to the B_1 receptor, which is not normally present, but can be upregulated by a wide variety of noxious stimuli, inflammatory cytokines, and bacterial lipopolysaccharides (LPS) [6]. The kallikrein-kinin system components, HF, PK, and the kininogens are present in plasma and leak into the extravascular space physiologically as interstitial fluid or as exudate in inflamed lesions. Therefore, the activation of either HF or PK results in BK production via the proteolytic cascade pathway, whenever these zymogens are activated by microbial proteases. Bacterial proteases can also release kinins directly from kininogens. The targets of microbial proteases in the cascade pathway of the kallikrein-kinin system are illustrated schematically in Fig. 1. As a result of the cascade activation, released kinins induce plasma leakage, which provides nutrients for invading/colonizing bacteria and although extravasated plasma is also a source of complement and antibodies, in general VL is beneficial for pathogens.

1.1.1 Activation of HF or PK

Porphyromonas gingivalis is a major causative agent of adult periodontal disease and secretes large amounts of cysteine proteases: two molecular-mass variants of an arginine-specific proteinase, the gingipains R, of 50 and 95 kDa [7, 8], and a lysine-specific proteinase, gingipain K, of 105 kDa [8]. Gingipain R causes VL via the activation of human PK, even directly in plasma [9], which is probably associated with the production of gingival crevicular fluid and the induction of gingival edema at sites of periodontal disease. *Pseudomonas aeruginosa*, *Serratia marcescens*, and *Vibrio vulnificus* are Gram-negative bacteria that are etiological factors in opportunistic infections. *Serratia* 56K, 60K, and 73K proteases, alkaline proteinase and elastase from *P. aeruginosa*, subtilisin from *Bacillus subtilis*, thermolysin from *B. thermophiles*, and a metalloprotease from *V. vulnificus* have also been shown to activate guinea pig HF and PK [10]. However, only the latter proteinase activates human HF and PK [11]. Clostripain from *Clostridium histolyticum* generates kinin from human plasma in a manner sensitive to inhibition by soybean trypsin inhibitor

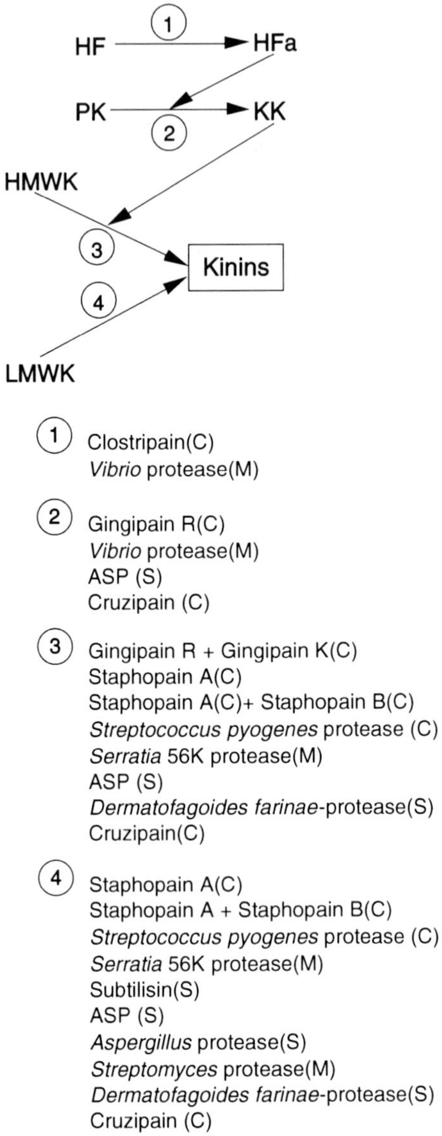

Fig. 1 Sites of action of microbial proteases in the cascade pathway of the kallikrein-kinin system. *HF* latent Hageman factor, *HFa* activated Hageman factor, *PK* prekallikrein, *KK* kallikrein, *HMWK* high-molecular-weight kininogen, *LMWK* low-molecular-weight kininogen. (S), (C), and (M) denote serine proteinases, cysteine proteinases, and metalloproteinases, respectively

(specific for plasma kallikrein) or lima bean trypsin inhibitor (specific for activated HF). This suggests that HF is a major target for proteolytic activation by clostripain [12]. *Aeromonas sobria* is a ubiquitous, waterborne, facultative anaerobic Gram-negative rod that can cause gastroenteritis and skin infections [13]. The serine protease of this pathogen (ASP) has been shown to generate VL activity in human plasma in a manner inhibited (50–60% inhibition) by HOE140, a BK B_2 receptor antagonist, and soybean trypsin inhibitor [14]. Consistent with this result, ASP was found to activate human PK, but not to activate HF at all [14]. *Trypanosoma cruzi* is a parasitic protozoan that causes Chagas' disease [15]. Cruzipain is the major cysteine protease isoform of this parasite and activates human PK [16].

1.1.2 Release from Kininogens

Staphylococcus aureus and *Streptococcus pyogenes* are common pathogens and major causative agents of Gram-positive sepsis. Cysteine proteinases from these pathogens release kinin from human kininogens [17, 18]. A cysteine proteinase from *S. aureus* (staphopain A) causes VL by releasing BK from kininogens. Moreover, when VL was induced by staphopain A, the leaked plasma area spread, whereas the VL area elicited by BK remained almost unchanged [17]. This may be attributable to connective tissue damage caused by the degradation of elastin [19], and possibly other extracellular matrix proteins by staphopain A. *Serratia* 56K protease, but not *Vibrio* protease, releases kinin from both LMWK and HMWK [20]. This is in contrast to subtilisin, a *Streptomyces caespitosus* proteinase, and an *Aspergillus melleus* proteinase, which release BK exclusively from human LMWK [20]. The mite *Dermatophagoides farinae* is one of the most important sources of allergens in house dust [21]. A *Dermatophagoides farinae* serine proteinase with a molecular mass of 30 kDa (Df protease) causes VL when injected into guinea pig skin and generates kinin from human plasma [22]. This protease also causes the release of BK from both human kininogens, with LMWK being a considerably better source of BK than HMWK [20]. In contrast, cruzipain liberates kallidin, but not BK, from human HMWK more efficiently than from LMWK [16]. ASP from *Aeromonas sobria* also induces the VL activity of human kininogens, more markedly from LMWK than from HMWK at their normal plasma concentrations, and this VL activity is predominantly inhibited by the B_2 receptor antagonist HOE140 [14]. Analysis of the kinins liberated by ASP revealed that this enzyme produces far more desArg-BK than BK from kininogens. DesArg-BK binds to the B_1 receptor and responses mediated by this receptor are upregulated by LPS or inflammatory cytokines in animal and human tissues [23]. Therefore, it is intriguing that ASP induces enhanced VL at sites of *Aeromonas* infection.

Because the release of kinin requires the cleavage of kininogen at two peptide bonds, it is possible that a bacterial protease or proteases, working in concert, can liberate kinin directly from kininogen without the activation of PK or HF. In fact, the combination of gingipain K and gingipains R releases BK directly from HMWK, cleaving the substrate at the amino and carboxyl termini of the BK sequence, respectively, whereas each gingipain alone cannot release kinin from HMWK [24]. Similarly, the VL activity of staphopain A is augmented by the addition of another cysteine proteinase (staphopain B) from this bacterium, although the latter enzyme has no VL activity by itself [17]. The enhancing effect of staphopain B is attributable to the release of a new kinin, in which the amino terminus is extended by three amino acid residues, whereas the carboxyl terminus is identical to that of BK (LMK-BK). The finding that this kinin has VL activity equivalent to that of BK is consistent with the fact that the carboxyl terminal

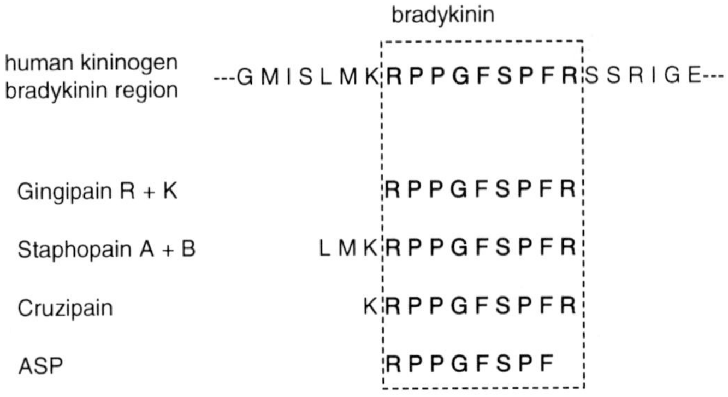

Fig. 2 Kinins released from kininogen by a microbial protease or by the synergistic action of two proteases

sequence of BK is important in its binding to the B_2 receptor [25]. Kinins released from kininogen by a microbial protease or by the synergistic action of two proteases are shown in Fig. 2.

Thus, various microbial proteinases, including those from the serine protease, cysteine protease, and metalloproteinase catalytic classes are capable of either activating the HF–PK system, or directly releasing kinin from HMWK and/or LMWK. However, many researchers still consider that the main function of bacterial proteases is in the digestion of host proteins as a nutrient source. This contention must be revised in the light of the data discussed here, which clearly show that bacterial proteases can efficiently contribute to the generation of kinin through the limited proteolysis of HF, PK, or kininogens. Released kinins are probably associated with the pathophysiology of infectious diseases. At the site of infection/inflammation, they are responsible for pain and local extravasation leading to edema, and can facilitate systemic dissemination of a pathogen from an initial site of colonization [26]. In addition to VL, the intra-arterial injection of staphopain A or ASP lowers the blood pressure of guinea pigs, as a function of kinin release [14, 17]. Sepsis is a serious medical condition, in which living bacteria are present and release proteases into the bloodstream. Therefore, the ability of bacterial proteases to cause hypotension can contribute to the onset of shock, a common and deadly consequence of sepsis.

1.2 Histamine

The anaphylatoxins, C5a and C3a, are fragments of complement factors C5 and C3, respectively, [27]. By binding to the corresponding receptors on the mast cells present in connective tissues and the mucosa of the airways and intestines, anaphylatoxins trigger the degranulation of mast cells and the release of histamine. C5 and C3 are present in the plasma that leaks into interstitial tissues, even under physiological conditions, and mast cells occur widely in the body. Therefore, it is plausible that microbial proteases liberated at an infection site readily produce anaphylatoxins, either directly or indirectly from the components of complement.

The VL caused in guinea pig skin by ASP was partially inhibited by the histamine H_1-receptor antagonist, diphenhydramine, indicating that histamine contributes to ASP-induced VL [14]. ASP mixed with C5, but not with C3, elicited VL in guinea pig skin in a dose- and incubation-time-dependent manner. The VL activity was completely inhibited by diphenhydramine, consistent with the ability of ASP to release the C5a molecule from C5 directly in human plasma [28]. This finding strongly suggests that ASP activity is relevant to the pathology of the infectious diseases caused by *Aeromonas*.

The oral intake of histamine-rich foods induces gastrointestinal symptoms in humans, such as diarrhea and flatulence [29]. Intestinal mast cells occur in increased numbers in a mouse model of diarrhea, induced by the oral ingestion of *S. aureus* peptidoglycan and histamine, and the activation and subsequent degranulation of these cells are involved in the induction of diarrhea [30]. Therefore, the activation of mast cells by ASP-induced C5a, causing histamine-dependent VL, could be associated with the gastroenteritis caused by *A. sobria* infection, a common cause of the disease [31, 32]. Moreover, the release of histamine from mast cells by ASP-generated C5a can act in concert with kinin, another VL-inducing factor generated by this protease [14], thus causing edema in aeromonad-infected wounds and lungs [33], which can lead to acute respiratory distress syndrome [31].

The proteolytic digestion of human C5 or C3 by gingipain R releases biologically active C5a, but not C3a [34]. The oxidation of C5 augments, its degradation and the production of C5a activity [35], suggests that the cooperative actions of oxidation and proteolysis can amplify both the severity and duration of the inflammatory reaction. It is likely that such reactions can promote the periodontal disease caused by *P. gingivalis* infection.

Df protease generates anaphylatoxins from the complement system components C3 and C5, and induces VL [36]. These data provide compelling evidence that this mite serine protease, as well as being an allergen, is strongly involved in the pathophysiology of allergic diseases via a proteolytic activity that generates complement-derived anaphylatoxins.

2 Leukocyte Accumulation

Leukocyte accumulation is a prominent feature of inflammation and follows VL. Generally, neutrophils accumulate at the infection sites of pyogenic bacteria, such as *S. aureus*. Conversely, macrophages and lymphocytes are involved in infections of bacteria that grow intracellularly, such as *Mycobacterium tuberculosis* and *Listeria monocytogenes*. Chemoattractants, including C5a released from the complement component C5 and chemokines secreted from various cells control the recruitment of leukocytes to inflammatory lesions. Therefore, the production or degradation of chemoattractants by microbial proteases may affect the severity and duration of inflammation.

2.1 Chemoattractant Release

C5a is a potent chemotactic factor for both neutrophils and macrophages [27]. As described above, the incubation of human C5 with either gingipain R, ASP, or Df protease leads to the release of biologically active C5a, which induces neutrophil migration [28, 35, 36]. ASP-treated C5 also induced neutrophil accumulation at the injected sites in guinea pig skin. The activation of the accumulated neutrophils, measured as the respiratory burst activity, was suppressed by a C5a receptor antagonist [28]. The biological relevance of this observation is supported by the finding that neither C5a nor the neutrophil C5a receptor activity was affected by ASP. Therefore, by generating C5a, ASP induced neutrophil accumulation leading to pus formation, as is clearly seen in the cellulitis and furuncles caused by local infections of *Aeromonas*.

Interleukin 8 (IL8) is a CXC chemokine that specifically attract neutrophils. Monocyte chemotactic protein 1 (MCP1), a CC chemokine, is chemotactic for monocytes, T cells, and eosinophils. Gingipains induce the secretion of IL8 and MCP1 from human monocytic THP-1 cells through the activation of proteinase-activated receptors (PARs) with synergistic signaling via Toll-like receptors or NOD1/2 [37, 38]. Gingipain R also enhances the production of IL8, but not MCP1, by human gingival fibroblasts in response to T-cell contact [38, 39]. The secretion of these chemokines in the gingiva in response to the gingipain-dependent activation of PARs would promote leukocyte infiltration from the bloodstream to the lesion, resulting in increased leukocyte accumulation, which is implicated in the aggravation of periodontal disease.

S. aureus is detected in the nasal lavage and biopsies of patients with chronic rhinosinusitis [40], and the culture supernatants of this pathogen induce IL8 production in nasal epithelial cells, which is dependent on serine proteases [41], probably via PAR2 activation. Neutrophil accumulation in the nasal cavity driven by the IL8 released from nasal epithelial cells upon stimulation with the *S. aureus* serine protease may be linked to the pathophysiology of chronic rhinosinusitis, including abscess formation in the paranasal sinus, and nasal secretions with pus.

P. aeruginosa produces a large exoprotease (LepA), an enzyme distinct from its other well-characterized proteases, including AprA, LasA, LasB, and protease IV, which induces the secretion of IL8 from human bronchiolar epithelial cells through the activation of PAR2 [42]. It has also been reported that Der p 1, a cysteine protease that is a clinically important house dust mite allergen, can induce IL8 release from respiratory epithelial cells via the activation of PAR2 [43, 44]. In addition to eliciting allergic reactions in the airways, causing asthma or rhinitis, this mite protease can induce inflammation through neutrophil recruitment.

Proteases are also important in the pathogenicity of *Blastocystis*, a ubiquitous enteric protozoan found in the intestinal tracts of humans and a wide range of animals. The presence of this protozoan is associated with gastrointestinal disorders involving diarrhea, abdominal pain, constipation, nausea, and fatigue. Cysteine proteases found in the central vacuole of *Blastocystis ratti* were shown to induce IL8 production in human colonic epithelial cells [45]. A similar effect has been described for the cysteine proteases of *Entamoeba histolytica* [46], another protozoan and the causative agent of amebiasis. Cumulatively, these protozoan cysteine proteases may initiate IL8-mediated inflammatory processes that ultimately lead to gastrointestinal symptoms.

2.2 Chemoattractant Degradation

Neutrophils constitute a formidable cellular line of defense against pathogenic bacteria and can kill bacteria efficiently using a variety of different mechanisms. Nevertheless, *P. gingivalis* actively promotes the accumulation of these phagocytes at its infection site by stimulating cells in the oral cavity to release IL8, as mentioned above. Neutrophil infiltration is further augmented by proteolytic trimming of the IL8 derived from keratinocytes, fibroblasts, and epithelial cells by gingipains [47]. Soluble gingipains R and gingipain K can convert $IL8_{1-77}$ very efficiently to $IL8_{6-77}$ and $IL8_{9-77}$ by cleaving it at the Arg_5-Ser_6 and Lys_8-Glu_9 peptide bonds, respectively, enhancing the biological activity of this cytokine two- to three-fold. In stark contrast, very low concentrations of cell-surface- and vesicle-associated gingipains can totally destroy IL8 activity. Because soluble enzymes can diffuse far from bacterial plaques and penetrate the periodontal tissues, they can enhance the biological activity of $IL8_{1-77}$, stimulating neutrophil migration towards the bacteria. At the same time, vesicle-bound gingipains, with only a limited ability to diffuse from the plaque surface, can destroy the chemotactic gradient by the rapid proteolytic inactivation of IL8-retaining neutrophils at a safe distance from the subgingival bacterial plaque. Thus, gingipains may protect *P. gingivalis* from leukocyte attack and secure the infection of this periodontal disease pathogen.

Group A streptococci (including *Streptococcus pyogenes*) are major Gram-positive bacterial pathogens associated with a wide spectrum of human diseases, ranging from superficial throat and skin infections to life-threatening invasive conditions, such as necrotizing fasciitis. Like *P. gingivalis*, this pathogen attenuates neutrophil infiltration when IL8 is degraded by ScpC, a serine protease [48]. ScpC specifically cleaves IL8 between Glu_{59} and Arg_{60} within the C-terminal α helix, reducing its chemokine activity for neutrophil activation and migration [49]. In a murine model of human necrotizing fasciitis in a ScpC-deficient mutant, ScpC was shown to degrade CXC chemokines at the site of *S. pyogenes* infection, thus impairing the recruitment of neutrophils, which play a major role in eradicating this bacterium [48].

Virulent strains of *S. pyogenes* and other β-hemolytic species of streptococci of human origin rapidly destroy the C5a chemotactic activity for phagocytes [50]. The bacterial protease responsible for C5a inactivation, referred to as streptococcal C5a peptidase (SCP, also called SpyCEP), cleaves C5a at the His_{67}-Lys_{68} peptide bond [51]. SCP is bound to the surface of this pathogen [52] and is related to the catalytic domain of subtilisin [53]. Mutant strains that lack SCP are cleared more rapidly than the parental strain from subcutaneous sites of infection and from the nasopharynx of mice. SCP influences the spread of streptococci from the infection site to the lymph nodes and spleen [54]. SCP also reduces neutrophil infiltration and their bactericidal activity by the cleavage and inactivation of two other human chemokines, granulocyte chemotactic protein 2 (GCP2, CXCL6), and growth-related oncogene α (GROα, CXCL1) [55]. It is likely that SCP supports sustained infections by streptococci, allowing their invasion and dissemination in the host.

3 Modulation of Leukocyte Functions

Leukocytes are essential mediators of inflammation and provide formidable bactericidal activity to curb and then eliminate invading microorganisms. In healthy tissues and the circulation, the antibacterial potential of these cells is dormant, and they become activated and "battle ready" only in response to microbial components and cytokines, including chemokines. The activating stimuli are sensed by receptors on the cell surface, so it is not surprising that many pathogenic microorganisms target these receptors and attenuate the antibacterial potential of phagocytes, as described below.

Gingipain K cleaves the N-terminal region of the C5a receptor (CD88) [56]. This Kgp-induced cleavage of the C5a receptor reduces its binding to C5a and suppresses neutrophil chemotaxis, phagocytosis, and bactericidal activity, all functions strongly stimulated by C5a. Gingipains preferentially degrade monocyte CD14, a major receptor for bacterial LPS, rendering the leukocytes hyporesponsive to LPS [57]. Moreover, it is likely that the LPS of *P. gingivalis* does not elicit signaling, because it is insufficient to stimulate monocytes/macrophages to secrete various proinflammatory cytokines (e.g., tumor necrosis factor α), which in turn stimulate other immune cells and cause the autocrine activation of phagocytes. This may explain, in part, the defective eradication of *P. gingivalis* from periodontal sites and the consequent persistent infection, despite the accumulation of large numbers of immune cells. Gingipains also cleave CD4 and CD8 on human T cells, with CD4 more susceptible than CD8 [58]. This is consistent with the fact that the CD4:CD8 ratio is reduced at sites of active periodontitis [59, 60]. Because CD4 and CD8 are important molecules in the recognition of antigens by helper T cells and killer T cells, respectively. The degradation of these molecules by gingipains could impair T-cell activation by *P. gingivalis* antigens. Therefore, this activity of gingipains must facilitate the bacterial evasion of the adaptive immune system and contribute further to bacterial survival and proliferation.

Another example of this type of strategy is the use of elastase by *P. aeruginosa* to inhibit human monocyte chemotaxis, and chemiluminescence in response to formyl-Met-Leu-Phe and zymosan-activated serum [44, 61, 62]. The inhibition of monocyte function by elastase may constitute an escape mechanism from immune cells for this bacterium.

4 Other Inflammatory Effects and Evasion of the Host Defense System

RgpB was shown to cause an increase in intracellular calcium in an oral epithelial cell line (KB) through the activation of PAR1 and PAR2, and induced the secretion of IL6, an important mediator of inflammation and humoral immunity [63]. Calcitonin gene-related peptide and substance P cause vasodilation, increased local blood flow, VL, and pain. The expression of these proinflammatory neuropeptides in human dental pulp cells is enhanced by RgpB signaling through PAR2 [64]. A similar signaling mechanism via PARs increases the production of hepatocyte growth factor, a broad-spectrum multifunctional cytokine involved in a variety of physiological processes, including tissue development, regeneration, and wound healing, in human gingival fibroblasts. These PAR-activation-mediated effects of RgpB on cells promote inflammation, leading to gingival tissue destruction. Taking all that into account, it is not surprising that activation of PAR-2 by gingipains plays an essential role in pathogenesis of *P. gingivalis* infections [65–67]. Furthermore, gingipain K was shown to promote LPS- and active vitamin-D_3-induced osteoclast differentiation by degrading osteoprotegerin [68], which is implicated in alveolar bone absorption, resulting in the loss of teeth, the ultimate consequence of periodontal disease.

Oral epithelial cells express intercellular adhesion molecule 1 (ICAM1, CD54), which mediates their interaction with neutrophils. Gingipains efficiently reduce ICAM1 expression by its proteolytic degradation, disturbing the interaction between epithelial cells and neutrophils [69]. RgpB cleaves CD14 on human gingival fibroblasts, leading to the downregulation of LPS-induced IL8 production [70]. These gingipain activities help *P. gingivalis* to evade the host defense system and, at the same time, exacerbate periodontal disease.

Periodontal disease is considered a possible contributor to systemic diseases, such as ischemic heart disease, and diabetes. Gingipain R induces the expression of the HIV-1 coreceptor CCR5 on oral keratinocytes by cleaving PAR1 and PAR2 [71], so *P. gingivalis* coinfection could promote the selective HIV-1 infection of these cells. Gingipain R may be an example of the microbial proteases that are involved in the development of other diseases.

5 Conclusion

The modulation of inflammation by microbial proteases is summarized in Fig. 3. These proteases can induce excessive inflammatory responses by upregulating the production or secretion of proinflammatory mediators, including cytokines, and by

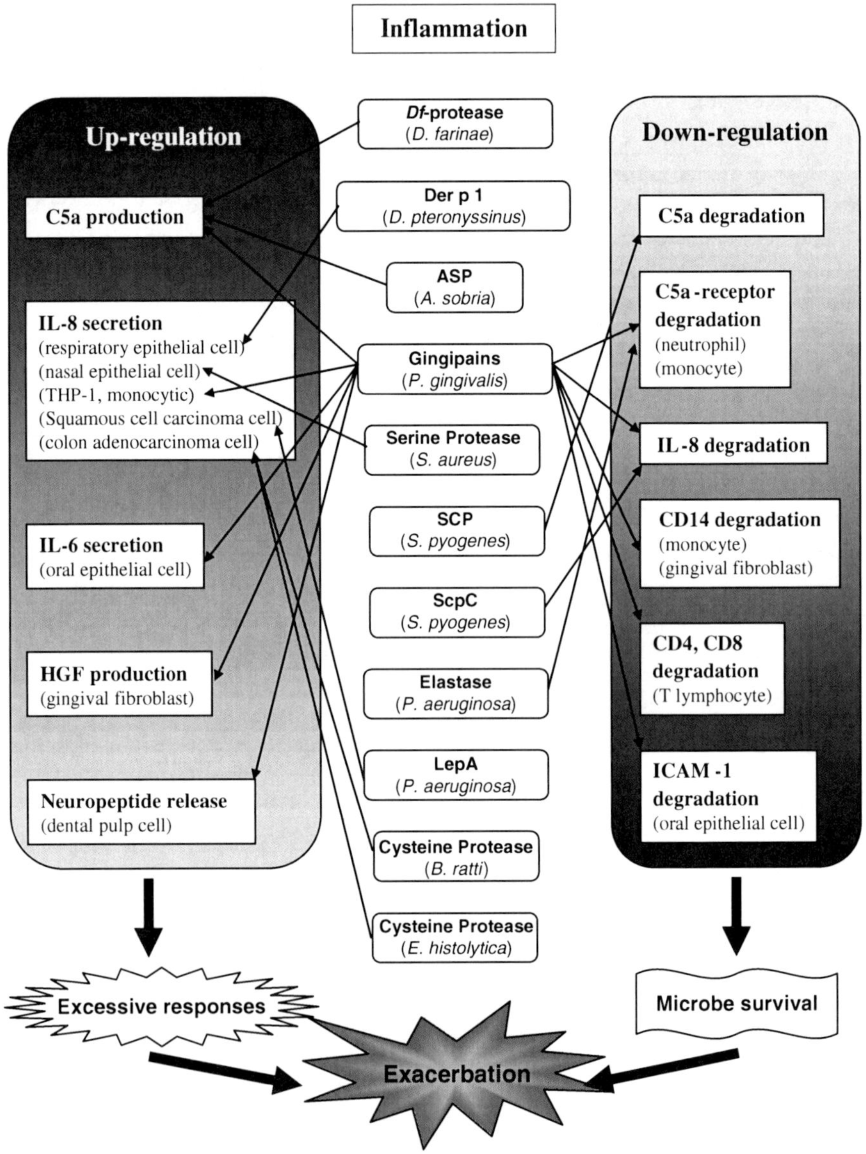

Fig. 3 Modulation of inflammation by microbial proteases

enhancing tissue-damaging reactions. Conversely, the same proteases can down-regulate leukocyte activities, thereby evading bacterial killing by the host defense system and supporting pathogen survival and proliferation in the hostile environment of inflamed tissue. The proinflammatory and immune-suppressing functions of microbial proteases together potentially result in the exacerbation of infectious diseases, in some cases leading to sepsis and shock. It is clear that microbial proteases are virulence factors, so the inhibition of their activity is a sound therapeutic strategy for infectious diseases. Inhibitors of these enzymes could be developed as drugs, which can be particularly directed against antibiotic-resistant strains.

References

1. Guo Y, Nguyen KA, Potempa J (2010) Dichotomy of gingipains action as virulence factors: from cleaving substrates with the precision of a surgeon's knife to a meat chopper-like brutal degradation of proteins. Periodontol 2000 54:15–44
2. Potempa J, Pike RN (2009) Corruption of innate immunity by bacterial proteases. J Innate Immun 1:70–87
3. Cochrane CG, Griffin JH (1982) The biochemistry and pathophysiology of the contact system of plasma. Adv Immunol 33:241–306
4. Hugli TE (1984) Structure and function of the anaphylatoxins. Springer Semin Immunopathol 7:193–219
5. Bhoola KD, Figueroa CD, Worthy K (1992) Bioregulation of kinins: Kallikreins, kininogens, and kininases. Pharmacol Rev 44:1–80
6. Leeb-Lundberg LMF, Marceau F, Müller-Ester W, Pettibone D, Zuraw B (2005) International union of pharmacology XLV. Classification of the kinin receptor family: from molecular mechanisms to pathophysiological consequences. Pharmacol Rev 57:27–77
7. Chen Z, Potempa J, Polanowski A, Wikström M, Travis J (1992) Purification of and characterization of a 50-kDa cysteine proteinase (gingipain) from *Porphyromonas gingivalis*. J Biol Chem 267:18896–18901
8. Pike R, McGraw W, Potempa J, Travis J (1994) Lysine- and arginine-specific proteinases from *Porphyromonas gingivalis*: isolation, characterization, and evidence for the existence of complexes with hemoagglutinins. J Biol Chem 269:406–411
9. Imamura T, Pike RN, Potempa J, Travis J (1994) Pathogenesis of periodontitis: a major arginine-specific cysteine proteinase from *Porphyromonas gingivalis* induces vascular permeability enhancement through activation of the kallikrein/kinin pathway. J Clin Invest 94:361–367
10. Molla A, Yamamoto T, Akaike T, Miyoshi S, Maeda H (1989) Activation of Hageman factor and prekallikrein and generation of kinin by various microbial proteinases. J Biol Chem 264:10589–10594
11. Miyoshi S, Shinoda S (1992) Activation mechanism of human Hageman factor-plasma kallikrein-kinin system by *Vibrio vulnificus* metalloproteases. FEBS Lett 308:515–519
12. Vargaftig BB, Giroux EL (1976) Mechanism of clostripain-induced kinin release from human, rat, and canine plasma. Adv Exp Med Biol 70:157–175
13. Jones BL, Wilcox MH (1995) *Aeromonas* infection and their treatment. J Antimicrob Chemother 35:453–461
14. Imamura T, Kobayashi H, Khan R, Nitta H, Okamoto K (2006) Induction of vascular leakage and blood pressure lowering through kinin release by a serine protease from *Aeromonas sobria*. J Immunol 177:8723–8729
15. Brenner Z (1973) Biology of *Trypanosoma Cruzi*. Annu Rev Microbiol 27:347–382

16. Del Nery E, Juliano MA, Lima APCA, Scharfstein JL (1997) Kininogenase activity by the major cysteinyl proteinase (Cruzipain) from *Tripanosoma cruzi*. J Biol Chem 272:25713–25718
17. Imamura T, Tanase S, Szmyd G, Kozik A, Travis J, Potempa J (2005) Induction of vascular leakage through bradykinin and a novel kinin by cysteine proteinases from *Staphylococcus aureus*. J Exp Med 201:1669–1676
18. Herwald H, Collin M, Müller-Ester W, Björck L (1996) Streptococcal cysteine proteinase releases kinins: a novel virulence mechanism. J Exp Med 184:665–673
19. Potempa J, Dubin A, Korzus G, Travis J (1988) Degradation of elastin by a cysteine proteinase from *Staphylococcus aureus*. J Biol Chem 263:2664–2667
20. Maruo K, Akaike T, Inada Y, Ohkubo I, Ono T, Maeda H (1993) Effect of microbial and mite proteases on low and high molecular weight kininogens. Generation of kinin and inactivation of thiol protease inhibitory activity. J Biol Chem 268:17711–17715
21. Miyamoto T, Oshima S, Ishizaki T, Sato S (1968) Allergenic identity between the common floor mite (*Dermatophagoides farinae*, Hughes, 1961) and house dust as a causative antigen in bronchial asthma. J Allergy 42:14–28
22. Maruo K, Akaike T, Matsumura Y, Kohmoto S, Inada Y, Ono T, Arao T, Maeda H (1991) Triggering of the vascular permeability reaction by activation of the Hageman factor-prekallikrein system by house dust mite proteinase. Biochim Biophys Acta 1074:62–68
23. Marceau F, Hess JF, Bachvarov DR (1998) The B_1 receptors for kinins. Pharmacol Rev 50:357–386
24. Imamura T, Pike RN, Potempa J, Travis J (1995) Dependence of vascular permeability enhancement on cysteine proteinases in vesicles of *Porphyromonas gingivalis*. Infect Immun 63:1999–2003
25. Regoli D, Barabé J (1980) Pharmacology of bradykinin and related kinins. Pharmacol Rev 32:1–46
26. Hu SW, Huang CH, Huang HC, Lai YY, Lin YY (2006) Transvascular dissemination of *Porphyromonas gingivalis* from a sequestered site is dependent upon activation of the kallikrein/kinin pathway. J Periodontal Res 32:200–207
27. Hugli TE (1986) Biochemistry and biology of anaphylatoxins. Complement 3:111–127
28. Nitta H, Imamura T, Wada Y, Irie A, Kobayashi H, Okamoto K, Baba H (2008) Production of C5a by ASP, a serine protease released from *Aeromonas sobria*. J Immunol 181:3602–3608
29. Wöhrl S, Hemmer W, Focke M, Rappersberger K, Jarisch R (2004) Histamine intolerance-like symptoms in healthy volunteers after oral provocation with liquid histamine. Allergy Asthma Proc 25:305–311
30. Feng BS, He SH, Zhang PY, Wu L, Yang PC (2007) Mast cells play a crucial role in *Staphylococcus aureus* peptidoglycan-induced diarrhea. Am J Pathol 171:537–547
31. Janda JM, Duffey PS (1988) Mesophilic aeromonads in human disease: current toxanomy, laboratory identification, and infectious disease spectrum. Rev Infect Dis 10:980–997
32. Deutsch SF, Wedzina W (1997) *Aeromonas sobria*-associated left-sided segmental colitis. Am J Gastroenterol 92:2104–2106
33. Janda JM, Abbott SL (1998) Evoling concepts regarding the genus *Aeromonas*: an expanding panorama of species, disease presentations, and unanswered questions. Clin Infect Dis 27:332–344
34. Wingrove JA, DiScipio RG, Chen Z, Potempa J, Travis J, Hugli TE (1992) Activation of complement components C3 and C5 by a cysteine proteinase (gingipain-1) from *Porphyromonas* (*Bacteroides*) *gingivalis*. J Biol Chem 267:18902–18907
35. DiScipio RG, Daffern PJ, Kawahara M, Pike R, Travis J, Hugli TE (1996) Cleavage of human complement component C5 by cysteine proteinases from *Porphyromonas* (*Bacteroides*) *gingivalis*. Prior oxidation of C5 augments proteinase digestion of C5. Immunology 87:660–667
36. Maruo K, Akaike T, Ono T, Okamoto T, Maeda H (1997) Generation of anaphylatoxins through proteolytic processing of C3 and C5 by house dust mite protease. J Allergy Clin Immunol 100:253–260

37. Uehara A, Naito M, Imamura T, Potempa J, Travis J, Nakayama K, Takada H (2008) Dual regulation of interleukin-8 production in human oral epithelial cells upon stimulation with gingipains from *Porphyromonas gingivalis*. J Med Microbiol 57:500–507
38. Uehara A, Imamura T, Potempa J, Travis J, Takada H (2008) Gingipains from *Prophyromonas gingivalis* synergistically induce the production of proinflammatory cytokines through protease-activated receptors with Toll-like and NOD1/2 ligands in human monocytic cells. Cell Microbiol 10:1181–1189
39. Oido-Mori M, Rezzonico R, Wang P-L, Kowashi Y, Dayer J-M, Baehni PC, Chizzolini C (2001) *Porphyromonas gingivalis* gingipain-R enhances interleukin-8 but decreases gamma interferon-inducible protein 10 production by human gingival fibroblasts in response to T-cell contact. Infect Immun 69:4493–4501
40. Niederfuhr A, Kirsche H, Deutschle T, Poppert S, Riechelmann H, Wellinghausen N (2008) *Staphylococcus aureus* in nasal lavage and biopsy of patients with chronic rhinosinusitis. Allergy 63:1359–1367
41. Rudack C, Sachse F, Albert N, Becker K, von Eiff C (2009) Immunomodulation of nasal epithelial cells by *Staphylococcus aureus*-derived serine proteases. J Immunol 183:7592–7601
42. Kida Y, Higashimoto Y, Inoue H, Shimizu T, Kuwano K (2008) A novel protease from *Pseudomonas aeruginosa* activates NF-κB through protease-activated receptors. Cell Microbiol 10:1491–1504
43. Asokananthan N, Graham PT, Stewart DJ, Bakker AJ, Eidne KA, Thompson PJ, Stewart GA (2002) House dust mite allergens induce proinflammatory cytokines from respiratory epithelial cells: the cysteine protease allergen, Der p 1, activates protease-activated receptor (PAR)-2 and inactivates PAR-1. J Immunol 169:4572–4578
44. Dulon S, Leduc D, Cottrell GS, D'Alayer J, Hansen KK, Bunnett NW, Hollenberg MD, Pidard D, Chignard M (2005) *Pseudomonas aeruginosa* elastase disables proteinase-activated receptor 2 in respiratory epithelial cells. Am J Respir Cell Mol Biol 32:411–419
45. Puthia MK, Lu J, Tan KSW (2008) *Blastocystis ratti* contains cysteine proteases that mediate interleukin-8 response from human intestinal epithelial cells in an NF-kB-dependent manner. Eukaryotic Cell 7:435–443
46. Yu Y, Chadee K (1997) *Entamoeba histolytica* stimulates interleukin 8 from human colonic epithelial cells without parasite-enterocyte contact. Gastroenterology 112:1536–1547
47. Mikolajczyk-Pawlinska J, Travis J, Potempa J (1998) Modulation of interleukin-8 activity by gingipains from *Porphyromonas gingivalis*: implications for pathogenicity of periodontal disease. FEBS Lett 440:282–286
48. Hidalgo-Grass C, Mishalian I, Dan-Goor M, Belotserkovsky I, Eran Y, Nizet V, Peled A, Hanski E (2006) A streptococcal protease that degrades CXC chemokines and impairs bacterial clearance from infected tissues. EMBO J 25:4028–4037
49. Edwards RJ, Taylor GW, Ferguson M, Murray S, Rendell N, Wrigley A, Bai Z, Boyle J, Finny SJ, Jones A et al (2005) Specific C-terminal cleavage and inactivation of interleukin-8 by invasive isolates of *Streptococcus pyogenes*. J Infect Dis 192:782–790
50. Wexler DE, Cleary PP (1985) Purification and characteristics of the streptococcal chemotactic factor inactivator. Infect Immun 50:757–764
51. Bohnsack JF, Mollison KW, Buko AM, Ashworth JC, Hill HR (1991) Group B streptococci inactivate C5a by enzymatic cleavage at the carboxy terminus. Biochem J 273:635–640
52. O'Conner SP, Cleary PP (1986) Localization of the streptococcal C5a peptidase to the surface of group A streptococci. Infect Immun 53:432–434
53. Chen C, Cleary P (1990) Complete nucleotide sequence of the streptococcal C5a peptidase gene of *Streptococcus pyogenes*. J Biol Chem 265:3161–3167
54. Ji Y, McKandsborough L, Kondagunta A, Cleary PP (1996) C5a peptidase alters clearance and trafficking of group A streptococcus by infected mice. Infect Immun 64:503–510
55. Sumby P, Zhang S, Whitney A, Falugi F, Grandi G, Graviss EA, DeLeo FR, Musser JM (2008) A chemokine-degrading extracellular protease made by group A *Streptococcus* alters pathogenesis by enhancing evasion of the innate immune response. Infect Immun 76:978–985

56. Jagels MA, Travis J, Potempa J, Pike R, Hugli TE (1996) Proteolytic inactivation of the leukocyte C5a receptor by proteinases derived from *Porphyromonas gingivalis*. Infect Immun 64:1984–1991
57. Sugawara S, Nemoto E, Tada H, Miyake K, Imamura T, Takada H (2000) Proteolysis of human monocyte CD14 by cysteine proteinases (Gingipains) from *Porphyromonas gingivalis* leading to LPS-hyporesponsiveness. J Immunol 165:411–418
58. Kitamura Y, Yoneda M, Imamura T, Matono S, Aida Y, Hirofuji T, Maeda K (2002) Gingipains in the culture supernatant of *Porphyromonas gingivalis* cleave CD4 and CD8 on human T cells. J Periodont Res 37:464–468
59. Okada H, Kasai Y, Kida T (1984) T lymphocyte subsets in the inflamed gingiva of human adult periodontitis. J Periodont Res 19:595–598
60. Stoufi ED, Taubman MA, Ebersole JL, Smith DJ, Stashenko PP (1987) Phenotype analyses of mononuclear cells recovered from healthy and diseased human periodontal tissues. J Clin Immunol 7:235–245
61. Kharazmi A, Nielsen H (1991) Inhibition of human monocyte chemotaxis and chemiluminescence by *Pseudomonas aeruginosa* elastase. Acta Pathol Microbiol Immunol Scand 99:93–95
62. Leduc D, Beaufort N, de Bentzmann S, Rousselle JC, Namane A, Chignard M, Pidard D (2007) The *Pseudomonas aeruginosa* LasB metalloproteinase regulates the human urokinase-type plasminogen activator receptor through domain-specific endoproteolysis. Infect Immun 75:3848–3858
63. Lourbakos A, Potempa J, Travis J, D'Andrea MR, Andrade-Gordon P, Santulli R, Mackie E, Pike RN (2001) Arginine-specific protease from *Porphyromonas gingivalis* activates protease-activated receptors on human oral epithelial cells and induces interleukin-6 secretion. Infect Immun 69:5121–5130
64. Tancharoen S, Sarker KP, Imamura T, Biswas KK, Matsushita K, Tatsuyama S, Travis J, Potempa J, Torii M, Maruyama I (2005) Neuropeptide release from dental pulp cells by RgpB via proteinase-activated receptor-2 signaling. J Immunol 174:5796–5804
65. Holzhausen M, Spolidorio LC, Ellen RP, Jobin MC, Steinhoff M, Andrade-Gordon P, Vergnolle N (2006) Protease-activated receptor-2 activation: a major role in the pathogenesis of *Porphyromonas gingivali*s infection. Am J Pathol 168:1189–1199
66. Wong DM, Tam V, Lam R, Walsh KA, Tatarczuch L, Pagel CN, Reynolds EC, O'Brien-Simpson NM, Mackie EJ, Pike RN (2010) Protease-activated receptor 2 has pivotal roles in cellular mechanisms involved in experimental periodontitis. Infect Immun 78:629–638
67. Holzhausen M, Cortelli JR, da Silva VA, Franco GC, Cortelli SC, Vergnolle N (2010) Protease-activated receptor-2 (PAR-2) in human periodontitis. J Dent Res 89:948–953
68. Yasuhara R, Miyamoto Y, Imamura T, Potempa J, Yoshimura K, Kamijo K (2009) Lysine-specific gingipain promotes lipopolysaccharide- and active vitamin D3-induced osteoclast differentiation by degrading osteoprotegerin. Biochem J 419:159–166
69. Tada H, Sugawara S, Nemoto E, Imamura T, Potempa J, Travis J, Shimauchi H, Takada H (2003) Proteplysis of ICAM-1 on human oral epithelial cells by gingipains. J Dent Res 82:796–801
70. Tada H, Sugawara S, Nemoto E, Takahashi N, Imamura T, Potempa J, Travis J, Shimauchi H, Takada H (2002) Proteolysis of CD14 on human gingival fibroblasts by arginine-specific cysteine proteinases from *Porphyromonas gingivalis* leading to down-regulation of lipopolysaccharide-induced interleukin-8 production. Infect Immun 70:3304–3307
71. Giacaman RA, Nobbs AH, Ross KF, Herzberg MC (2007) *Porphyromonas gingivalis* selectively up-regulates the HIV-1 coreceptor CCR5 in oral keratinocytes. J Immunol 179:2542–2550

Terminating Protease Receptor Signaling

Kathryn A. DeFea

Abstract Signals generated by proteolytic activation of Protease-activated receptors (PARs) must be terminated to avoid uncontrolled cellular events, such as proliferation, cell migration, and inflammation. Because PARs are irreversibly activated, this is of paramount importance because, unlike most other GPCRs, the ligand cannot dissociate or diffuse away. Signal termination within the cell consists of a multi-step process involving desensitization and internalization of receptors. Furthermore, given the wealth of available proteases to activate these receptors, mechanisms exist to disarm these receptors extracellularly, rendering them insensitive to activation. In this chapter, the process of signal termination will be separated into: (1) Receptor uncoupling, in which we discuss intracellular and extracellular mechanisms of receptor desensitization, (2) Receptor Endocytosis, and (3) Termination of downstream signaling pathways. Discussion will focus on PAR_1, PAR_2, and PAR_4, as PAR_3 is thought to signal primarily through activation of either PAR_1 or PAR_4.

Keywords AP-2 • Arrestin • Bicaudal D1 • Clathrin • Desensitization • Endocytosis • Endosome • Internalization • PARs • Protease-activated-receptor-1 • Protease-activated-receptor-2 • Signaling

1 Introduction

Termination of signaling through Protease-activated-receptors (PARs) shares some of the classical mechanisms utilized by many other G-protein coupled receptors (GPCRs), involving desensitization of the receptor such that it no longer responds to extracellular agonist and removal of the receptor from the cell surface via

K.A. DeFea (✉)
Biomedical Sciences Division, University of California, Riverside, CA 92521, USA
e-mail: katied@ucr.edu

N. Vergnolle and M. Chignard (eds.), *Proteases and Their Receptors in Inflammation*, Progress in Inflammation Research, DOI 10.1007/978-3-0348-0157-7_13,

endocytosis. Additionally, downstream signals, set in motion by the initial ligand binding event, are terminated intracellularly. However, while most GPCRs are reversibly activated by soluble agonists, proteolytic activation of PARs is irreversible, leading to some important distinctions in the processes governing their signal termination. The first is that because the tethered agonist cannot diffuse away, proteolytic activation of PARs could result in sustained signaling, necessitating a precise temporal control of the downstream signals. The second is that PARs are not typically recycled to the plasma membrane from sorting endosomes but are rather targeted for degradation. This necessitates regulated trafficking of PARs to the membrane in response to their depletion by proteolytic activation. The third distinction is that, because the actual agonist for PARs is a sequence within their own N-terminus, proteolytic cleavage at sites other than the activating sites can render the receptor unresponsive to activation by other proteases. Thus termination of one signal leads to activation of another. An overview of these mechanisms is diagrammed in Fig. 1. The concept of signal termination by PARs is complicated by the fact that some of the mechanisms terminating the classical G-protein signal are capable of generating their own signal.

2 Mechanisms of Uncoupling PARs from G-Proteins

2.1 Desensitization Through Steric Inhibition

Receptor Phosphorylation and β-arrestin binding: The classical model for GPCR desensitization involves rapid phosphorylation of the cytoplasmic C-terminus by downstream kinases such as G-protein receptor kinases (GRKs), PKC, and PKA, leading to increased recruitment of β-arrestins and uncoupling of the receptor from the bound Gα-protein [1–3]. Additionally, there is evidence that phosphorylation can reduce the efficiency of G-protein coupling even in the absence of β-arrestins. Indeed both PAR_1 and PAR_2 are heavily phosphorylated upon agonist treatment, and phosphorylation is important for their desensitization [4–6]. In contrast, PAR_4 has not been reported to be phosphorylated, and termination of its signaling may occur primarily through internalization [7]. Typically, these events are thought to occur as a result of the original G-protein signal, serving as a feedback inhibition loop to avoid prolonged activation of G-proteins. However, one cannot ignore the fact that the kinases responsible for phosphorylating PARs are activated by numerous other cellular pathways and that, under certain physiological conditions, PARs may be phosphorylated prior to exposure to agonist, which would result in decreased efficacy of receptor activation. Furthermore, studies on PARs 1 and 2, in which C-terminal phosphorylation sites have been removed, suggest both phosphorylation-dependent and independent mechanisms for desensitization [4]. There is some degree of diversity within the PAR family in terms of how phosphorylation occurs and what role it plays in desensitization. Phosphorylation of PAR_1 by

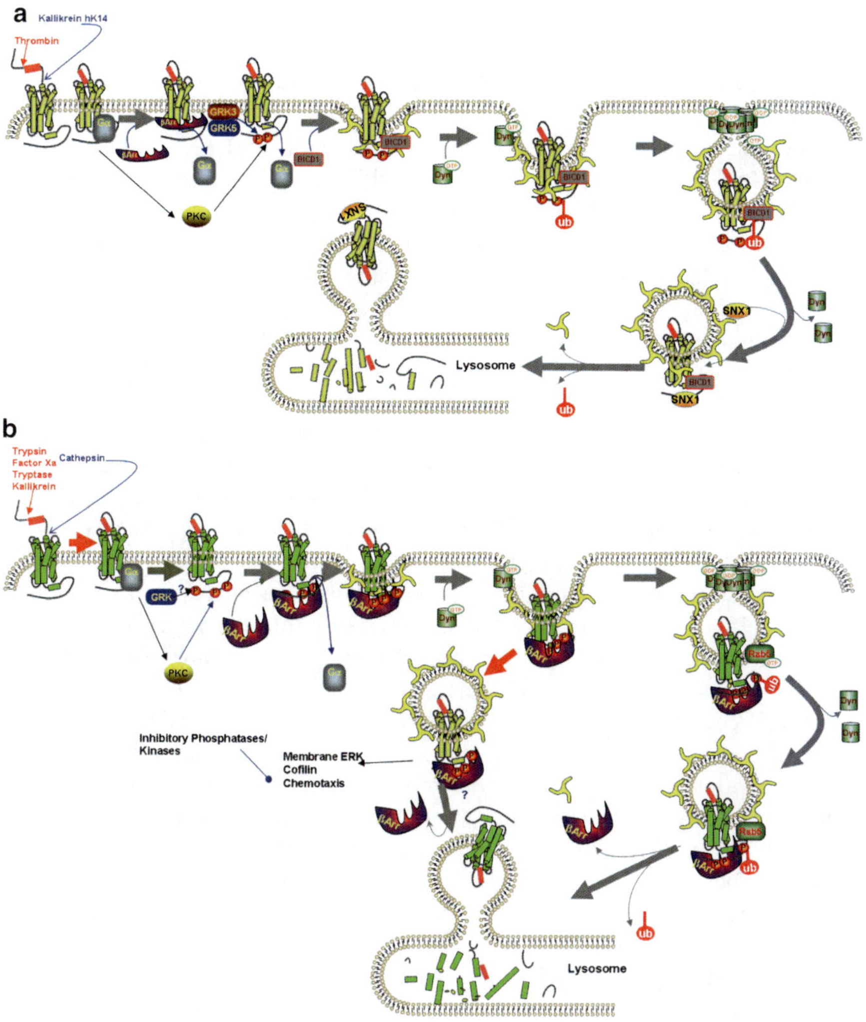

Fig. 1 Termination of Protease activated Receptor-1 and 2 signaling via desensitization, internalization and degradation. (**a**) PAR_1 is activated by thrombin and can be inactivated extracellularly by the Kallikrein, hK14. Activated PAR_1 is phosphorylated by PKC and GRKs 3 and 5 which decreases interaction with Gα subunits. Recruitment of β-arrestin-1 and Bicaudal D1 (BICD1) further contribute to Gα uncoupling rendering the receptor desensitized. Activated PAR_1 is ubiquitinated and BICD1 facilitates clathrin and dynamin-dependent endocytosis. In the early endosome, binding of SNX1 and deubiquitination target the receptor for lysosomal degradation. (**b**) PAR_2 is activated by multiple serine proteases and can be inactivated extracellularly by neutrophil Cathepsin G. Activated PAR_2 is phosphorylated by PKC and possibly GRK and β-arrestins are recruited to the phosphorylated receptor to mediate uncoupling from Gα and facilitate clathrin, dynamin and Rab5-dependent endocytosis. β-arrestin bound receptor can generate a novel set of signals from the endosome which must be terminated by downstream pathways or can be targeted to lysosomes for degradation. Ubiquitination is essential for targeting to early endosomes and subsequent deubiquitination is necessary for lysosomal targeting and degradation. β-arrestin signaling endosomes may also be targeted for degradation but this is not yet substantiated experimentally

GRK-3 and 5 is important for mediating its desensitization, but GRKs have not been directly shown to play a role in desensitization of PAR_2 [8]. Several studies implicate PKC in both PAR_1 and PAR_2 desensitization; however, this mechanism does not appear to account for all ligand-induced desensitization [9–11]. Presumably multiple kinases are required for full desensitization of PAR/G-protein signaling. GRKs also have multiple means of desensitizing signaling, as GRK3 possesses an RGS domain and can bind Gαq, while simultaneously uncoupling the receptor (which acts as a guanine exchange factor in the G-protein activation cycle) [12]. On the other hand, kinases such as PKC can only promote desensitization by phosphorylation.

Recruitment of β-arrestins to PARs appears to result in decreased G-protein signaling, presumably through the ability of β-arrestins to sterically block G-protein association [2]. This uncoupling event is associated with a decreased sensitivity to additional agonist addition as well as a termination in the production of G-protein associated intracellular messengers. In the case of Gαq, which couples to both PAR_1 and PAR_2, studies have shown enhanced accumulation of intracellular Ca^{2+} and IP_3, both in the absence of β-arrestins and with heterologous activation of downstream kinases [4, 9, 11, 13, 14]. These studies suggest that while phosphorylation alone may contribute to receptor/G-protein uncoupling, recruitment of β-arrestins may be important for desensitization of both phosphorylated and unphosphorylated receptors. A number of differences exist in how PAR_1 and PAR_2 utilize β-arrestins. First, phosphorylation clearly plays an important role in β-arrestin recruitment to PAR2, as mutations of putative phosphorylation sites reduce β-arrestin recruitment [4, 9]. In contrast, PAR_1 appears to have independent phosphorylation-dependent and β-arrestin-dependent mechanisms for desensitization. Second, PARs differ in their "β-arrestin preference": while PAR_2 recruits both β-arrestins, PAR_1 seems to preferentially recruit β-arrestin-1 [4, 10, 14–16]. However, these distinctions are not completely understood. Over-expression of either β-arrestin-1 or 2 mutants that bind tightly to unphosphorylated receptors enhances PAR1 desensitization, suggesting a possible role for β-arrestin-2 in terminating PAR_1 signaling. Conversely, since recruitment of β-arrestin-2 to PAR_1 has not been reported, β-arrestin-2 may be involved only in terminating thrombin signals that are elicited through transactivation of PAR_2 [8, 16]. Third, PAR_1 is not reported to signal through β-arrestins, while PAR_2 promotes several well-defined β-arrestin-dependent signaling events [8, 17–21]. Whether β-arrestin recruitment can truly be considered a mechanism for signal termination is a more complicated question, as it is now known that PAR_2 can also generate intracellular signals through β-arrestins, in response to proteolytic cleavage, independent of G-protein engagement (discussed in subsequent sections). Thus, recruitment of β-arrestin to the receptor can no longer be viewed as a read out of signal termination.

Other PAR interacting proteins and signal termination: Identification of novel PAR binding partners has suggested a more complicated layer of signaling and signal termination than was originally thought. Recently, Bicaudal D1 (BICD) was identified in a yeast two hybrid screen for PAR_1 interacting proteins. BICD was originally identified in *Drosophila Melanogaster* as a microtubule-associated protein essential for mRNA localization during development. These new studies indicate that BICD associates with PAR1, upon receptor activation (both proteolytic and non-proteolytic), via its cytoplasmic C-terminus, and that it decreases accumulation of IP3 [22]. Interestingly, BICD-mediated desensitization was specific for PAR_1 and did not affect PAR_2-induced IP3 accumulation. The site of interaction on PAR_1 overlaps that of Gαq, suggesting that BICD sterically disrupts association with the heterotrimeric G-protein. BICD, like β-arrestin, also plays a role in PAR_1 internalization (discussed in the next section). Although β-arrestins have been predicted to promote G-protein uncoupling from PARs through the same mechanism as BICD, the precise sites of interaction between β-arrestin and PAR_1 and PAR_2 have not been elucidated. This may be due to the fact that there may be multiple PAR/β-arrestin interactions associated with distinct cellular events.

There are other identified PAR binding partners that may interact with similar domains on the receptor as either G-proteins or β-arrestins, some of which represent additional PAR signaling pathways. Termination of these additional signaling pathways remains somewhat of a mystery. Furthermore, one must consider the possibility that association of any of these novel PAR binding partners might sterically inhibit the canonical G-protein signaling. One recently identified PAR_1 binding partner is the LIM-domain containing protein, zyxin, which has been shown to play a role in PAR_1-stimulated actin reorganization [23]. Unlike β-arrestins and BICD, zyxin knockout did not enhance G-protein signaling, suggesting it does not play a role in G-protein desensitization. However, only Gαi and Gα12 were assayed in this case so the effect of zyxin on Gαq signaling remains to be determined. PAR_2 has been shown to constitutively interact with the Jun activation domain binding protein 1 (Jab1). Proteolytic activation of PAR_2 causes Jab1 to dissociate from the receptor and promote gene transcription, suggesting some PAR_2-mediated events are due to release of Jab 1 [24]. Whether Jab1 association with PAR_2 can render the receptor less able to couple to G-proteins, and whether Jab1 requires additional modifications such as receptor phosphorylation, remain unknown.

2.2 *Preemptive Desensitization of PARs*

While the typical GPCR must be desensitized, not just as a means of terminating the signal already generated but also to render the receptor less sensitive to continued agonist stimulation, the irreversible nature of PAR activation means that receptors remaining at the surface cannot be reactivated by the continued presence of protease. However, studies on PAR_2 have suggested multiple mechanisms preemptively desensitizing the receptors. As mentioned earlier, heterologous activation of downstream kinase such as PKC renders PAR_1 and PAR_2 less susceptible to activation by proteolytic cleavage, presumably due to decreased G-protein coupling [13]. However, whether activation of PKC alone can lead to β-arrestin-dependent receptor internalization and signaling has not been demonstrated. Alternate proteolytic cleavage, such as by neutrophil serine proteases, removes the canonical cleavage site on PAR_2 and renders it insensitive to activation by trypsin [25–27]. Additionally, many Kallikreins are reported to activate PAR_1, PAR_2, and PAR_4 but the Kallikrein hK14 can inactivate PAR_1 by the same mechanism [26]. Thus, a number of cellular mechanisms exist both for termination of PAR signaling, as well as limiting the pool of receptor that is capable of generating an initial signal in response to extracellular proteases. More recently, non-canonical cleavage of PAR_2 has been shown to activate alternate signaling pathways [26] (also discussed in Chapter "Proteolytic enzymes and cell signaling: pharmacological lessons"). Thus, once again, turning off one PAR signaling pathway potentially activates another pathway that must in turn be terminated.

3 Mechanisms of PAR Endocytosis

While uncoupling of GPCRs from their upstream effectors is important for termination of their signaling, typically signal termination is not complete until the receptor is removed from the cell surface. This results in termination of original signal and prevents reactivation of receptor even if the agonist remains present. Because the PARs are irreversibly cleaved, additional proteolytic cleavage cannot induce a second wave of activation, but the receptors could potentially continue to signal. Thus, by endocytosis and lysosomal targeting of activated PARs, prolonged activation is prevented even in the face of continued protease production. However, in light of evidence that endocytosed PAR_2 can continue to signal through β-arrestins, the ability of endocytosis to terminate the protease-initiated signal is dependent upon the eventual targeting of the vesicle contents for degradation. Additionally, there are multiple mechanisms by which PARs are internalized, leading to a complex temporal control over their signaling duration. Both PAR_1 and PAR_2 undergo agonist-induced endocytosis via a clathrin-dependent mechanism, but additional mechanisms for internalization exist as well. Less is known about the internalization of PAR_4 and it appears to be removed from the cell surface by a clathrin-dependent process less rapidly than either PAR_1 or PAR_2. Clathrin-dependent endocytosis requires a mechanism by which the receptor is targeted to clathrin coated pits. β-arrestins, in addition to uncoupling receptors from G-proteins, can also serve as clathrin adaptor proteins. As mentioned above, phosphorylation is involved in the recruitment of β-arrestins to most receptors, and so it is not surprising that mutations of C-terminal phosphorylation sites in both PAR_1 and PAR_2 inhibit agonist-induced endocytosis. However, while PAR_2 internalization is severely impaired in the absence of both β-arrestins, PAR_1 internalization is β-arrestin-independent. The recently identified microtubule-associated protein, BicD1, is also important for mediating agonist-induced PAR_1 internalization, which could provide the missing link in this pathway, given the fact that β-arrestins are not involved. Studies suggest that BicD1 colocalizes with only a portion of total cellular PAR_1 in cytoplasmic vesicles [22]. PAR_2 requires both β-arrestins-1 and

2 for internalization, but studies suggest specific temporal roles for each. While β-arrestin-2 is required for prolonged internalization of PAR_2, β-arrestin-1 appears to mediate rapid, transient internalization. In studies in MEFs expressing β-arrestin-1 alone, PAR_2 was rapidly internalized and colocalized with the lysosomal marker LAMP-1, but additional stores of PAR_2 were mobilized from the Golgi apparatus to the cell membrane to restore the response to agonist. In MEFs expressing β-arrestin-2 alone, PAR_2 was removed from the surface less rapidly but remained internalized and no visible mobilization of internal stores was observed. Thus, even in the absence of one β-arrestin, some internalization still occurs. In addition to clathrin and β-arrestins, the GTPases dynamin and Rab5 are required for the early steps in endocytosis of PAR_2 and disruption of their function prevents PAR_2 desensitization, as well as resensitization [4, 28]. PAR_1 also requires dynamin for trafficking to early endosomes, but the role of Rab5 has not been investigated [29].

PARs also undergo a certain amount of tonic internalization (Fig. 2), mediated by distinct mechanisms for each receptor family member. In the case of PAR_1, tonic internalization requires clathrin and dynamin, and the clathrin adaptor protein, AP-2, which binds to a specific tyrosine within the C-terminus to facilitate recruitment of receptor into clathrin-coated pits. In contrast to agonist-induced

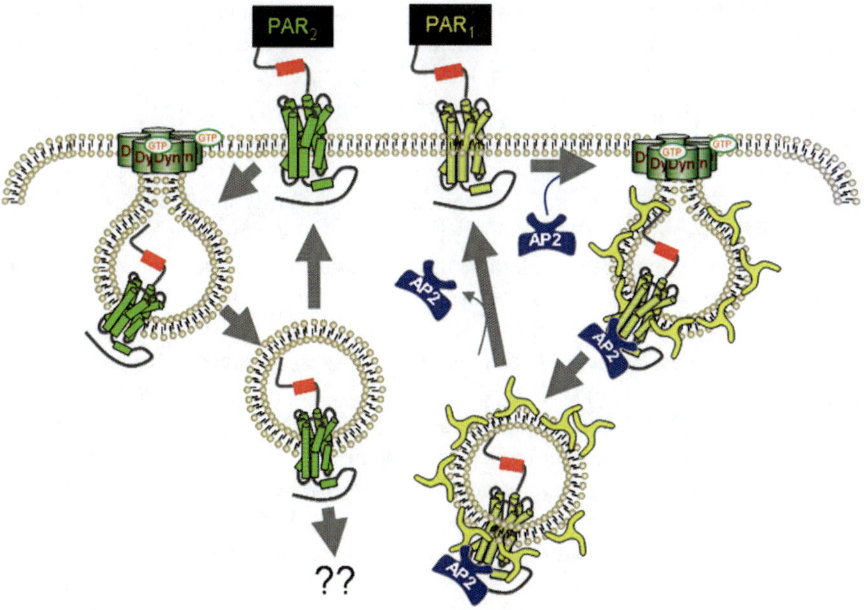

Fig. 2 Different mechanisms for tonic cycling of PARs regulate the amount of receptor at the surface. PAR_2 (*green*) is constitutively removed from the cell surface by a β-arrestin-independent, clathrin-independent mechanism, while PAR_1 (*yellow*) is constitutively removed by a clathrin and dynamin-dependent process. Both processes occur independent of receptor phosphorylation and serve to control the amount of receptor that is irreversibly activated. The fate of all of the tonically internalized PAR_2 is unclear; some receptor may be degraded and replaced with receptor from Golgi stores. Most of the PAR_1 removed by this mechanism is recycled back to the cell surface

endocytosis, tonic internalization of PAR_1 does not require serine/threonine phosphorylation. Similarly a phosphorylation-deficient mutant of PAR_2 was still able to undergo tonic internalization, but this mechanism appeared to be independent of clathrin and dynamin. Tonic internalization of PAR_1 provides a means of maintaining receptor at the cell surface, as the internalized receptor is protected from protease cleavage and is constitutively returned to the plasma membrane. Tonic internalization of PAR_2 may serve the opposite purpose, i.e., to ensure that any unactivated receptor does not remain at the cell surface for prolonged periods of time.

4 Receptor Degradation

The fate of endocytosed PARs is only partially understood as different internalization mechanisms appear to be associated with different post-internalization events. The final stage of PAR endocytosis is degradation of the receptor. In the case of PAR signal termination, this is a highly important step, as the receptor is irreversibly activated and cannot be recycled to the surface. Additionally, because endocytosis is often the beginning of β-arrestin-dependent signals, receptor degradation is important for termination of these signals as well. Although in some cell types, the majority of endocytosed PAR_1 and PAR_2 appear co-localized with lysosomal markers, suggesting they are targeted for degradation, there are also reports for prolonged signaling of PAR_2 within endosomal vesicles [9, 10, 14, 30, 31]. Over the past decade, identification of various mechanisms for tagging proteins for specific degradation pathways has raised the possibility that not all activated PARs follow the same fate. A number of factors determine the degradation of PARs, beyond endocytosis, including ubiquitination and association with sorting proteins. Endocytosed PAR_1 associates with the sorting nexin, SNX1, which is important for lysosomal degradation [32, 33]. In the absence of SNX1 association, PAR_1 accumulates in early endosomes and is poorly degraded. Both PAR_1 and PAR_2 are ubiquitinated and this modification is crucial for agonist-mediated receptor degradation [34]. Interestingly, ubiquitination of PAR_1 inhibits tonic receptor internalization but not receptor recycling, which results in an overall increase in the amount of receptor at the cell surface. A mutant PAR_1 lacking all ubiquitination sites showed enhanced constitutive internalization but normal agonist-induced degradation, suggesting ubiquitin is not required for degradation of PAR_1. As the ubiquitinated sites overlap with a putative AP2 binding site on PAR_1 which is required for constitutive but not agonist-induced internalization, it has been proposed that binding of this clathrin adaptor protein is prevented by addition of ubiquitin. Thus, deubiquitination would increase tonic cycling of PAR_1. In contrast, upon activation PAR_2 associates with the E3 ubiquitin ligase, c-cbl, and ubiquitination is essential for its sorting to lysosomes and eventual degradation [35]. PAR_2 is deubiquitinated after targeting to the lysosomes by endosomal deubiquitinating proteases AMSH and UBPY. Just as ubiquitination is essential for the initial tagging of PAR_2 to send it to the lysosome, deubiquitination is required for sorting to late endosomes [36]. Deubiquitination is not required for dissociation of PAR_2 from β-arrestin-2 or for ERK1/2 signaling, but it is not yet known whether cbl-mediated ubiquitination is involved in endosomal PAR_2 signaling. In summary, a number of modifications and protein/protein interactions are important for removing PARs from the cell surface and targeting them for degradation, which is the final mechanism for termination of protease signaling.

5 Termination of Downstream Signals

As mentioned previously, PARs, being physically tethered to their own ligands, have the potential for prolonged signaling and emerging paradigms on their signaling suggest they can signal via multiple adaptor proteins. Many of the signaling pathways downstream of PARs are associated with inflammation and proliferation; thus, termination of each of these signaling events is of paramount importance. As mentioned above, β-arrestins play an important role in terminating G-protein stimulated signals via uncoupling, but they can also directly inhibit signals downstream of the activated G-proteins (Fig. 3). This phenomenon was first observed for termination of PAR_2-stimulated PI3K activity. In the absence of β-arrestin engagement, PAR_2 promotes Gαq/Ca2+/PKC-dependent activation of classical PI3K isoforms and recruitment of β-arrestins to PAR_2 results in the sequestration of PI3K and direct inhibition of PI3K activity [17, 19]. This mechanism for termination of PI3K signaling is not shared by all PARs. Interestingly, in the case of PAR_1, β-arrestins appear to enhance rather than inhibit PI3K, via its ability to associate with Src. Silencing of β-arrestins also inhibits PAR_2-stimulated LIMK activity and PAR_2-stimulated AMPK activity, which are also dependent on Ca^{2+-}-dependent signaling [20, 37]. PAR_1 can also activate LIMK to inhibit cell migration and the possibility that β-arrestins also play a role in negatively regulating LIMK activity downstream of PAR_1 has not yet been investigated. There are other pathways directly inhibited by β-arrestins; thus it is formally possible that additional PAR signaling pathways are subject to this type of regulation downstream of the initial G-protein signal.

In the context of signal termination, it is also important to remember that many of these β-arrestin signals (both inhibitory and stimulatory) potentiate their own signals downstream of PAR_2, raising the question of how these are terminated. For example, inhibition of LIMK is associated with increased activity of its downstream target, cofilin, resulting in increased actin filament severing and cell motility [20]. Downstream of PAR_2, β-arrestins can directly facilitate cofilin activity via mechanisms other than LIMK inhibition. β-arrestins also activate the ERK1/2 pathway via the formation of a complex containing the entire MAPK module (ERK1/2, MEK1/2 and Raf). Complexes containing both β-arrestin-1 and β-arrestin-2 are associated with activation of ERK1/2. In the case of β-arrestin-1, the sequestered ERK1/2 may eventually phosphorylate it resulting in its dissociation from the PAR_2

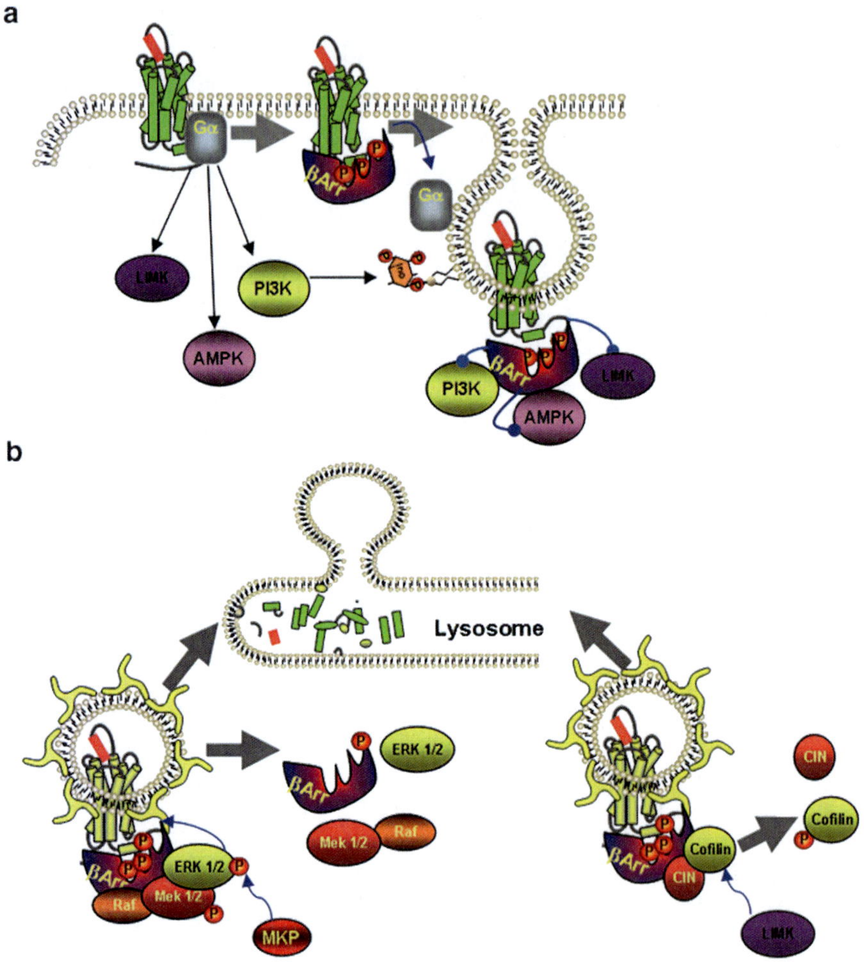

Fig. 3 Downstream mechanisms for terminating G-protein and β-arrestin-dependent PAR_2 signaling pathways. (**a**) G-protein-dependent pathways activated by PAR_2 are terminated by β-arrestins at multiple levels. As depicted in Fig. 1b, β-arrestins uncouple PAR_2 from its cognate G-protein. β-arrestins bind to and inhibit PI3K, LIMK and AMPK activated by the G-protein signaling arm. (**b**) β-arrestin-dependent signaling complexes on endosomes promote prolonged ERK1/2 and cofilin activation by scaffolding them with their upstream activators. Termination of these signaling pathways is not well defined and may involve inactivation by downstream inhibitors (MKPs for ERK1/2 and LIMK for cofilin) and complexes may subsequently disassemble. Additionally these complexes may eventually be targeted for lysosomal degradation. In the case of ERK1/2/β-arrestin-1 complexes, the sequestered ERK1/2 may phosphorylate β-arrestin-1 leading to its dissociation from the receptor and subsequent disassembly of the complex

and disassembly of the complex. MAP Kinase Phosphatases (MKPs) may also play an important role in terminating this signaling arm. PAR_1 does not appear to promote β-arrestin-dependent signals, but whether other binding partners such as the recently identified BICD1 can signal is not known. Similarly, the signaling pathways elicited by zyxin are not yet well understood; dissociation of zyxin from PAR_1 may be sufficient tot terminate this signal. To date, little is known about how any of these β-arrestin signaling events, and other G-protein-independent PAR signaling events, are terminated; presumably dissociation of these signaling complexes, and possibly lysosomal degradation of signaling endosomes, is the mechanism here as well.

6 Concluding Remarks

Signals emanating from PARs are involved in numerous physiological events, including proliferation, cell migration, production of inflammatory mediators, stimulation of pain and tissue edema, ion transport, immune cell degranulation, and changes in vascular and epithelial permeability. There are a number of mechanisms in place for terminating these signals, ranging from transient uncoupling of heterotrimeric G-proteins from their cognate receptors and modifications rendering the receptor non-responsive to subsequent agonist stimulation to long-term removal of PARs from the cell surface (summarized in Table 1). Additionally, there are numerous mechanisms for terminating distal signals set in motion by the initial G-protein coupling event. Many of these signal termination mechanisms are part of a built-in feedback loop stimulated by the initial proteolytic activation. One unresolved paradox is the dual role of β-arrestins in both termination of some and activation of other PAR_2-stimulated signals. However, it is clear that defects in termination of any of these PAR-mediated signals can result in a myriad of pathological events including tumorigenesis, metastasis, chronic inflammation, and dysregulation of mucosal functions.

Table 1 Mechanisms for desensitizing, internalizing, degrading and preemptively inhibiting (disarming) PARs

Receptor	Desensitization	Endocytosis	Degradation	Dis-arming
PAR_1	Phosphorylation, β-arrestin-1, BICD1	Phosphorylation, clathrin, AP-1, dynamin, ubiquitin, BICD1	SNX1	Kallikrein, PKC phosphorylation
PAR_2	Phosphorylation, β-arrestins	Phosphorylation, β-arrestins, dynamin, clathrin, Rab5, cbl, ubiquitin	Ubiquitination/ deubiquitination, AMSH, UBPY	Cathepsin, elastase, tryptase, PKC phosphorylation
PAR_3	??	?? mediated by association other PARs??	??	??
PAR_4	??	clathrin	??	??

?? mechanism unknown

References

1. Gehret AU, Hinkle PM (2010) Importance of regions outside the cytoplasmic tail of G protein-coupled receptors for phosphorylation and dephosphorylation. Biochem J 428:235–245
2. Shenoy SK, Lefkowitz RJ (2003) Multifaceted roles of beta-arrestins in the regulation of seven-membrane-spanning receptor trafficking and signalling. Biochem J 375:503–515
3. Tobin AB, Butcher AJ, Kong KC (2008) Location, location, location. . .site-specific GPCR phosphorylation offers a mechanism for cell-type-specific signalling. Trends Pharmacol Sci 29:413–420
4. Ricks TK, Trejo J (2009) Phosphorylation of protease-activated receptor-2 differentially regulates desensitization and internalization. J Biol Chem 284:34444–57
5. Ishii K, Chen J, Ishii M, Koch WJ, Freedman NJ, Lefkowitz RJ, Coughlin SR (1994) Inhibition of thrombin receptor signaling by a G-protein coupled receptor kinase. Functional specificity among G-protein coupled receptor kinases. J Biol Chem 269:1125–1130
6. Tiruppathi C, Yan W, Sandoval R, Naqvi T, Pronin AN, Benovic JL, Malik AB (2000) G protein-coupled receptor kinase-5 regulates thrombin-activated signaling in endothelial cells. Proc Natl Acad Sci USA 97:7440–7445
7. Shapiro MJ, Weiss EJ, Faruqi TR, Coughlin SR (2000) Protease-activated receptors 1 and 4 are shut off with distinct kinetics after activation by thrombin. J Biol Chem 275:25216–25221
8. Soh UJ, Dores MR, Chen B, Trejo J (2010) Signal transduction by protease-activated receptors. Br J Pharmacol 160:191–203
9. DeFea KA, Zalevsky J, Thoma MS, Dery O, Mullins RD, Bunnett NW (2000) Beta-arrestin-dependent endocytosis of proteinase-activated receptor 2 is required for intracellular targeting of activated ERK1/2. J Cell Biol 148:1267–1281
10. Kumar P, Lau C, Wang P, Mathur M, DeFea KA (2007) Differential effects of {beta}-arrestins on internalization, desensitization and ERK1/2 activation downstream of Protease Activated Receptor-2. Am J Physiol Cell Physiol 293:C346–C367
11. Yan W, Tiruppathi C, Lum H, Qiao R, Malik AB (1998) Protein kinase Cbeta regulates heterologous desensitization of thrombin receptor (PAR-1) in endothelial cells. Am J Physiol Cell Physiol 274:C387–C395
12. Ferguson SSG (2007) Phosphorylation-independent attenuation of GPCR signalling. Trends Pharmacol Sci 28:173–179
13. Bohm SK, Khitin LM, Grady EF, Aponte G, Payan DG, Bunnett NW (1996) Mechanisms of desensitization and resensitization of proteinase-activated receptor-2. J Biol Chem 271:22003–22016
14. Stalheim L, Ding Y, Gullapalli A, Paing MM, Wolfe BL, Morris DR, Trejo J (2005) Multiple independent functions of arrestins in the regulation of protease-activated receptor-2 signaling and trafficking. Mol Pharmacol 67:78–87
15. Chen CH, Paing MM, Trejo J (2004) Termination of protease-activated receptor-1 signaling by + -arrestins is independent of receptor phosphorylation. J Biol Chem 279:10020–10031
16. Paing MM, Stutts AB, Kohout TA, Lefkowitz RJ, Trejo J (2002) +-Arrestins regulate protease-activated receptor-1 desensitization but not internalization or down-regulation. J Biol Chem 277:1292–1300
17. Wang P, DeFea K (2006) Protease-activated-receptor-2 simultaneously directs beta-arrestin-dependent inhibition and Gaq-dependent activation of PI3K. Biochemistry 45:9374–9385
18. Ge L, Ly Y, Hollenberg M, DeFea K (2003) A beta-arrestin-dependent scaffold is associated with prolonged MAPK activation in pseudopodia during protease-activated receptor-2-induced chemotaxis. J Biol Chem 278:34418–34426
19. Wang P, Kumar P, Wang C, DeFea K (2007) Differential regulation of Class IA Phosphatidylinositol 3-Kinase catalytic subunits p110a and ß by protease-activated-receptor-2 and ß-arrestins. Biochem J 428:221–230

20. Zoudilova M, Kumar P, Ge L, Wang P, Bokoch GM, DeFea KA (2007) beta-arrestin-dependent regulation of the cofilin pathway downstream of protease-activated receptor-2. J Biol Chem 282:20634–20646
21. Zoudilova M, Min J, Richards HL, Carter D, Huang T, DeFea KA (2010) {beta}-Arrestins scaffold cofilin with chronophin to direct localized actin filament severing and membrane protrusions downstream of protease-activated-receptor-2. J Biol Chem 2010 May 7;285 (19):14318–14329
22. Swift S, Xu J, Trivedi V, Austin KM, Tressel SL, Zhang L, Covic L, Kuliopulos A (2010) A novel protease-activated receptor-1 Interactor, Bicaudal D1, regulates G protein signaling and internalization. J Biol Chem 285:11402–11410
23. Han J, Liu G, Profirovic J, Niu J, Voyno-Yasenetskaya T (2009) Zyxin is involved in thrombin signaling via interaction with PAR-1 receptor. FASEB J 23:4193–4206
24. Luo W, Wang Y, Hanck T, Stricker R, Reiser G (2006) Jab1, a novel protease-activated receptor-2 (PAR-2)-interacting protein, is involved in PAR-2-induced activation of activator protein-1. J Biol Chem 281:7927–7936
25. Ramachandran R, Sadofsky LR, Xiao Y, Botham A, Cowen M, Morice AH, Compton SJ (2007) Inflammatory mediators modulate thrombin and cathepsin-G signaling in human bronchial fibroblasts by inducing expression of proteinase-activated receptor-4. Am J Physiol Lung Cell Mol Physiol 292:L788–L798
26. Oikonomopoulou K, Hansen KK, Saifeddine M, Tea I, Blaber M, Blaber SI, Scarisbrick I, Andrade-Gordon P, Cottrell GS, Bunnett NW, Diamandis EP, Hollenberg MD (2006) Proteinase-activated receptors, targets for kallikrein signaling. J Biol Chem 281:32095–32112
27. Dulon S, Cande C, Bunnett NW, Hollenberg MD, Chignard M, Pidard D (2003) Proteinase-activated receptor-2 and human lung epithelial cells: disarming by neutrophil serine proteinases. Am J Respir Cell Mol Biol 28:339–346
28. Roosterman D, Schmidlin F, Bunnett NW (2003) Rab5a and rab11a mediate agonist-induced trafficking of protease-activated receptor 2. Am J Physiol Cell Physiol 284:C1319–C1329
29. Trejo J, Altschuler Y, Fu HW, Mostov KE, Coughlin SR (2000) Protease-activated receptor-1 down-regulation. J Biol Chem 275:31255–31265
30. Dery O, Thoma MS, Wong H, Grady EF, Bunnett NW (1999) Trafficking of proteinase-activated receptor-2 and beta-arrestin-1 tagged with green fluorescent protein. beta-Arrestin-dependent endocytosis of a proteinase receptor. J Biol Chem 274:18524–18535
31. Trejo J, Hammes SR, Coughlin SR (1998) Termination of signaling by protease-activated receptor-1 is linked to lysosomal sorting. Proc Natl Acad Sci USA 95:13698–13702
32. Gullapalli A, Wolfe BL, Griffin CT, Magnuson T, Trejo J (2006) An essential role for SNX1 in lysosomal sorting of protease-activated receptor-1: evidence for retromer-, Hrs-, and Tsg101-independent functions of sorting nexins. Mol Biol Cell 17:1228–1238
33. Wang Y, Zhou Y, Szabo K, Haft CR, Trejo J (2002) Down-regulation of protease-activated receptor-1 is regulated by sorting nexin 1. Mol Biol Cell 13:1965–1976
34. Wolfe BL, Marchese A, Trejo J (2007) Ubiquitination differentially regulates clathrin-dependent internalization of protease-activated receptor-1. J Cell Biol 177:905–916
35. Jacob C, Cottrell GS, Gehringer D, Schmidlin F, Grady EF, Bunnett NW (2005) c-Cbl mediates ubiquitination, degradation, and down-regulation of human protease-activated receptor 2. J Biol Chem 280:16076–16087
36. Hasdemir B, Murphy JE, Cottrell GS, Bunnett NW (2009) Endosomal deubiquitinating enzymes control ubiquitination and down-regulation of protease-activated receptor 2. J Biol Chem 284:28453–28466
37. Wang P, Jiang YWY, Shyy JY, DeFea KA (2010) Beta-arrestin inhibits CAMKKbeta-dependent AMPK activation downstream of protease-activated-receptor-2. BMC Biochem 2010 Sep 21;11:36

Index

N. Vergnolle and M. Chignard (eds.), *Proteases and Their Receptors in Inflammation*, Progress in Inflammation Research, DOI 10.1007/978-3-0348-0157-7,